STANDARD PLUMBING ENGINEERING DESIGN

STANDARD PLUMBING ENGINEERING DESIGN

SECOND EDITION

Louis S. Nielsen, B.S., P.E.

Director, New York State Housing and Building Codes Bureau

Associate Plumbing Engineer, New York State Building Code Commission

Instructor, Water Systems and Plumbing Design, New York University

Instructor, Plumbing Theory, Law and Design, Delehanty Institute and New York Structural Institute

McGraw-Hill Book Company

New York St. Louis San Francisco Auckland
Bogotá Hamburg Johannesburg London Madrid Mexico
Montreal New Delhi Panama Paris São Paulo
Singapore Sydney Tokyo Toronto

Library of Congress Cataloging in Publication Data

Nielsen, Louis S.
Standard plumbing engineering design.

Includes index.
1. Plumbing. I. Title.
TH6122.N5 1981 696'.1 81-2451
AACR2

ISBN 0-07-046541-X

3 4 5 6 7 8 9 0 KPKP 8 9 8 7 6

The editors for this book were Robert L. Davidson, Frank J. Cerra, Charles P. Ray, and Celia Knight, the designer was Elliot Epstein, and the production supervisor was Sally Fliess. It was set in Baskerville by The Kingsport Press.

Printed and bound by The Kingsport Press.

CONTENTS

Preface xi

1 HISTORICAL DEVELOPMENT OF PLUMBING ENGINEERING 1

History of Plumbing, 1
Ancient Plumbing, 2
Early American Sanitation Standards, 4
Plumbing Systems in Buildings, 6
Fixture Development, 13
Standards for Plumbing Materials, 20
Plumbing System Regulations and Standardization, 25

2 SYSTEMS, MATERIALS, AND FIXTURES FOR OCCUPANCIES 37

Performance Required of Building Systems, 37
Performance Required of Materials for Plumbing Systems, 37
Building Occupancy Classifications, 38
Fixtures Required for Building Occupancies, 41
Facilities for the Physically Handicapped, 41
Standard Code Regulations, 44

3 FIXTURES, FIXTURE TRAPS, AND INTERCEPTING DEVICES 53

Standards for Quality of Fixtures, 53
Water Closets and Urinals, 54
Flush Tanks and Flush Valves (Flushometers), 55
Lavatories, Bathtubs, and Showers, 56
Sinks and Laundry Trays, 57
Drinking Fountains, 58
Dishwashing Machines and Fixtures, 59
Floor Drains, 60

Fixture Overflows and Strainers, 60
Special Use Fixtures, 61
Intercepting Strainers, 61
Swimming Pools, 62
Religious, Ornamental, and Aquarium Fixtures, 63
Fixture Installation, 63
Fixture Traps, 66
Use of Intercepting Fixture Traps, 69
Vapor Relief Pipes or Local Vents for Fixtures, 70
Standard Code Regulations, 71

4 WATER SUPPLY SYSTEMS 81

General Design Considerations, 81
Source of Water Supply, 82
Quality of Water Supply for Fixtures, 82
Water Services and Water Meters, 84
Water Treatment for Corrosion Control, 85
Protection of Potable Water Supply, 86
Protection against Contamination by Backflow and Back Siphonage, 89
Backflow Prevention Devices for Special Fixtures and Equipment, 93
Water Supply Requirements, 94
Multizone Water Supply Systems for Tower Buildings, 95
Water Supply Tanks, 97
Piping Installation, 100
Expansion and Contraction of Piping, 102
Standard Pipe Thread, 107
Offset Calculations, 108
Valving, 110
Water Supply Piping Tests, 112
Disinfection Methods, 112
Standard Code Regulations, 113

5 FLOW IN WATER SUPPLY PIPING 123

Principles of Hydraulics, 123
Physical Properties of Water, 123
Streamline and Turbulent Flow, 125
Velocity, 128
Quantity Rate of Flow through Piping, 128
Potential and Kinetic Energy, 129
Static Head, 130
Velocity Head, 131
Bernoulli's Theorem, 132
Friction Head, 133
Torricelli's Theorem, 134
Jet Flow, 134
Quantity Rate of Flow from Outlets, 136
Quantity Rate of Flow through Submerged Orifices, 138
Pressure Loss through Meters, 138
Pressure Loss Due to Friction in Piping, 139
Pipe Friction Formulas for Various Service Conditions, 143
Equivalent Length of Fittings and Valves, 145
Pressure Loss through Miscellaneous Equipment, 150
Division of Flow in Parallel Pipe Circuits, 151
Water Hammer, 151
Velocity Effects in Piping, 155
Suction Lift of Pumps, 155
Work Done in Pumping, 157

6 SIZING THE WATER SUPPLY SYSTEM 159

Design Objectives, 159
Minimum Available Pressure, 159
Demand at Individual Water Outlets, 160
Estimating Demand, 163
Preliminary Information, 167
Limitation of Friction, 169
Accurate Pipe Friction Charts for Various Service Conditions, 172
Limitation of Velocity, 176
Sizing Tables Based on Velocity Limitation, 181
Simplified Method for Sizing Systems, 184
Step-by-Step Procedure for the Simplified Sizing Method, 197
Application of Simplified Method to Illustrative Problems, 198
Detailed Method for Sizing Systems in Buildings of Any Height, 208
Application of Detailed Method to Illustrative Problems, 213
Standard Code Regulations, 226

7 HOT WATER SUPPLY SYSTEMS 231

Objectives of Standard Design, 231
Safety Devices, 233
Water-Heating Methods and Heater Types, 238
Direct Heating Equipment Installations, 239
Indirect Tankless Heater Installations, 244
Indirect Storage Tank Heater Installations, 248
Demand and Heater Capacities, 256
Return Circulation Systems, 261
Procedure for Determining Circulation Rates and Sizing Return Piping, 273
Standard Code Regulations, 280

8 DRAINAGE AND VENT SYSTEMS 283

General Design Considerations, 283
Sewage Disposal, 284
Storm Water Disposal, 285
Building Sewers, 285
Combined Systems in Buildings, 286
Storm Water Drains, 287
Controlled Flow Storm Water Drainage System, 287
Fixture and Equipment Connections to Sanitary Drainage System, 290
Venting of Sanitary Drainage System, 292
Objectionable Wastes, 293
Drainage Systems below Sewer Level, 295
Subsoil Drainage, 296
Backwater Valves, 296
Building Traps and Fresh Air Inlets, 297
Connections to Sanitary Building Drains, 301
Branch Connections to Drainage Stack Offsets, 302
Drainage Stack Vent Extensions and Vent Stacks, 302
Air Pressure Relief Vents, 304
Suds Pressure Zones and Suds Pressure Relief Vents, 306
Air Pressure Relief for Pneumatic Ejectors, 312
Piping Installation, 312

Expansion and Contraction of Piping, 316
Cleanouts in Drainage Piping, 319
Fixture Trap Vents, 320
Wet Venting, 323
Stack Venting, 326
Circuit and Loop Venting, 327
Combination Waste and Vent System, 329
Indirect Waste Piping, 331
Special Wastes, 332
Testing of Drainage and Vent Systems, 333
European Single Stack Drainage Systems, 335
Standard Code Regulations, 338

9 FLOW OF WATER IN DRAINAGE PIPING 355

Principles of Hydraulics, 355
Gravity Flow in Sloping Drains, 355
Velocity of Flow for Scouring Action, 359
Gravity Flow in Vertical Drains, 360
Hydraulic Jump at Base of Vertical Drains, 362
Quantity Rate of Flow in Fixture Drains, 363
Quantity Rate of Flow in Branch Drains, 364

10 SIZING DRAINAGE SYSTEMS 365

Sizing for Adequacy of Performance, 365
Fixture Drains, 365
Sanitary Drainage Fixture Units, 366
Design Load for Sanitary Drainage System, 369
Drainage Stacks and Branches, 369
Sanitary Building Drains, 370
Storm Drainage System, 372
Combined Storm and Sanitary Building Drains, 373
Standard Code Regulations, 373

11 FLOW OF AIR IN VENT PIPING 377

Principles of Fluid Mechanics and Pneumatics, 377
Physical Properties of Air, 377
Equivalent Static Head of Water, Air, and Suds, 380
Conditions of Flow, 380
Pneumatic Effects in and Venting Design Criterion for Sanitary Drainage Systems, 381
Quantity Rate of Flow from Outlets, 382
Pressure Loss Due to Friction in Piping, 384
Permissible Length of Vent Piping, 385
Equivalent Length of Fittings in Vent Piping, 386
Value of Coefficient of Friction f, 387
Quantity Rate of Flow in Vent Stacks, 388
Quantity Rate of Flow in Individual Vents, 389
Quantity Rate of Flow in Branch Vents and Vent Headers, 390
Gravity Circulation of Air by Induced Head or Draft, 391

12 SIZING VENT PIPING 393

Sizing for Adequacy of Performance in Extended Service, 393

Vent Stacks, 393

Vent Extensions and Terminals of Stacks, 394

Vent Headers, Their Vent Extensions and Terminals, 395

Individual Vents, 396

Branch Vent Drops to Lower Floors, 397

Branch Vents, 399

Circuit and Loop Vents, 400

Relief and Yoke Vents for Soil and Waste Stacks, 401

Suds Pressure Relief Vents, 401

Vents for Building Sewage Sumps and Receiving Tanks, 402

Standard Code Regulations, 402

Index 405

ABOUT THE AUTHOR

LOUIS S. NIELSEN holds a B.S. degree from Carnegie-Mellon University and licenses to practice engineering in the states of New York and New Jersey. From 1954 to 1970 he was plumbing technical editor of *Plumbing-Heating-Cooling Business* magazine. In 1963 he participated with other leading plumbing engineers in founding the Plumbing Designers Educational Foundation and served as its first executive vice president. A former research fellow in plumbing at Virginia Polytechnic Institute, he has taught at the New York University School of Continuing Education. He is also the author of more than 300 articles on a variety of subjects relating to plumbing design and installation. Mr. Nielsen recently retired from service with the State of New York, where he served as Director of the State Housing and Building Codes Bureau from 1973 to 1979.

PREFACE

This second edition of *Standard Plumbing Engineering Design* has been written to present significant additions and modifications which have become necessary since the first edition was published in 1963. The changes are relatively broad. They cover changes in design objectives, new standards for materials, simplified methods for sizing water supply systems, design provisions for energy conservation, facilities for the physically handicapped, application to tower-type buildings with multizoned systems, new provisions for suds pressure relief in drainage systems, application of innovative designs and methods in housing construction, and the introduction of metrication terminology.

The updated content of this text is especially appropriate as it is the standard text used in the first year of the 2-year certificate course, Water Systems and Plumbing Design, given by New York University, School of Continuing Education, in New York City. This course for training plumbing designers has been sponsored by the Plumbing Designers Educational Foundation, composed of recognized plumbing engineers from the offices of registered architects, consulting engineers, government agencies, manufacturers, and the New York chapter of the American Society of Sanitary Engineering. The plumbing designers training program was started in 1964 in New York City to meet the need for certified designers and has continued to date. Training programs have also been conducted in a number of other large metropolitan areas by various chapters of the American Society of Sanitary Engineering and the American Society of Plumbing Engineers.

Adequate performance and economical installation can be achieved in the design of plumbing systems for buildings by applying rational engineering principles and generally accepted standard code regulations.

Lack of knowledge regarding the pertinent engineering principles and how to apply them in design has left most plumbing system designers with little alternative other than to rely upon and be guided by the specific, detailed requirements of existing plumbing codes. Many such codes currently are recognized as being substandard, obsolete, inadequate, or arbitrarily and excessively restrictive and cannot be considered suitable as a guide to satisfactory design of plumbing systems for modern buildings.

The purpose of this text is to present modern, standardized engineering design of plumbing systems for buildings of all occupancy classifications in conformity with standard requirements of codes, noting specific items where significant variations may occur in numerous individual codes, so that standard design based upon recognized engineering principles may be applied regardless of building location. A special aim is to present usable information and illustrations on standard design and economical arrangements and the theoretical basis upon which the design standards and code requirements were developed.

This book provides standard engineering methods to satisfy the needs of plumbing designers, architects, engineers, contractors, builders, plumbing officials, and students of plumbing design. It has been arranged so as to be suitable as a reference book, including charts, tables, and step-by-step procedures for sizing piping systems, and their applications in numerous illustrative problems.

A clear understanding of the theories involved in plumbing design should permit them to be extended and applied advantageously to new ideas of design and to better and simpler regulations. For this reason, the hydraulics and pneumatics of plumbing systems have been specially treated in the text and have been written so as to be understandable to the average individual having a significant interest in plumbing design.

The standard code regulations cited in this text have been excerpted from the parts of the New York State Building Construction Code applicable to plumbing, the first performance-type plumbing code enacted into law. It should be recognized that almost all of these regulations have been presented in the past in many model plumbing codes recommended by authoritative standards-writing bodies, associations, societies, and government agencies. In addition, the same regulations are currently in effect in most up-to-date plumbing codes, although there may be some slight differences in wording.

Standard plumbing engineering design is not the product of any one mind, code committee, organization, or official regulatory body, but is rather the evolutionary result of advances in the field of designing and installing plumbing systems in buildings over a period of approximately 140 years. Its historical development is discussed in the text. Standard plumbing engineering design may be recognized as being the best way we know of currently to protect the health, safety, and welfare of people through sanitary plumbing installations designed for adequate performance and suited to the needs of building occupants. When a better way is found to do this, the design standards should be changed accordingly.

Recognition must be granted to all who have had a hand in the development of standard plumbing engineering design as we know it today. The National Association of Plumbing-Heating-Cooling Contractors (NAPHCC) has had a leading role in this work, as witnessed by its organizational objectives, standardization committee activities, scholarship programs, and research sponsorship. The United Association of Journeymen Plumbers and Apprentices of the Plumbing and Pipe Fitting Industry of the United States and Canada (UAJPAPPFI) collaborated very closely in this development and maintained training programs for the people who had to install new plumbing designs and make the designs work satisfactorily in service. The American Society of Sanitary Engineering (ASSE) has been a leader in the evaluation of design methods, establishment of performance requirements for plumbing products, and discussion of ways and means to improve performance of plumbing systems. During the 1960s, the newly formed American Society of Plumbing Engineers (ASPE) assumed an active leadership role in this continuing development. In 1973, they became cosponsors of the National Standard Plumbing Code, published by the National Association of Plumbing-Heating-Cooling Contractors (NAPHCC).

The federal government has played a major role. The National Bureau of Standards, U.S. Department of Commerce, provided a firm basis for establishing scientific data from research on plumbing systems and by advising how such data should be applied in order to formulate necessary regulations. During the 1970s, the U.S. Department of Housing and Urban Development (HUD) contributed greatly by developing the design criteria for residential building construction in its major experimental Operation Breakthrough housing construction program. HUD has been a leader in the introduction of innovative designs and methods of construction, has sponsored many new plumbing research programs including in-depth investigation of European single stack drainage systems at the Davidson Laboratory of Stevens Institute, and has been a driving force toward the objective of achieving national uniformity of code requirements.

Recognition is due three regional associations of code enforcement officials, the Building Officials and Code Administrators International (BOCA), the International Association of Plumbing and Mechanical Officials (IAPMO), and the Southern Building Code Congress International

(SBCC), for the publication of model plumbing codes containing plumbing requirements which have been deemed to meet generally accepted standard status. As a practical matter, uniformity nationally has been reached except for insignificant details.

Health agencies throughout the nation have collaborated to set up and maintain effective regulations of plumbing installations to protect the health, safety, and welfare of the people within their jurisdictions. Manufacturers of plumbing equipment have kept pace with this development, sometimes far in advance of codes and practice.

These are but a few of the many groups to whom respectful acknowledgment should be accorded for their contributions to the development of standard plumbing engineering design. To all of them, we express our sincere appreciation.

LOUIS S. NIELSEN

1

HISTORICAL DEVELOPMENT OF PLUMBING ENGINEERING

HISTORY OF PLUMBING

The history of plumbing has special significance to all who are involved in the design and installation of plumbing systems. It provides depth of knowledge, broad objectivity, helpful guidance, needed cautions, and informative records of plumbing performance and adverse experiences. Recognition of past mistakes and learning from them provides an elevated basis for plumbing system design and installation.

The progressive development of sanitary standards in America evolved from very primitive and rude beginnings. Intolerable health conditions and epidemics of waterborne diseases caused strong health protection measures to be adopted in highly populated metropolitan areas. Extensive disastrous fires in congested city regions led to construction of large public waterworks systems used for both fire-fighting purposes and for potable water supply to buildings.

Introduction of plumbing systems in buildings brought with it unique problems related to public health, personal hygiene, building design, plumbing materials, advanced techniques, and governmental regulations. As these problems evolved during a revolutionary period of industry, the solutions developed were intimately related to new materials, methods, standards, and standardization.

History provides a clear record of many mistakes, bad practices, shoddy materials, and insanitary installations which were made in the introduction of plumbing systems into buildings. In each case, appropriate corrections had to be made and precautions prescribed for the future.

Performance required of building plumbing systems gradually became a recognized subject, and a long list of plumbing principles was developed and published. The broad performance objective is to provide reasonable safeguards for sanitation in and adjacent to buildings to protect the public health, safety, and welfare against the hazards of inadequate or insanitary plumbing installations.

ANCIENT PLUMBING

In ancient times, plumbing and sanitation were not always primitive. Human beings elevated them to significant levels in past ages. History reveals that one of the fundamental differences between civilization and barbarism is related to the installation of piping systems for providing an adequate, pressurized supply of safe drinking water, sanitary disposal of sewage, and efficient, unobjectionable disposal of storm water. This is evidenced by the fact that those peoples who enjoyed high civilizations in the past had developed plumbing systems for protecting health.

Confirmation on this matter is provided in the reports of discoveries by archeologists while digging in various parts of the world where ancient civilizations were known to have flourished. For example, the ruins of a plumbing system estimated to be from 3000 to 6000 years old were found in excavations in the Indus River valley in India. In Egypt, sections of copper water pipe estimated to be about 5500 years old were unearthed along with palace apartments in which each bedroom apparently had been provided with a bathroom.

In the ancient empire of Babylonia, a nation centered in the general area between the Euphrates and the Tigris Rivers, the science of hydraulic engineering seems to have had its beginning. A network of canals, all skillfully planned and regulated, covered the area. Large brick drainage sewers with access holes were installed in Babylon. Greek writers told of the Hanging Gardens of Babylon; from this, it may be inferred that some means for pumping water to considerable heights had been developed at that time.

From Babylonia came the Hammurabi Code, a code of laws regulating business and custom. It is reported to have been drawn up probably by Shulgi, second king of the third Ur dynasty, in the period between 2400 and 2150 B.C. Included in this code were regulations governing the construction of buildings. This period evidently was a very formative one in the organization of society and the progress of civilization in Babylonia.

On the island of Crete, the remains of a plumbing system at least 3000 years old were unearthed in excavations on the site of an ancient palace at Knossos. Evidence was found of plumbing fixtures, a water supply system, a sanitary drainage system, and a heating system. One of the fixtures was a bathtub made of hard pottery and 5 ft (1.5 m) in length. It was a floor-standing model with an integral base, resembling in shape the cast-iron bathtub-on-base widely installed in America in the latter part of the nineteenth century. Another fixture was a water closet, also of hard pottery. It showed evidence of having been equipped with a water closet seat and a flushing device. Found intact were long sections of clay drain pipe of the bell-and-spigot type. Pipe lengths were short, and branch fittings were provided with T and Y connections adjacent to the bells or hubs.

Of all the ancient peoples, the Romans carried sanitation to the highest and broadest degree of development. From their language, Latin, have come such words as *sanitation* and *plumber,* the latter being derived from *artifex plumbarius,* meaning a worker in lead. Roman aqueducts still grace the Italian countryside and rank among the world's engineering triumphs. Extensive large underground sewer systems, public and private baths, lead and bronze water piping systems, and marble fixtures with gold and silver fittings have come to be symbolic of the civilization of Ancient Rome. An especially significant feature of progress may be cited as being the fact that much of the underground public water supply system was constructed of standardized cast lead sections.

Public bathing colonies dotted the Roman Empire. Some covered as much as a square mile. One of them, the baths of Diocletian, accommodated 3200 bathers. Baths and bathing pools were lined with ceramic glazed tile. In residences, bathtubs often occupied an entire room and were supplied with both hot and cold water. Hot water was provided by means of lead or bronze piping which conveyed water across open fires. Bathtubs often were carved from solid marble or lined with ceramic glazed tile and equipped with gold or silver fittings.

After almost a thousand years of world rule, the empire of Ancient Rome crumbled. In the fifth century, it was subjected to successive invasions by Goths and Vandals, barbaric tribes from the north of Europe. In 455, Vandals swept south through Rome, sacked it of all things of value including any metals that could be removed, and destroyed its public works. With the destruction of Rome, its civilization rapidly decayed, and sanitary standards regressed almost to the vanishing point.

The following 10 centuries have been historically termed the *Dark Ages.* For many centuries, people in general paid little attention to personal cleanliness and other domestic sanitary needs involving the use of water. Bathing was frowned upon by persons of influence and not taken seriously even by members of the ruling class, many of whom preferred to use perfume. Plumbing fixtures fell into disuse, including water closets which had been developed and widely used during the fourth and fifth centuries in Rome. They were not used again until about the twelfth century, and even then their use was extremely limited.

During the fourteenth century, Europe was ravaged by disease. Bubonic plague swept the continent and England reportedly killed 25 million people. To improve sanitary conditions in Paris in 1395, the authorities ordered a stop to the practice of throwing sewage out of building windows onto the streets below. But this was a common practice that continued unabated in other cities.

As late as the early part of the eighteenth century, European cities had not been equipped with sanitary sewage disposal facilities. The mortality rate in many cities exceeded the birth rate. When building owners were ordered to install domestic sewage vaults, considerable opposition was raised. It was not until the latter part of the eighteenth and early part of the nineteenth centuries that European cities started to provide public sewer systems beneath city streets. Slowly people began to use the convenient public sewer facilities for the disposal of sewage from buildings and to develop progressively higher sanitary standards.

EARLY AMERICAN SANITATION STANDARDS

Although America has become a symbol of high standards in plumbing and sanitation, these evolved from very primitive and rude beginnings. Along the Atlantic Coast, firmly established settlements developed local industries and conducted trade with Europe. Among the numerous early settlements were several which later became major port cities, such as Boston, New York, Philadelphia, and Baltimore. Each faced the same general sanitation problems and progressed in developing sanitary standards almost simultaneously.

Available reports of the progressive development of sanitary standards in New York may be cited as typical. Following settlement of the port area in 1626, houses were built. None had within them any water supply or sewage disposal facilities. Drinking water was used sparingly as it had to be carried from springs or wells, or purchased by the bucket from water peddlers who traveled through the streets selling water from wooden barrels on horse-drawn trucks. Outdoor earth-pit privies were used as toilet facilities. Wastes from dishwashing, clothes washing, and bathing were disposed of outdoors by dumping them onto the ground adjacent to buildings. Rainwater from roofs also was disposed of onto the ground. As the population of the settlement increased with the arrival of new immigrants, conditions deteriorated. Shallow wells became polluted by seepage from earth-pit privies, areas around homes became excessively fouled from sewage and refuse dumped onto the ground, and streets were quagmires of mud long after rainstorms ended.

Health conditions became intolerable in time and forced organization of a Common Council in 1675. The council appointed a health officer in charge of sewage and refuse disposal and other health matters. Watertight privy vaults began to be installed instead of earth-pit privies as toilet facilities. Scavenging regulations governing the disposal of privy-vault wastes were put into effect in 1676. Public wells were projected in 1677 and completed in 1686. Street gutters were installed in built-up areas in 1687, and homeowners were ordered to pave sidewalks. In 1700, a sanitary ordinance was adopted prohibiting the dumping of scavengers' barrels of vault wastes into the street gutters. In 1703, an open-ditch public sewer or sewage canal was constructed, and city surveyors were appointed to establish street and sewer grades. Complaints arose about the unsanitary conditions created by the open-ditch public sewer, and in 1717 the sewer was extended to empty into New York Bay. In 1728, the first underground sewer was laid under the streets of New York. The first water supply reservoir was constructed in 1776. It collected water from wells and ponds and distributed water through a supply system consisting of hollow wooden logs laid under principal streets.

Epidemics of waterborne diseases occurred in New York, Philadelphia, Baltimore, and other population centers along the Atlantic Coast. Public pressure developed as complaints to authorities mounted regarding the unsanitary disposal of sewage and the lack of an adequate, available supply of safe drinking water. To improve conditions, boards of health were established, Philadelphia organizing a board in 1794, and Boston in 1797.

As a health protection measure, communities began to install all public sewers underground and to extend them to buildings, although many people considered the sewers merely as a means of eliminating unsightly conditions. These early underground sewers were constructed with flat stone tops and bottoms and brick masonry sidewalls. They were intended to serve just for storm water drainage from streets and buildings. But they soon became foul and odorous from sewage and garbage dumped into street gutters. In 1831, catch-basin traps were installed in street gutters to intercept solids conveyed by storm water draining into the public sewer.

In 1830, after numerous serious fires had demonstrated the need for an adequate, available supply of water for fire fighting, New York City installed its first public waterworks. This consisted of a large above-ground water supply tank into which water was pumped from shallow wells, and from which water was supplied through two 12-in cast-iron underground water mains to fire hydrants installed along several of the main streets where business buildings were located. But this system proved to be totally inadequate when a severe fire broke out on December 16, 1835. A total of 530 buildings were destroyed overnight.

The disastrous fire of 1835 in New York City awoke the people to action and led to developments of great significance and benefit. People became aware of the necessity for having an adequate pressurized water supply system readily and constantly available for fire fighting in built-up areas. They also realized that there was great need, both as a sanitary measure and as a laborsaving convenience, for having an adequate pressurized water supply system from which safe drinking water could be piped directly to buildings. Soon after the fire, plans were projected for providing a large public water supply system which would satisfy both of these needs.

This project was completed seven years later, in 1842, at which time the original Croton Aqueduct System was placed in operation. In this system, water from the Croton River was collected in Croton Reservoir, 40 mi north of the city, and supplied therefrom through an underground piping system to two distribution reservoirs in the city, one at 42d Street and another in Central Park. From these reservoirs, water was distributed through a system of cast-iron water mains installed underground in city streets, and fire hydrants were installed on sidewalks at appropriate locations along the curb. Building owners were permitted to have water service connections made to the public main, and water service piping extended therefrom to supply faucets or hydrants in building cellars or yards. At that time, the population of the city of New York was about 300,000.

PLUMBING SYSTEMS IN BUILDINGS

The installation of pressurized water services into building cellars and yards in New York City in 1842, upon completion of the Croton Aqueduct System, marked the start of a radical change in building construction—the installation of plumbing systems in buildings. Pressurized water supply piping systems made it possible to satisfy, at the turn of a faucet, the need of building occupants for a safe and abundant supply of water for all domestic purposes and to eliminate the drudgery, labor, and inconvenience of having to carry water from the source. No plumbing fixtures had been installed within buildings prior to this time, except for a few crude sink installations reportedly made in kitchens and provided with water supply by means of adjacent hand pumps that drew water from shallow wells.

As late as 1845, records indicate that buildings were not provided with interior drainage piping systems. Most buildings were equipped with exterior leaders which conveyed storm water from roofs to pavements and sidewalks from which the water ran into street gutters. In some cases where branches had been installed from the public sewer to buildings, the exterior leaders discharged directly into such branches or building sewers. Before fixtures could be installed with water supply and drainage piping systems, building sewers had to be installed first so as to convey sewage away from the buildings to a suitable disposal terminal, such as a public sewer system. To satisfy this need in New York City in 1845, sanitary building sewers were permitted to be connected to the existing public sewer system which originally had been provided just for storm water disposal. These building sewers, and the main drains installed underground in buildings at the time, were constructed with flat stone tops and bottoms and brick masonry sidewalls.

By 1850, plumbing fixtures had been installed in a number of New York City homes. These were principally private residences owned by wealthy people who could afford to alter their buildings to accommodate such facilities. Provision had to be made to protect the fixtures and piping against frost damage by means of heating equipment, or insulation, or both. Earliest installations consisted of wooden and sheet-metal sinks in kitchens, wooden washtubs in kitchens or in cellar or basement laundry rooms, and sheet-metal bathtubs in special bathrooms or closets.

For these early installations, water supply and drainage piping were attached to building walls and either left exposed in rooms or concealed in boxwork. A handmade trap was installed in the drain of each individual fixture to prevent escape of obnoxious odors and sewer gases from fixture waste outlets. However, these traps often lost their water seals owing to siphonage and back-pressure conditions in the drainage system, and this caused fouling of the atmosphere of rooms in which fixtures were placed. Check valves and many specially designed traps were installed in efforts to prevent loss of trap seal, but such devices were found to be totally ineffective. At that time, the principle of venting fixture drains to protect trap seals was unknown.

Nevertheless, progress was made in the installation of plumbing systems in buildings. Fixtures were placed in locations where they would not be too objectionable. Sinks and washtubs were put in kitchens and basements. Lavatories and bathtubs were located on various floors and connected to separate stacks. Long hopper water closets, so named because of their funnel or long hopper shape, were installed in toilet rooms or compartments accessible only from outdoors, because it was considered hazardous to health for rooms which housed such odorous fixtures to be directly accessible from the interior of buildings. This type of water closet was installed so as to be relatively frost-proof by placing the trap and water supply valve below the floor level.

In the late 1850s, people became more and more aware of the need for improving sanitary standards in and adjacent to buildings. Widening recognition was given to the fact that plumbing systems in buildings could provide adequate safe water for drinking, cooking, bathing, and for flushing fixtures and also could safely and efficiently dispose of sewage and other wastes from buildings. Extensions were built on many homes specifically to provide bathrooms at the upper stories of existing buildings. Lavatories, bathtubs, and water closets were installed in these extension bathrooms, many of which were also provided with heating equipment. Double doors were placed in passageways between extension bathrooms and the main building in order to prevent bathroom odors and sewer gases from entering the living quarters.

Directly following the Civil War, immigration swelled the populations of industrial cities in the eastern part of the country. In many cities, rows of attached three- and four-story tenement houses were built to take care of the additional population. These buildings were provided just with yard hydrants for drinking water supply, while toilet facilities consisted of rows of privies built above watertight privy vaults located in the backyards of the buildings. Extremely objectionable, unsanitary conditions soon developed under such circumstances. Health authorities had to take stringent action to halt the spread of disease. To protect the health of building occupants, the public was alerted to the necessity of equipping buildings with adequate means for supplying safe drinking water for domestic purposes and with adquate facilities for sanitary disposal of sewage. Health authorites advocated the installation of plumbing systems in buildings, and as a result this became a subject of regulation in sanitary codes.

In the early 1870s, water-supplied kitchen sinks came into general use in private homes and other small buildings. Fireboxes of coal-fired kitchen ranges were equipped with water backs and water fronts, and circulation piping was installed between these water-heating units and hot water storage tanks so as to make pressurized hot water available in volume at fixtures. The use of outdoor privies and privy vaults for private homes was discontinued gradually as indoor water closets, directly connected to building drains, were installed in toilet rooms accessible from backyards.

A major stymie to more rapid introduction of plumbing systems in buildings was the fact that, as late as 1874, no way was known for preventing fixture trap seals from being lost because of siphonage and backpressure conditions in the drainage system. Where fixture trap seals were lost, objectionable odors and sewer gases escaped from the system at fixture outlets and fouled the atmosphere of rooms in buildings. A significant instance of this occurred when a plumbing system was installed in a large new private dwelling in New York City in 1874. Soon after occupying the building, the owner complained to the plumbing contractor that the stench of sewer gas from fixtures in the building was unbearable.

After receiving this complaint, the plumbing contractor discussed it at a conference with other New York City master and journeymen plumbers. At this conference in 1874, the theory of protecting fixture trap seals by means of vent pipes was originally proposed. The theory was that air pressure in the drain at the outlet of a fixture trap had to be in relatively exact balance with the atmospheric pressure at the inlet of the trap, and this balance could be maintained by means of a vent pipe connected to the drain at the trap outlet and extended to atmospheric pressure outdoors so that air could flow freely into or out of the drain in response to pressure variations in the drain. This theory was tested by contractors and journeymen in the field on numerous installations, and it was proved to be correct. However, numerous details of vent-piping installation and sizing had to be determined by further testing and field experience before continuous, satisfactory performance of vent piping was assured. Nevertheless, the principle of venting sanitary drainage systems by means of attendant vent pipes, to protect fixture trap seals against loss by siphonage and back pressure, was established. The way had been found to prevent objectionable odors and sewer gases from escaping at fixture waste outlets and fouling the atmosphere in buildings.

News of the development of the principle of venting sanitary drainage systems spread rapidly to all parts of the country. Detailed information on vent-piping installation, test reports, and experience with systems in service were carried in trade publications, association reports, and newspapers at the time. A major breakthrough had been achieved in knowledge of the design of plumbing systems in buildings which made it possible to locate plumbing fixtures inside without fouling the atmosphere. Objections to installing plumbing systems in buildings rapidly vanished, and plumbing installations proceeded at a greatly accelerated pace.

Within a few years, kitchen sinks were installed in each dwelling unit in tenement houses. Owners of private homes began to have kitchen sinks put in, followed soon after by laundry trays, then bathtubs, and later by lavatories placed in appropriate locations for convenient use. About 1880, the use of privies and privy vaults in the backyards of tenement houses was discontinued. In their place batteries of hopper-type water closets, directly connected to building drains, were installed in either backyards or cellars. Similarly, at schools privies and privy vaults were removed. They were replaced by installations of trough-type water closets, known as *school sinks,* directly connected to building drains. These fixtures were provided in separate schoolyard toilet buildings.

By 1881, the health protection benefits of sanitary plumbing systems in buildings were clearly recognized by health officials in cities. Prior to this time, in New York City, 90 percent of all human wastes had to be disposed of by removing such wastes from privy vaults and transporting them through buildings, along city streets to docks, and then out to sea where they were dumped. This method of sewage disposal was a severe health hazard and had to be eliminated for this reason. Sanitary plumbing systems in buildings were the answer. People in cities knew this from hard experience. They began to rely upon plumbing facilities for improved sanitary conditions, and to reduce their daily work and increase their enjoyment of living. For economy in installation, sinks and laundry trays were grouped together in kitchens; and water closets, bathtubs, and lavatories were grouped together in bathrooms. This was possible to do in cities with public water supply and sewage disposal systems. But in rural areas, having no such public systems available for building connection, homes had no plumbing facilities. The only sanitary provisions for building occupants in such areas were an outdoor earth-pit privy and a well. Portable washtubs and bathtubs were used either indoors or under an outdoor shed in most such areas.

In the 1890s, two important fixture developments, combined with newly available gas and electric public utility systems laid under city streets, aided in further expanding the use of plumbing systems in buildings. The first water closet design considered to be really sanitary was introduced about 1890 with the development of the washdown water closet. Almost simultaneously, the free-standing, white-enameled cast-iron bathtub appeared. They were hailed as important new sanitary advances, as they were reasonably priced, mass-produced fixtures which homeowners desired. Doctors and health authorities advocated the expanded use of hot water as a sanitary measure and proclaimed the health benefits of bathing. The ready availability of public utility gas supply systems, which had been newly laid under city streets, aided in expanding the use of hot water supply systems in buildings and the installation of gas-fired water heaters. The availability of public utilty systems for supplying electricity for light and power in buildings made possible the installation of efficient electric pumps for pumping water to plumbing fixtures at any height. It was at this time that skyscraper-type office buildings were first erected in New York City, Chicago, Philadelphia, and other major cities. These buildings were equipped with plumbing systems that performed satisfactorily and unobjectionably, and suitable kinds and numbers of fixtures were provided in convenient locations for building occupants.

At the start of the twentieth century, laws had already been enacted in many areas of the country requiring the installation of plumbing systems in buildings and the provision of suitable kinds and numbers of fixtures in convenient locations for the use of building occupants. In general, such areas were large municipalities where public water supply and public sewer systems were available for building connections. In areas beyond the limits of public systems, it was deemed unreasonable to require installations of plumbing systems and fixtures. Nevertheless, people desired sanitary plumbing facilities and sought to equip their buildings with appropriate systems. Hot water supply was especially desired as manufacturers publicized their new developments in water heater equipment. Coal- and gas-fired sidearm water heaters appeared on the scene. Automatic controls were developed to eliminate the dangers associated with manual operation of water heaters, and range-boiler manufacturers introduced tanks made of several different kinds of materials with greater durability.

Many new tenements were erected in large industrial cities to house the swelling populations. These buildings had sinks and laundry trays in each dwelling unit, but water closets were provided in toilet compartments accessible from the public hallways on each floor. In many cases, more than one family used the same toilet facilities. It soon was apparent that such arrangements were inadequate and objectionable and fostered unsanitary conditions. Health authorities put new regulations into effect requiring that water closets be installed in toilet rooms or bathrooms in each dwelling unit, and strenuous efforts were made to bring existing building facilities up to the revised standards.

Following World War I and continuing through the early 1920s, the large industrial cities expanded tremendously. New housing developments were built on the fringes of cities, and public water supply, sewer, and utility systems were extended to serve the new buildings. All these were equipped with the most modern plumbing systems and fixtures of the day. Complete bathroom installations, consisting of a water closet, lavatory, and bathtub with an overhead shower were provided in each dwelling unit along with modern kitchen sinks and laundry trays. The growing importance of sanitary plumbing systems in buildings was shown by large-scale plumbing installations in hotels, office buildings, factories, food processing plants, and dairy buildings. Most buildings were provided with more plumbing equipment than was required by law. Multistory residential buildings in great numbers were erected in the central parts of cities where land values were relatively high. They too were fully equipped with complete bathroom, kitchen, and laundry fixtures of modern, sanitary design. Many were equipped with colored plumbing fixtures which were introduced in the middle 1920s. But this tremendous new building construction wave reached its peak in 1929 and came to a sudden halt in 1930 when the severe business depression occurred.

During the 1930s, relatively few new buildings were erected until the latter part of the decade. This period was devoted principally to the correction and modernization of plumbing systems and equipment in existing buildings. Many buildings with inadequate plumbing facilities were improved by the installation of additional, new plumbing fixtures and the replacement of old, obsolete types. Important corrections were made in the potable water supply systems of buildings to eliminate all water supply piping connections and fixture supply connections which were recognized as potential sources of contamination. This drive for correction of systems was led by health, water supply, and building officals so as to avoid a repetition of the amoebic dysentery epidemic which occurred in the city of Chicago during its World's Fair in 1933. Other important improvements were made in the hot water supply systems in existing buildings. Many were equipped with modern, automatically controlled hot water heaters designed for use with gas, oil, or electricity as the source of heat. During this period, the public utility systems around the country extended their electric supply lines into a great portion of the rural area. This provided a source of power for pumping water from wells and for supplying plumbing systems with all the water needed to maintain the same sanitary standards that were enjoyed in the cities. Private sewage disposal systems were provided by means of underground septic tank and leaching field installations in appropriate locations. In this way, modern sanitary plumbing systems and fixtures became available even in remote regions of the country.

In the latter 1940s, following World War II, and continuing through the 1950s, 1960s, and into the 1970s, there was a tremendous expansion of housing developments and industrial plant construction outside the central areas of cities in the United States. New buildings were erected along new principal highways, and public water, sewer, gas, and electric systems were provided for building service needs in most areas. Private systems were utilized in many areas where public systems were not available. All such buildings were equipped with modern plumbing systems conforming to sanitary standards elevated to a higher level than ever previously enjoyed by people. In the central areas of cities, many old buildings were removed, and in their places large skyscraper office buildings and residential buildings were erected. They too were equipped with modern plumbing systems designed in accordance with the highest sanitary standards in history in order to serve the greatest occupancy loads of all time.

Tower building construction accelerated in the late 1950s and early 1960s, and necessitated changes in design to meet changing conditions. Increased building heights and increased water usage, including water for air conditioning, required water supply tanks so large that they caused significant space problems and were uneconomical. To meet the changing conditions, design was changed to provide tankless, automatic constant-pressure booster-pump systems which required a minimum of valuable building space and which also provided a sealed-in supply of potable water from the source of supply to the plumbing fixture outlet.

In 1966, a critical shortage of copper occurred in the United States because of stoppage of shipments from foreign sources of supply. Inventories of copper DWV tube and fittings were rapidly exhausted. Large developments of single family residences were halted for most of 1966 because of the unavailability of copper DWV piping which had originally been planned to be installed. This urgent need was soon filled by nonmetallic, plastic DWV pipe and fittings, which were then introduced into use for building plumbing systems under carefully prescribed installation conditions.

A most significant change in the design of buildings used by the public began in 1961. The object of the change was to make all buildings and facilities, including plumbing, used by the public accessible to, and functional for, the physically handicapped, to, through, and within their doors, without loss of function, space, or facility where the general public is concerned. The changes were set forth in the American National Standards Institute standard, Specifications for Making Buildings and Facilities Accessible to and Usable by Physically Handicapped People, originally issued as A117.1-1961 (Reaffirmed 1971) and recently superseded by A117.1-1980. By 1971, governmental regulations were enacted mandating the necessary changes including many related to plumbing systems in buildings.

Since 1974, when the supply of foreign oil to the United States was interrupted and oil prices rose sharply, ways to conserve energy have been a constant concern. Some important conservation measures relate to the design of plumbing systems. Elimination of water waste, limitation of water use to a reasonable minimum, limitation of hot water supply temperature and rate of flow from hot water faucets, insulation of hot water heaters, tanks, and piping, and use of heat reclaiming systems and solar heating systems are some of the conservation measures to be applied in the design of plumbing systems for buildings.

FIXTURE DEVELOPMENT

The modern sink, laundry tray, lavatory, bathtub, water closet, and other fixtures did not evolve overnight. Their development extended over a period exceeding 130 years. Even after appropriate fixture designs were achieved, public demand for them had to be stimulated. Expressions such as "Cleanliness is next to Godliness," were adopted and popularized to induce people to practice sanitary habits. Fixtures were sold more on the basis of the comfort, convenience, and privacy they afforded to users than on health protection benefits.

Portable fixtures were used at first. In bedrooms, a wooden washstand and toilet set were provided. The top of the washstand was usually covered with a marble slab on which were placed a glazed pottery washbasin and large water pitcher. Other items included a glazed pottery slop jar and a chamber pot, which were generally concealed in a compartment in the lower part of the stand. Towels were hung from bars attached to the sides of the stand. Portable wooden washtubs and wooden and sheet-metal bathtubs were commonly used. One of the early sheet-metal tubs used in France was shaped like a slipper. The bather sat upon a seat in the "heel" and extended his feet into the "toe" of the tub. Beneath the heel was a grill upon which charcoal was burned to heat water in the tub.

For all such facilities, water had to be carried to them, and wastes and sewage had to be carried away to an appropriate place for disposal. The labor and inconvenience involved thereby were factors which influenced many people against adopting and practicing sanitary habits. It gradually became evident that in order to raise sanitary standards and protect health, it was necessary to provide pressurized water supply piping systems to convey an adequate, safe supply of water directly to fixtures and to provide sanitary drainage piping systems to convey sewage directly from fixtures to an unobjectionable terminal for disposal.

In the 1840s, pressurized water supply systems and sanitary drainage systems were first introduced into buildings in the United States. Thereafter, plumbing fixtures began to be installed with connections to such systems, and the development of plumbing fixtures proceeded at a rapid pace to satisfy a constantly increasing demand. The first fixtures to be installed in buildings reportedly were kitchen sinks and water closets. Shortly afterward, washtubs, bathtubs, and lavatories were installed.

Early washtubs were made of plain, bare wood while sinks and bathtubs were merely wooden frames or wooden boxes lined with sheet metal. Sheet lead was used at first for linings, then sheet zinc, and later sheet copper. Showers were provided above some of the early bathtubs. Although these fixtures were usable, they had many objectionable features. Linings were easily dented and damaged; they developed leaks at seams and at waste and overflow connections and became unsightly owing to corrosion. To improve the appearance of bathtubs, it was common practice to paint the exposed surface of zinc and copper linings with white or cream-colored paint and to repaint them again and again as the paint chipped or peeled off.

More durable sinks of black cast iron were developed. They soon gained popularity over sheet-metal-lined wooden-frame sinks. The cast-iron sink was of the rectangular flat-rim type and was installed on a supporting wooden frame or was placed against the wall with the rear rim resting on a wall cleat while the front was supported from the floor by means of two cast-iron legs inserted into slots on the front of the sink. Often a cast-iron splash back was fastened to the wall above the rear rim of the sink. A single hole was provided in the center of the splash back so as to permit a faucet to be connected there and firmly fastened above the sink. Later models of splash backs were provided with two holes to permit installation of both hot and cold water faucets.

Early lavatory installations evolved from the old wooden washstand and pottery toilet set. A glazed pottery washbasin, either round or oval in shape, was installed beneath a large opening in a marble slab and attached to the underside of the slab by means of bolts. Holes were drilled through the slab to permit long-shank faucets to be attached thereto with spouts projecting above the basin. An opening in the bottom of the basin was equipped with a waste plug outlet fitting for direct connection to the drainage piping system. The joint between the top of the basin and underside of the marble slab was sealed by means of plaster of paris. Usually the marble slab and washbasin assembly was installed against a wall and supported by wooden framework. Drip trays often were installed on the floor beneath washbasins because of the incidence of leakage development at the plaster joint between basin and slab.

One early improvement in washbasin design was the provision of an overflow fitting on the side of the bowl so that an overflow pipe could be installed between the overflow fitting and the drain just below the waste-outlet plug. This was followed by another improvement, an integral overflow built into one side of the glazed pottery washbasin.

The first of the early water closets was known as the *valve closet.* It was developed by Joseph Bramah, an English inventor, about 1788. As the original model did not work too well, it was improved later by the addition of a flushing rim. The deep bowl of this closet was flushed and refilled with water by means of a valve controlled by an air cylinder adjusted in accordance with the water pressure provided by a Bramah pump. This type of water closet was used for many years in toilet compartments of railroad cars.

About 1833, the *pan closet* appeared on the scene, after having originated in England. This type of water closet soon gained preference, as it was much cheaper than the Bramah valve closet, and continued to be in common use for more than 40 years. The pan closet consisted of a deep lead bowl with a hinged copper pan that held water in the bowl to form a water seal. The hinged pan was dumped by means of a hand crank. The bowl was flushed by manual operation of a valve in the waterline, supplied directly from an elevated water-storage tank which often was located in an attic. Elevated flush tanks, installed just about 5 ft (1.5 m) above the fixture, were later used to flush pan water closets.

Around 1850, long hopper water closet bowls came into common use. They were made of glazed pottery and shaped like a long funnel or hopper, after which they were named. This type of water closet was installed so as to be relatively frost-proof. It discharged into a trap located below the floor and was flushed by means of a valve in the water supply piping which was directly connected to the bowl of the fixture. Exposed water supply piping was covered with insulation, and the water supply valve was located below the floor. This valve was operated by a rod connected to the underside of the water closet seat, so that the bowl was flushed continuously throughout the period a user sat on the seat.

Almost coincidentally, short hopper water closet bowls were produced for installations where frost protection was not a problem. These bowls were of glazed pottery and shaped like a short hopper. This type of water closet was designed to be installed on, and attached to, the top flange of a cast-iron P trap equipped with a floor standard. The joint between the bowl and the trap flange was made with putty and secured by means of clamps. The bowl was flushed by water from an elevated flush tank.

About 1870, the *plunger closet* was introduced and gained popularity. It too originated in England. For about 20 years, it was widely installed in buildings where it was not subject to frost conditions. But it required frequent maintenance and repairs to keep it functioning properly.

In the 1870s bathing became much more popular. This was partly due to the fact that in 1872 the ancient arts of founding and enameling were united in the production of the first enameled cast-iron bathtub which featured durable, smooth white-enameled surfaces. Two years later, mass production of such bathtubs was started by a New York manufacturer. This was the beginning of modern enameled cast-iron plumbing fixtures.

Soon thereafter, solid procelain bathtubs were imported from England. They had smooth white hard-glazed surfaces which made them easy to maintain in sanitary condition. However, they were prone to crazing and were heavier and more expensive than enameled cast-iron bathtubs. The popularity of porcelain tubs was relatively limited, the enameled tubs being both lower-priced and reasonably durable.

Two-compartment and three-compartment washtubs, made of mill-cut soapstone slabs, were marketed. At first, the installer had to assemble the fixture at the building site, install it on standards, and seal the joints between slab sections with a paste mixture of litharge and glycerin.

Completely assembled soapstone washtubs were later manufactured to meet the competition of solid porcelain and solid concrete washtubs.

About 1880, the first all earthenware water closet, known as the *washout closet,* was developed in England. An integral trap was built into its design, and it had provision for attaching a toilet seat directly to the top of the bowl. This latter feature eliminated any need for installing framework and legs to support a toilet seat above the bowl, as was the case with all the earlier designs. Since this water closet was made completely of earthenware, it was easier to maintain in sanitary condition. In addition, it had better flushing characteristics than any of the earlier water closets. It was flushed by means of an elevated flush tank located on the wall about 5 ft (1.5 m) above the fixture.

Up until 1880, the design of plumbing fixtures originated principally in England. But, thereafter, developments in plumbing fixture design proceeded independently and at an accelerated pace in the United States. Much of this may be attributed to the completion of new railroads which opened up the western part of the continent, the formation of large industrial corporations to exploit natural resources of the undeveloped areas, the continuous increase in population due to waves of immigration, and the tremendous demand for new homes and buildings to house the swelling numbers in industrial centers all over the country.

By 1890, all earlier designs of water closets were made obsolete and relegated to the category of unsanitary fixtures with the development of the washdown water closet, which originated in America. This was an all earthenware water closet having an integral S trap and provision for attaching a seat directly to the top of the bowl, features similar to the washout closet. But the washdown water closet showed such design advances as siphonic action, greater depth of water in the bowl, greater water coverage of interior bowl surfaces, elimination of unventilated spaces, and complete scouring of all interior bowl surfaces with each flushing. These advances prevented progressive fouling of interior fixture surfaces and odorous conditions after extended service. When first introduced, the washdown water closet was flushed by water from an elevated flush tank located on the wall about 5 ft (1.5 m) above the fixture. This flush tank was designed to hold 8 gal (30 L) of water and was equipped with a siphon-type flush valve which siphoned from the tank at least 6 gal (22.7 L) of water at each flushing. Several years later, flushing was also accomplished by means of a Flushometer, the first automatic flush valve introduced on the market.

In the 1890s, the free-standing, white-enameled cast-iron bathtub on legs enjoyed great popularity as a replacement for earlier models made of sheet metal and wooden framework. The free-standing bathtub was much more sanitary and durable. However, it was difficult to clean under the fixture and between the fixture and the adjacent wall. As a result, many free-standing bathtubs were later provided with cast-iron bases, rather than short cast-iron legs, in order to keep the floor under the bathtub clean.

By 1900, American pottery manufacturers had developed glazed vitreous chinaware with smooth, impervious surfaces. This material was so well suited to plumbing fixtures that it became a standard for water closet bowls and was preferred by many individuals for various other fixtures. In view of this, many plumbing-fixture manufacturing firms combined so as to unite the arts of pottery, founding, and enameling under single firms.

Bathtubs, sinks, washtrays, and lavatories made of glazed terra cotta were manufactured and proved popular for a time. These sinks, washtrays, and lavatories were relatively heavy and had to be provided with sturdy legs or bases for support. The bathtubs were designed to be built into wall and floor construction, a feature which was hailed as an advance of considerable merit from a sanitary view. The popularity of these terra cotta fixtures gradually diminished with the development of similar designs in enameled cast iron, which were lighter and more economical. However, fixtures made of glazed vitreous china continued to compete in terms of weight and cost with those made of enameled cast iron.

The beginning of the twentieth century saw the development of the water closet as we know it today. Water closets designed for wall-hung installation appeared about 1905. In 1915, manufacturers introduced the lowdown flush tank and water closet as a combination unit, consisting of a floor-outlet type washdown water closet and a porcelain flush tank designed for installation on the wall just above the top of the water closet. From 1916 to 1920, other advances in water closet design included a reverse-trap model and the use of siphon jets for stronger siphonic action and a reduction of noise in operation.

During the 1920s, improved design features appeared, such as priming jets in washdown-type and reverse trap water closets, increased surface area of water in closet bowls, and further reduction of noise in operation. One-piece water closet bowl and flush tank units were introduced in the 1930s. They provided relatively silent operation owing to a number of design improvements, including the use of a quiet-action ball cock in the flush tank compartment and a bowl design which produced rotary or vortex movement of water in the closet bowl sufficient to afford adequate scouring action and complete siphonage of the contents of the bowl.

Improvements have continued to be made in the design of water closets. The most recent has been the introduction in 1960 of a wall-hung water closet and lowdown flush tank combination of simplified design, weighing less than the other wall-hung models and provided with a lightweight concealed metal fixture carrier of simplified design by which it can be attached to structural elements of walls.

Great improvements have also been made in the design of sinks and laundry trays. Prior to 1900, the flat-rim and roll-rim sinks were equipped with separate splash backs on which separate faucets were mounted. But, about that time, the need for improved sanitation in kitchens of dwelling units resulted in a trend away from the use of flat-rim sinks, wooden enclosures beneath sinks, and separate splash backs. One-piece roll-rim and apron-type sinks with integral splash backs appeared about 1910, and faucets were mounted on the vertical back wall of sinks. These one-piece sinks were designed for installation on walls by means of metal brackets securely attached to the structural elements of the wall and were not provided with legs for support from the floor. Larger one-piece roll-rim and apron-type sink and drainboard combination fixtures were introduced about 1920. They had integral splash backs and were designed to be installed on walls with metal brackets. But owing to the greater weight and size of the combination, they also had to be supported from the floor by means of two legs set beneath the front rim of the fixture.

One-piece sink and washtray combination fixtures and two-compartment sink combination fixtures were introduced about 1930. They were designed with roll rims or aprons and integral splash backs on which combination faucets were mounted. Because of their weight and size, these fixtures were not only securely attached to the wall, but also supported from the floor by means of two legs set beneath the front rim of the fixture or by two pedestals, one beneath the sink compartment and the other beneath the washtray compartment.

In 1940, a short integral splash back and ledge was introduced into the design of sink and washtray combinations and two-compartment sink combinations. A combination faucet was specially designed for installation on the ledge of the fixture, which was attached to the wall by means of metal brackets and additionally supported by a strong metallic cabinet with an adjustable base set beneath the fixture rim.

Soon thereafter, one-piece flat-rim style sink and washtray combinations and two-compartment sink combination fixtures for counter-top installation were introduced. These flat-rim fixtures were installed in waterproof counter tops by means of a metal frame with a watertight seal. The counter top was provided with a short splash back and installed on top of a floor-standing cabinet securely attached to the wall by brackets. A deck-type combination faucet was specially designed for installation on the counter top adjacent to the fixture.

These improvements in the design of sinks and laundry trays were made principally in enameled cast-iron fixtures. However, they were also included in enameled pressed steel fixtures following their introduction in the late 1920s. In 1950, one-piece flat-rim counter-top sinks made of stainless steel were introduced. Thus, improved sanitary design has now been built into sinks and laundry trays in several different kinds of durable materials.

In the early 1950s, the design of kitchens was changed so as to provide extended counter-top space, with cabinets above and below. This change was utilized to permit the under-counter installation of two household plumbing appliances, the domestic dishwashing machine, and the domestic automatic laundry washing machine. In large multistory residential buildings, automatic laundry washing machines were installed in general laundry rooms on each floor or at basement or cellar levels.

In 1952, plastic bathtubs, plastic shower stalls, and plastic wall enclosures for above bathtubs and shower receptors were introduced by fixture manufacturers, and were utilized in many large, new multistory residential buildings.

STANDARDS FOR PLUMBING MATERIALS

Plumbing systems in buildings are designed and constructed using the materials currently available in our highly industrialized society. Each system is composed of many different individual parts, pipes, fittings, valves, fixtures, and numerous other items, which are assembled to function and provide the performance required to satisfy the needs of building occupants and to protect the health, safety, and welfare of the people. To satisfy these needs economically, most of the many parts of plumbing systems are mass-produced by industry.

Since the start of the industrial revolution in England in the mid-1700s, the great change from an agricultural and handicraft economy to a modern industrial economy has progressed hand in hand with new discoveries and inventions and the exploitation of new sources of power and raw materials. Their utilization through expanding mass production of goods to satisfy public demand has wrought untold benefits to the people and has revolutionized their way of life.

Standards are at the base of all mass production. Before starting to mass-produce any given item, industry must first establish a standard for it, incorporating every feature necessary to satisfy the public demand or need for it, and then proceed to make it at a cost that will satisfy the consumer. Consequently, standards are everybody's business in the broad sense.

The earliest standards for plumbing materials were those devised

Table 1-1
STANDARDS FOR PLUMBING MATERIALS*

Materials	Standards	Source†
Plumbing fixtures and fittings:		
Vitreous china	A112.19.2-1973	ANSI
Enameled cast iron	A112.19.1-1979	ANSI
Stainless steel, residential use	A112.19.3-1976	ANSI
Porcelain enameled formed steel	A112.19.4-1977	ANSI
Gel-coated glass-fiber reinforced polyester resin bathtub units	Z124.1-1974	ANSI
Gel-coated glass-fiber reinforced polyester resin shower receptor and shower stall units	Z124.2-1974	ANSI
Laundry equipment, household	A197.2-1973	ANSI
Dishwashers, household	A197.1-1973	ANSI
Dishwashers, commercial	A197.3-1973	ANSI
Drinking fountains and drinking water coolers, self-contained, mechanically refrigerated	A112.11.3-1973	ANSI
Floor drains	A112.21.1-1974	ANSI
Finished and rough brass plumbing fixture fittings	A112.18.1M-1979	ANSI
Shower head, ball joint (integral volume control)	WW-S-001913-1975	FS
Supports for off-the-floor plumbing fixtures for public use	A112.6.1-1978	ANSI
Ferrous pipe and fittings:		
Cast-iron soil pipe and fittings, extra heavy and service weight	A112.5.1-1973	ANSI
Cast-iron threaded drainage fittings	B16.12-1971	ANSI
Hubless cast-iron sanitary system pipe and fittings	301-75	CISPI
Hubless stainless-steel couplings	310-78	CISPI
Cast-iron water pipe, cast-in-metal molds	A21.6-1975	ANSI
Cast-iron water pipe, cast-in-sand lined molds	A21.8-1975	ANSI
Cast-iron water pipe (2″)	A21.12-1971	ANSI
Cast-iron water pipe fittings	A21.10-1977	ANSI
Steel pipe, seamless and welded, zinc coated	A120-77	ASTM
Malleable iron fittings, threaded, 150 lb	B16.3-1971	ANSI
Pipe fittings, threaded (bushings, locknuts, and plugs)	B16.14-1971	ANSI
Roof drains	A112.21.2-1971	ANSI

Table 1-1 (*Continued*)

Materials	Standards	Source†
Ferrous pipe and fittings (*continued*)		
Valves, backwater	A112.14.1-1975	ANSI
Valves, gate, cast iron, threaded and flanged, 125 and 250 lb	WW-V-58b-1971	FS
Nipples, pipe, threaded	WW-N-351B(1)1970	FS
Unions, pipe, steel or malleable iron	WW-U-531D-1973	FS
Nonferrous pipe and fittings:		
Brass pipe	B43-76	ASTM
Brass tube	B135-74	ASTM
Copper pipe, standard pipe size	B42-78	ASTM
Copper pipe, threadless	H26.2-76	ANSI
Copper water tube, types K, L, and M	B88-78	ASTM
Copper drainage tube, type DWV	B306-78	ASTM
Cast-bronze screwed fittings, 125 and 250 lb	B16.15-78	ANSI
Cast-bronze solder joint pressure fittings	B16.18-78	ANSI
Cast-bronze solder joint drainage fittings	B16.23-76	ANSI
Brass or bronze flanges and flanged fittings, 150 and 300 lb	B16.24-71	ANSI
Cast-bronze fittings for flared copper tube	B16.26-75	ANSI
Wrought-copper and bronze solder joint pressure fittings	B16.22-73	ANSI
Wrought-copper and wrought-copper-alloy solder joint drainage fittings	B16.29-73	ANSI
Lead pipe, bends and traps	WW-P-325A-1967	FS
Unions, pipe; brass or bronze, 250 lb	WW-U-516A-1967	FS
Valves, ball	WW-V-35B-1973	FS
Valves, bronze, gate	WW-V-54D-1974	FS
Valves, bronze; angle, check and globe; screwed flanges, solder; 125, 150, and 200 lb	WW-V-51E-1974	FS
Valves, water pressure reducing	A112.26.2-1975	ANSI
Nonmetallic pipe and fittings:		
Asbestos-cement pipe, nonpressure (sewer)	C428-78	ASTM
Asbestos-cement pipe, pressure (water)	C400-77	AWWA
Asbestos-cement pipe, perforated	C508-76	ASTM
Bituminized-fiber pipe, homogeneous (sewer)	D1861-73	ASTM
Bituminized-fiber pipe, homogeneous, perforated	D2312-73	ASTM

Table 1-1 (*Continued*)

Materials	Standards	Source†
Nonmetallic pipe and fittings (*continued*)		
Bituminized-fiber pipe, laminated wall (sewer)	A176.2-72	ANSI
Bituminized-fiber pipe, laminated wall, perforated	A176.5-71	ANSI
Clay pipe, perforated, standard and extra strength	A106.8-78	ANSI
Compression joints for vitrified clay bell and spigot pipe	C425-77	ASTM
Concrete pipe, sewer, nonreinforced	C14-78	ASTM
Acrylonitrile-butadiene-styrene (ABS) plastic pipe, schedules 40 and 80	D1527-77	ASTM
Acrylonitrile-butadiene-styrene (ABS) plastic pipe fittings, socket type, schedule 40	D2468-76	ASTM
Acrylonitrile-butadiene-styrene (ABS) plastic drain, waste and vent pipe and fittings	D2661-78	ASTM
Solvent cement for acrylonitrile-butadiene-styrene (ABS) plastic pipe and fittings	D2235-76a	ASTM
Polyethylene (PE) plastic pipe, schedule 40	D2104-74	ASTM
Plastic insert fittings for polyethylene (PE) plastic pipe	D2609-74	ASTM
Polyvinyl chloride (PVC) plastic pipe, schedules 40, 80, and 120	D1785-76	ASTM
Polyvinyl chloride (PVC) plastic pipe fittings, socket type, schedule 40	D2466-78	ASTM
Polyvinyl chloride (PVC) plastic drain, waste and vent pipe and fittings	D2665-78	ASTM
Solvent cement for polyvinyl chloride (PVC) plastic pipe and fittings	D2564-78a	ASTM
Backflow prevention devices:		
Air gaps in plumbing systems	A112.1.2-1979	ANSI
Vacuum breakers, antisiphon	A112.1.1-1971	ANSI
Vacuum breakers, hose connection	A112.1.3-1976	ANSI
Vacuum breakers, pressure type	A112.1.7-1976	ANSI
Double check with atmospheric vent	1012-1978	ASSE
Reduced pressure principle back pressure, backflow preventer	1013-1971	ASSE
Double check valve back pressure, backflow assembly	1015-1978	ASSE

Table 1-1 (*Continued*)

Materials	Standards	Source†
Miscellaneous materials:		
Cleanouts, metallic	A112.36.2	ANSI
Calking lead, type 1	QQ-C-40(2)-1970	FS
Cement lining	A21.4-1974	ANSI
Coal-tar enamel, protective coatings for steel water pipe	C203-78	AWWA
Fixture setting compound	TT-P-001536-1968	FS
Grease interceptors	G101	PDI
Hose clamps	WW-C-440B(2)1973	FS
Hydrants for utility and maintenance use	A112.21.3-1976	ANSI
Pipe hangers and supports	WW-H-171D-1970	FS
Relief valves, pressure and temperature, and automatic gas shutoff devices for hot water supply systems	Z21.22-1979	ANSI
Rubber gaskets for asbestos-cement pipe	D1869-78	ASTM
Rubber gaskets for cast-iron soil pipe and fittings	C564-76	ASTM
Rubber gaskets for concrete sewer pipe	C443-78	ASTM
Water hammer arresters	A112.26.1-1975	ANSI
Water heaters, automatic storage type	Z21.10.1-1975	ANSI
Water heaters, electric, storage tank	WW-H-196H-1971	FS
Water heaters, instantaneous	WW-H-191B-1970	FS
Water meters, cold, displacement type	C700-71	AWWA
Water meters, cold, current type	C701-70	AWWA
Water meters, cold, compound type	C702-70	AWWA
Sheet copper	B152-76	ASTM
Sheet lead, grade A	QQ-L-201F(2)1970	FS
Soft solder	QQ-S-571E-1972	FS

* Standards listed in this table are the latest available at publication of this book. As standards are revised and updated regularly, reference to standards in contract specifications should be made to the latest edition in each case.

† Abbreviations used in this table to indicate the source of each particular standard refer to the following issuing organizations:

ANSI American National Standards Institute, 1430 Broadway, New York, NY 10018.

ASSE American Society of Sanitary Engineering, 960 Illuminating Building, Cleveland, OH 44113.

ASTM American Society for Testing and Materials, 1916 Race Street, Philadelphia, PA 19103.

AWWA American Water Works Association, 6666 W. Quincy Avenue, Denver, CO 80235.

CISPI Cast Iron Soil Pipe Institute, 2029 K Street, N.W., Washington, DC 20405.

PDI Plumbing and Drainage Institute, 5342 Boulevard Pl., Indianapolis, IN 46208.

FS Federal Supply Service, Standards Division, General Services Administration (standards are obtainable from the Superintendent of Documents, Government Printing Office, Washington, DC 20402).

by individual manufacturers for their products in England in the early 1800s. Similar standards were applied in the United States by manufacturers in the production of pipes, fittings, and fixtures, which reportedly began about 1842. Prior to this time, plumbers had to make their own fixtures, traps, and fittings. The trend toward manufactured piping and fixtures gained momentum coincident with the installation of public water supply systems in large cities. Clay, cast iron, and lead were the earliest kinds of materials manufactured for plumbing systems, followed shortly thereafter by wrought iron, brass, and copper in the 1850s. Cast or factory-made traps became available about 1871.

The development of standards for plumbing materials may be classified generally into four stages: (1) by individual companies; (2) by industrial associations, technical societies, and government bureaus; (3) on a national scale; and (4) on an international scale. From one stage to the next, the importance of, the difficulties involved in, and the number of organizations interested in a given standard increase greatly. Usually, two or more stages develop simultaneously as the result of a significant and apparent need to resolve many problems which affect numerous diverse industries and require a common solution. In this way, standards are vitally necessary to establish the best way known for producing any given item. Generally accepted standards facilitate the integrating processes necessary for large-scale production and distribution and for satisfying the demand of the ultimate consumer or user.

National standards for plumbing materials began to appear during and shortly after World War I. In the 1920s, such standards were developed rapidly to embrace most of the range of plumbing materials. Since then, these standards have been changed in accordance with evident needs, and new standards have been developed coincident with the introduction of new materials and new methods and with changing conditions. Since World War II, the use of new materials, methods, and techniques for plumbing systems has resulted in a broad advance in the development of new standards and the updating of old standards to meet current needs. Standards for the principal kinds of plumbing materials currently used in building construction are listed in Table 1-1.

PLUMBING SYSTEM REGULATIONS AND STANDARDIZATION

Every state has police power to protect the health, safety, and welfare of its people. Many states exercise such power directly through state agencies, while some states delegate specific powers, duties, and responsibilities to municipalities established under state law. Regulations to protect drinking water supplies against the hazards of pollution and contamination and to provide for safe, sanitary disposal of sewage are necessary to protect health. The design, installation, and maintenance of plumbing systems are subjects within the category of regulations necessary to protect health.

When plumbing fixtures were first introduced into buildings in America, about 1842, no plumbing regulations existed other than those dealing with the maintenance of privy vaults. But as the populations of the industrial cities swelled in size and people were crowded into new tenement houses with very meager unsanitary facilities located in the backyards in close proximity to the buildings, health conditions deteriorated and epidemics occurred. The situation became so bad that in 1866, after urgent appeals were made, the legislature of the State of New York granted power to the Metropolitan Board of Health of New York City to control and regulate plumbing in the city. Soon thereafter, the New York City Sanitary Code included plumbing regulations.

These regulations required that care be taken to prevent pollution of the public water supply and that adequate drains and soil pipes be provided in buildings. Fixtures connected to sewers had to have approved means, i.e., fixture traps with water seals, for preventing gases or odors from passing out through the fixtures into rooms. Ventilation and means of cleansing water closets and privies were required, ventilation in this case meaning a local ventilation pipe connected beneath the seat of each water closet bowl or privy vault. It was prohibited for the contents of privy vaults and cesspools to run off into the ground or street. Toilets were required to be kept clean, and no offensive gases or odors were to be permitted to pass into the house or any other house or building. So far as is known, these were the earliest regulations dealing with plumbing systems in buildings. They were prepared by health authorities after numerous conferences with plumbing contractors.

In 1875, soon after the theory of venting was originally proposed, tested, and proved to be correct, the New York City Board of Health declared unvented fixture traps to be useless and unsafe and required all fixture traps to be protected by means of adequate vent pipes. In addition, all soil and waste stacks were to be extended to the atmosphere above building roofs. In 1877, soil and waste stack terminals had to be located at least 2 ft (0.6 m) above building roofs; privy vaults were required to have an 8-in (200-mm) ventilation pipe extending above the roof of the building; and owners, tenants, and lessees were made liable for violations. In 1880, after conducting many tests on venting, using glass pipe and fixture traps, minimum sizes of vent pipes for various sizes of traps were established by the Board of Health. In 1881, the installation of water closets in rooms used for living or sleeping purposes was prohibited.

The enactment of plumbing system regulations in other densely populated cities of the country proceeded almost simultaneously with those put into effect in New York City, as cited in the foregoing discussion. All major cities adopted plumbing system regulations as part of their sanitary codes, but Washington, DC, in the 1870s, put its plumbing regulations into a separate code, which became known as the first plumbing code in the nation.

Many cities followed the example of Washington, DC, and established separate plumbing codes. States began to authorize the establishment of examining boards of plumbers in cities and to empower such boards to write plumbing regulations in cooperation with local health boards. As plumbing practices in different municipalities varied considerably, their code requirements soon reflected these variations, although they were usually of minor importance.

A most significant regulation was put into effect in New York City by the Board of Health in 1883. All water supply connections made to fixtures below rim level were ordered removed; their future use was discontinued, and it was strictly prohibited to make them. This regulation was the result of an investigation and series of tests conducted by the board after receiving a report from both master plumber and journeymen plumber associations that many existing water supply connections to plumbing fixtures were potential sources of contamination to the potable water supply system, as water could be drained or siphoned back into the system through supply connections made to fixtures below rim level. One of the most hazardous of these connections was the direct water supply valve for flushing hopper-type water closets. Flush tanks equipped with bottom supplied, submerged ball cocks and bathtubs equipped with either bottom or sidewall bell supply inlets were among the fixtures from which water could drain back into the potable water supply system.

However, this regulation was not strictly enforced once the power over plumbing standards was transferred from the Board of Health in 1890 to an Examining Board of Plumbers and a Department of Buildings. These agencies did not have the same strict means of enforcement as did the Board of Health. This was of special importance with regard to fixtures which had been originally approved in existing buildings. Health boards had power to order correction of unsanitary conditions and health hazards regardless of prior approvals. The attitude of building department officials became relaxed on the subject of this regulation. In 1896, they permitted installation of a patented flush valve, known as a Flushometer, for flushing water closets and urinals by means of a direct connection to the potable water supply system. At first, these valves were permitted to be installed only when supplied from a separate water supply system for water closets and urinals. But this was later relaxed further to permit flush valves to be supplied simply from separate risers for water closets and urinals.

By 1913, the plumbing codes of cities throughout the nation had become comprehensive documents which specified how almost every detail of plumbing systems in buildings was to be installed; what fixtures had to be installed for the use of building occupants; minimum permissible sizes for drains and vents based upon established methods of determining loads on such piping; minimum permissible sizes for water supply piping; the various types and kinds of materials which had been approved for various uses; and administrative procedures which had to be followed in securing permits to do plumbing work, inspection and testing of installations, and other related details. Plumbing codes had become lengthy, detailed, and very complicated. Nevertheless, the codes of different cities very closely paralleled each other in most respects. Differences in the various codes related principally to items such as kinds of venting methods permitted, kinds of materials recognized as durable under service conditions in different areas, and numerous minor points of relatively little significance.

The need for standardization of plumbing system regulations had long been realized and voiced by master plumber associations, plumbing inspector associations, and plumbing equipment manufacturers associations. Experience with standardization in the mass production of products by industry, and the tremendous benefits that resulted therefrom, as especially evident by the end of World War I, further promoted the idea of trying to achieve standardization of plumbing code regulations. This was in tune with the trend toward accelerated standardization in every phase of industry from 1918 onward. Interested associations appealed to the United States government to initiate authoritative studies and develop model plumbing regulations which could serve as a recognized standard.

A comprehensive effort toward standardization of plumbing codes was made by the U.S. Department of Commerce in 1921. A building code committee of the department began to formulate rules for plumbing systems in small dwellings. To investigate and determine the facts regarding the hydraulics and pneumatics of plumbing systems, scientific experiments were conducted by the National Bureau of Standards. The findings of these experiments were applied as the basis for numerous plumbing requirements. The committee's report appeared in the publication Building and Housing Report No. 2, "Recommended Minimum Requirements for Plumbing in Dwellings and Similar Buildings," dated July 3, 1923, and published by the U.S. Department of Commerce, National Bureau of Standards.

The building code committee of the U.S. Department of Commerce reconvened in 1928 to review the results of 5 years of use of its 1923 model plumbing requirements. Several changes were made, chiefly in pipe sizes, and a revised report was issued. This report appeared in the publication Building and Housing Report No. 13, "Recommended Minimum Requirements for Plumbing," dated August 30, 1928, and published by the U.S. Department of Commerce, National Bureau of Standards. Supplemental revisions were made up to May 1931, and the final report combining the original and supplemental reports was published in 1932.

During the early 1930s, hot water storage tank explosions began to occur frequently and demonstrated the need for hot water supply systems to be equipped with positive means for preventing excessive pressure and temperature conditions. Pressure and temperature relief valves were developed to meet the need, and plumbing codes soon included regulations requiring such devices to be installed at appropriate locations in the hot water supply system.

In the city of Chicago, during its World's Fair in 1933, an amoebic dysentery epidemic occurred. It was of extensive proportions as shown by subsequent reports issued by the Chicago Board of Health and was directly attributed to contamination of water supply piping systems in several buildings. The report emphasized that all water supply connections made to fixtures below rim level were potential sources of contamination to the potable water supply system and should be eliminated as health hazards. Laboratory tests furnished ample confirmation of this fact, and public demonstrations were held to show how readily water supply systems could be contaminated by most of the water inlets to fixtures in common use at the time. These tests and demonstrations merely confirmed the correctness and properness of the sanitary code regulation adopted in 1883 by the New York City Board of Health, prohibiting all water supply connections made to fixtures below rim level and ordering discontinuance of such dangerous connections. The necessity for such regulations to protect potable water supply systems against contamination was amply demonstrated by the amoebic dysentery epidemic in Chicago 50 years later.

By 1935, regulations had been adopted in most plumbing codes to prohibit below-rim water supply connection to fixtures. Enforcement of these new regulations was pressed by health authorities and water supply authorities acting in close collaboration to protect health and to maintain the potable quality of the public water supply systems. Changes in the water supply connections to fixtures in existing buildings had to be made. In many cases, the changes were simple, while in others they were costly. In some cases, where the function of fixtures depended upon a below-rim water supply connection, changes were either impractical or impossible to make. But necessity was the mother of invention, for vacuum breakers were soon developed as satisfactory protective devices for use in cases where fixtures had to be equipped with below-rim potable water supply connections. In 1938, regulations dealing with permissible vacuum-breaker installations on fixture water supply connections appeared in many plumbing codes.

The U.S. Department of Commerce in 1935 established a Central Housing Committee to study ways of improving the housing situation in the nation. In 1938, a subcommittee was formed to study plumbing. This group picked up the work of the previous building code committees and proceeded to develop a standard plumbing manual for use in connection with low-cost housing where the special need was to take advantage of all legitimate economies. This committee's report appeared in the publication Building Materials and Structures Report No. 66, "Plumbing Manual," dated November 22, 1940, and published by the U.S. Department of Commerce, National Bureau of Standards.

From the day it was organized in 1883, the National Association of Master Plumbers was vitally concerned with plumbing codes and their improvement. Association members had to comply with such codes in their daily work and thus had intimate knowledge of the good and bad points of plumbing system regulations. Standardization committees of the association were continuously active in promoting development of standards for all types of plumbing equipment and materials. In 1933, the association's standardization committee developed and published a model plumbing code. It was recommended to code-writing authorities as a suitable standard. To resolve numerous code problems and to develop a scientific basis for code provisions, a research program was sponsored at the State University of Iowa, resulting in considerable scientific data on plumbing system design. Many of these findings were applied by the standardization committee in revising its 1933 model code. In 1942, the National Association of Master Plumbers published its new code, recommended to code-writing authorities as a modern standard.

During World War II, there was an extreme need to conserve critical metals, particularly those commonly used in plumbing systems. The Office for Emergency Management of the Executive Office of the President asked for and received the cooperation of the National Association of Master Plumbers and the United Association of Journeymen and Apprentices of the Plumbing and Pipe Fitting Industry of the United States and Canada. They collaborated closely with representatives of many federal agencies to develop an emergency plumbing code which limited the use of critical metals wherever possible while maintaining safe and sanitary plumbing standards. This code was published in 1942 by the

Office for Emergency Management as the Emergency Plumbing Standards for Defense Housing. It was based upon plumbing requirements contained in Building Materials and Structures Report No. 66, "Plumbing Manual," but appropriate changes were made as required for the conservation of critical metals. This emergency plumbing code was applicable throughout the nation for the balance of the war period.

In 1946, as a result of favorable experience with the Emergency Plumbing Standards during the war period, the United States Housing and Home Finance Agency sponsored a joint committee, known as the Uniform Plumbing Code Committee, to engage in research on the nation's plumbing needs and to draft a plumbing code suitable for adoption by code authorities throughout the nation. Participating with representatives of many federal agencies on this committee were representatives of the National Association of Master Plumbers and the United Association of Journeymen and Apprentices of the Plumbing and Pipe Fitting Industry of the United States and Canada. Research work at the National Bureau of Standards and at the Public Health Service Environmental Health Center provided the committee with scientific data to resolve some of the controversial matters in plumbing system regulations. The committee's work resulted in the publication "Report of the Uniform Plumbing Code Committee," dated July 1949, and published by the U.S. Department of Commerce and the Housing and Home Finance Agency.

For many years, the Western Plumbing Officials Association (WPOA) had also worked actively to develop uniformity of plumbing code regulations. This association produced a model plumbing code in 1938 and designated it as the Uniform Plumbing Code adopted by the WPOA. It was revised at 2-year intervals in order to keep abreast of new materials and methods of construction, the last edition appearing in 1948.

Another organization interested in standardization of plumbing system regulations was the American Society of Sanitary Engineering. In 1942, this organization published a set of standards for use as a guide in plumbing installation. A committee was established also to develop a model plumbing code. But, somewhat later, instead of pursuing this task independently, the society decided to collaborate with other organizations and agencies interested in developing a model plumbing code that would have broad sponsorship.

The American Standards Association, founded in 1918 by five great engineering societies and later broadened in membership to include all nationally recognized technical societies, trade associations, and government agencies having an interest in standards, became actively interested in the coordination of efforts to develop a single, generally accepted standard plumbing code. In 1936, the association's A40 sectional committee organized a subcommittee to undertake the task of establishing minimum requirements for plumbing, but little progress was made and the subcommittee was disbanded in 1939. A new subcommittee was organized in 1941 to develop an American Standard plumbing code. This new group was made up of official representatives from a wide range of interested organizations. The final report of the subcommittee was approved by the A40 sectional committee and its sponsors, the American Society of Mechanical Engineers and the American Public Health Association, and then was submitted to the American Standards Association for adoption and designation as an American Standard. This standard was adopted on February 17, 1949, and designated American Standard Plumbing Code, A40.7-1949.

In 1949, the existence of several different model plumbing codes recommended by various authoritative associations clearly indicated the desirability of reconciling differences between the various codes and developing a single standard plumbing code which would be generally accepted for adoption by code authorities throughout the nation. Under joint sponsorship of the U.S. Department of Commerce and the Housing and Home Finance Agency, a Coordinating Committee for a National Plumbing Code was formed. Represented on this committee were the American Public Health Association, American Society of Mechanical Engineers, American Society of Sanitary Engineering, Building Officials Conference of America, Conference of State Sanitary Engineers, National Association of Plumbing Contractors, Western Plumbing Officials Association, the Housing and Home Finance Agency, and the U.S. Department of Commerce. The committee was assisted by laboratory research at the National Bureau of Standards and the State University of Iowa, and by advisory committee work of representatives of many federal agencies. The report of this committee appeared in the publication "Report of the Coordinating Committee for a National Plumbing Code," dated June 1951, and published by the U.S. Department of Commerce and the Housing and Home Finance Agency.

Soon thereafter, the American Society of Mechanical Engineers and the American Public Health Association, sponsors of the American Standards Association A40 sectional committee project, conducted a survey of organizations interested in the plumbing code standardization project. Favorable response to the coordinating committee's report was received. Several suggestions for improvement of the report were considered and incorporated into the draft of a proposed American Standard National Plumbing Code. Following approval by the sponsor organizations, the proposed new standard was submitted to the American Standards Association for adoption. This new standard was adopted on January 25, 1955, and designated American Standard National Plumbing Code, A40.8-

1955. It soon became the recognized, generally accepted standard for the engineering design of plumbing systems in buildings.

The need to update the A40.8-1955 standard became evident with new developments in materials, methods, and technology. A new American National Standards Committee A40 was organized in 1964 and proceeded to update the A40.8 standard. In this revision project, the National Association of Plumbing–Heating–Cooling Contractors served as one of the sponsors.

In 1968, the proposed revision was submitted for approval to the United States of America Standards Institute, successor to the American Standards Association. (Since then, the United States of America Standards Institute has changed its name to the American National Standards Institute.) The proposed revision did not receive approval because the institute found that consensus had not been achieved. Lost by this action were years of committee efforts and many important revisions related to new materials, methods, and technical advances.

A serious void existed for a great number of parties of interest. There still remained the unresolved need for a modern, updated standard which both plumbing contractors and plumbing engineers deemed essential for their work.

To satisfy this need in 1971, the National Association of Plumbing–Heating–Cooling Contractors took the lead and published the National Standard Plumbing Code. Its stated purpose was to provide local and state governments, code administrative bodies, and industry with a modern, updated code following the format and sequence of A40.8-1955 so as to provide maximum convenience for users. The National Standard Plumbing Code presented significant revisions related to new materials, methods, and technology.

With the June 1973 revision, the American Society of Plumbing Engineers joined this effort by endorsing the National Standard Plumbing Code. Representatives of the American Society of Plumbing Engineers now serve as members of the code committee and participate in a continued code updating program.

A swing toward state plumbing codes, and away from municipal plumbing codes, began in the 1930s. The amoebic dysentery epidemic, which occurred in the city of Chicago during the period of its World's Fair in 1933, and subsequent reports on the epidemic issued by the Chicago Board of Health brought home to responsible state officers and state health officials the realization that there was need for more extensive and more adequate protection of the public health and welfare against the hazards of unsanitary and substandard plumbing installations. The trend toward state codes accelerated after World War II as the result of experience with the many unsanitary and substandard plumbing installations of the war period and because of the great number of new housing developments which mushroomed after the war period in suburban and rural areas, most of which had no effective protection in the form of plumbing system regulations. By 1956, twenty-six states had codes to regulate plumbing system installations. These codes varied in form and applicability. Of these codes, 8 were mandatory statewide, 13 were mandatory statewide with certain exceptions, 10 established minimum standards below which local municipal codes could not provide although they could provide higher standards, 3 were mandatory only where adopted or accepted by municipalities, and 5 were model codes recommended to muncipalities for adoption.

A significant change in plumbing system regulations, the establishment of performance requirements in codes, rather than specific requirements prescribing use of certain methods, devices, materials, and techniques, appeared in the State Building Construction Code of New York State as the various portions of this code were promulgated in 1951, 1953, 1956, 1958, and in subsequent amended editions. This was one of the three state codes which were mandatory only where municipalities voluntarily accepted applicability of the code. The reason for the establishment of performance requirements was given in the New York State Building Code Law, enacted in 1949, and appeared in the statement of legislative findings and purposes, part of which is as follows:

> Among the factors inducing high costs of construction are various laws, ordinances, rules, regulations and codes regulating the construction of buildings and the use of materials therein. They serve to increase cost, without providing correlative benefits or safety to owners, builders, tenants and users of buildings. It is the purpose of this act to institute the preparation of a state code of building construction to provide, so far as may be practicable, basic and uniform performance standards. Thus, while establishing reasonable safeguards for the security, welfare and safety of the occupants and users of buildings, the use of modern methods, devices, materials and techniques will be encouraged. This should be effective in lowering construction costs.

Performance regulations in plumbing codes have gained wider application in recent years. By making adequate performance of any given plumbing system installation the test of its acceptability, the ingenuity of the designer, installer, and manufacturer is permitted to be employed, rather than being overly restrained by the necessity for conforming strictly to specific requirements in codes. Such restrictions pose severe difficulties in meeting the many new and varied conditions for which systems must be designed in different areas. In addition, it has been found in numerous instances that conformity with specific requirements of generally accepted standard plumbing codes does not assure adequate performance of installations, especially where code requirements are inappropriate or inadequate for conditions of the installations.

Under a performance code, such as the New York State Building Construction Code, compliance with its plumbing performance requirements may be achieved in either of two ways. The first way is to comply with applicable provisions of a generally accepted standard recognized and listed by the code-writing authority. In New York State, the applicable standard is the Standard National Plumbing Code, dated 1978.

The second way to achieve compliance with performance requirements applies to installations which *deviate* from applicable provisions of generally accepted standards. In such cases, compliance may be deemed to be achieved only when it shall have been conclusively proved that the deviations meet the performance requirements of the code.

The second way has been proved to be of great value. New ideas in building design and technology, which are in advance of and deviate from recognized standards, may be approved and put into service upon conclusive proof of adequacy of performance. One example which may be cited is the installation of storm water drainage systems, sized in accordance with the rated discharge of weir-type roof drains. Such a system was approved for a single building having 30 acres of roof area. Another example is the introduction of the Sovent system, one of the European single stack drainage systems, into several new, high-rise residential buildings, one of which was a 20-story-tower type. In these cases, performance criteria and testing procedures had to be specially developed and then applied during the construction phase so as to prove adequacy of performance.

The durability of materials and equipment and the adequacy of their performance in service vary appreciably with the chemical characteristics of the water supplies they convey and of the soils in which they may be installed. Some materials which perform with acceptable durability in most water supply areas may have very limited durability and perform inadequately in certain areas where conditions are adverse to their use. Similarly, the performance and design adequacy of building drainage systems are affected by significant changes in the physical characteristics of the wastes conveyed. For example, in recent years, suds-backup conditions have occurred with considerable frequency at lower floor fixtures in many different types of buildings, both new and old. This condition resulted from the great increase in the volume of suds accompanying wastes in building drainage systems coincident with the introduction and widespread use of sudsy detergents since World War II, and the inadequacy for suds flow of vent pipes designed simply for airflow in certain zones of the drainage system. Satisfactory design for adequate performance under suds-flow conditions is discussed under appropriate subjects in other chapters.

Much still remains to be done before generally acceptable standardization of plumbing system regulations is achieved. In any event, regulations must keep pace with changing conditions and should not be allowed to restrict the development of new methods, devices, materials, and techniques. The present trend is toward performance requirements, making adequate performance the test of acceptability. This is a reasonable and objective basis upon which to establish regulations to protect the health, safety, and welfare of the people.

2

SYSTEMS, MATERIALS, AND FIXTURES FOR OCCUPANCIES

PERFORMANCE REQUIRED OF BUILDING SYSTEMS

Buildings intended for human habitation, occupancy, or use should have plumbing systems designed in accordance with modern standards of sanitation. A sanitary drainage system should be provided to convey sewage from all fixtures to an adequate, approved means of disposal. There should be a water supply system to furnish cold water to every water closet and urinal, and hot and cold water to every sink, laundry tray, automatic laundry washing machine, lavatory, bathtub, and shower required for the building occupancy, except as otherwise specifically noted for certain occupancies in other parts of this chapter.

Plumbing systems for buildings should be designed, installed, located, and maintained so that under normal conditions of use they will not become a potential danger to health or welfare, a danger because of defects, a source of ignition, or a radiation hazard and will not create excessive noise, or otherwise become a nuisance. Systems should be installed in such a manner so as not to weaken structural members or cause damage or deterioration to any part of the building through fixture usage. They should be designed and constructed so as to avoid fouling, clogging, and depositing of solids and should be maintained in sanitary and serviceable condition. Equipment requiring operation, inspection, or maintenance should be located so that easy access to it is provided. Moving parts of equipment which may be a potential hazard should have guards to protect against accidental contact.

PERFORMANCE REQUIRED OF MATERIALS FOR PLUMBING SYSTEMS

Plumbing systems and equipment should be made of approved materials, should be free from defective workmanship, and should be designed and installed so as to be durable, without need for frequent repairs or major replacements. The life of the average building in the United States is considered to be about 70 years. However, after 35 years of service, it is generally recognized that buildings usually require major alterations and replacement of equipment for modernization and efficient continued service. Knowing the characteristics of the water to be conveyed by a water supply system and of the sewage and gases to be conveyed by a drainage and vent system, appropriate selection of piping materials and equipment may be made to afford either 35 or 70 years of durable service. Where piping is to be concealed and relatively inaccessible in floors or walls, it is essential that the piping be of adequate durability from the economic view because the cost of replacing damaged sections of walls and floors in addition to the piping is relatively prohibitive.

Consequently, before proceeding with any given installation, the designer and installer should consult with the local authority having jurisdiction to determine the durability of materials and their joints used under local conditions. Local authorities have knowledge of the characteristics of the available water supply and of the sewage and gases to be conveyed by sanitary drainage and vent systems, and they can provide authoritative determinations regarding the durability of various kinds of materials under local conditions for the guidance of the designer and installer.

The installer and designer should also be guided by the good practice recommendations of the manufacturers of materials and equipment regarding care, handling installation, and adjustment of their products in order that the performance of such products will not be impaired by defects or damage during installation, or by bad installation practices. Materials should be free of manufacturing defects or damage, however occasioned, which would render or tend to render such materials defective, unsanitary, or otherwise improper for effective and efficient service. Used plumbing equipment or material should not be installed unless specially approved by the local authority having jurisdiction. Plumbing equipment or material which is potentially defective because of wear, damage, sanitary hazards, or prior use should not be reused for plumbing purposes.

BUILDING OCCUPANCY CLASSIFICATIONS

In codes, buildings are classified by groups in respect to their occupancy or use and in respect to the number and physical condition of the occupants in buildings of the institutional type. The various building occupancy classifications are defined below:

The terms *one-* and *two-family dwelling* mean a building arranged for one or two dwelling units, a dwelling unit being one or more rooms with provision for living, sanitary, and sleeping facilities arranged for the use of one family.

The term *multiple dwelling* includes a building containing three or more dwelling units; a building containing living, sanitary, and sleeping facilities occupied by one or two families and more than four lodgers residing with either one of such families; a building with one or more sleeping rooms, other than a one- or two-family dwelling, used or occupied by permanent or transient paying guests or tenants; a building with sleeping accommodations for more than five persons used or occupied as a club, dormitory, fraternity or sorority house, or for similar uses; and a building used or occupied as a convalescent, old-age, or nursing home, but not including private or public hospitals or public institutions.

The term *business occupancy* applies to buildings in which the primary or intended occupancy or use is the transaction of administrative business or civic or professional services, and to buildings where the handling of goods, wares, or merchandise, in limited quantities, is incidental to the primary occupancy or use. Newsstands, lunch counters, barber shops, beauty parlors, and similar service facilities are considered as incidental occupancies or uses. Business occupancy includes but is not limited to the following: broadcasting stations; civic administration buildings; laboratories, other than chemical; office buildings, parking lot offices; professional offices, incidental to other uses; service facilities in connection with the primary occupancies and uses; and telephone exchanges.

The term *mercantile occupancy* applies to buildings in which the primary or intended occupancy or use is the display and sale to the public of goods, wares, or merchandise. Mercantile occupancy includes but is not limited to the following: display rooms; gasoline service stations; markets; salesrooms; and stores, both retail and wholesale.

The term *industrial occupancy* applies to buildings in which the primary or intended occupancy or use is the manufacture or processing of products requiring operations such as making, altering, assembling, bottling, canning, finishing, handling, mixing, packaging, repairing, cleaning, and laundering. Industrial occupancy includes but is not limited to the following: assembly plants, factories of all kinds, garages with repair facilities, gas plants, industrial and chemical laboratories, loft buildings, manufacturing plants of all kinds, mills, power plants, processing plants, pumping stations, refineries, smokehouses, tenant factories, waterworks stations, and work or repair shops.

The term *storage occupancy* applies to buildings in which the primary or intended occupancy or use is the storage of, or shelter for, goods, merchandise, products, vehicles, or animals. Storage occupancy includes but is not limited to the following: aircraft hangars, bulk oil storage, coal pockets, cold storage, freight depots, garages without repair facilities, gasoline bulk stations, grain elevators, lumber yards, stockrooms, storehouses, tank farm buildings, transit sheds, truck and marine terminals, and warehouses.

The term *assembly occupancy* applies to buildings in which the primary or intended occupancy or use is the assembly for amusement, athletic, civic, dining, educational, entertainment, patriotic, political, recreational, religious, social, sports, or similar purposes. Assembly occupancy includes but is not limited to the following: amusement park buildings of all kinds; armories; art galleries; assembly halls; auditoriums; bath houses; bowling alleys; club rooms; coliseums and stadiums; courtrooms; dance halls; exhibition halls or buildings; grandstands; gymnasiums; lecture halls; lodge halls or rooms; mortuaries; motion picture theaters; museums; night clubs; passenger stations and terminals of air, surface, underground, and marine public transportation facilities, restaurants; skating rinks; tents and similar shelters; theaters; churches, synagogues, mosques, and other places of worship, schools, colleges, and similar places of education.

The term *institutional occupancy* applies to buildings in which the primary or intended occupancy or use is for persons domiciled or detained under supervision, subclassified according to the movement of the occupants as follows:

> Institutional occupancy, supervised—includes but is not limited to dormitories and quarters for employees and staff and for persons whose movements are not limited, domiciled under supervision
>
> Institutional occupancy, restricted—for persons whose movements are limited because of illness, age, physical or mental handicap, including but not limited to child caring institutions, clinics, day nurseries, homes for ill and infirm, hospitals, infirmaries, and sanitariums, but excluding convalescent, nursing, and old-age homes classified as multiple dwellings
>
> Institutional occupancy, detained or confined—for persons detained or confined in a mental hospital or for correctional or penal purposes, including but not limited to detention homes, houses of correction, jails, mental hospitals, penitentiaries, police lockups, prisons, and reformatories.

The term *miscellaneous occupancy* applies to nonresidential buildings in which the primary or intended occupancy or use is not included in any of the preceding occupancy classifications; to accessory structures attached to, part of, or supported by, the buildings; and to buildings temporary in character. Miscellaneous occupancy includes but is not limited to contractors' temporary buildings.

FIXTURES REQUIRED FOR BUILDING OCCUPANCIES

Over a period of more than 100 years, various laws have been enacted requiring the installation of adequate numbers and types of plumbing fixtures in various classes of building occupancies. Such regulations may be found in plumbing codes, sanitary codes, housing laws, building codes, multiple dwelling codes, labor laws, industrial codes, public building laws, codes for places of public assembly, standards for approval of school building construction, and others. These laws have sought to protect the health of building occupants by requiring that adequate sanitary facilities be provided to meet the basic needs of sanitation and personal hygiene. For each different class of building occupancy, the minimum number and types of plumbing fixtures required and the location of such fixtures have been clearly specified in law. Such required fixtures and the conditions of installation for each building occupancy classification are stated under the heading "Standard Code Regulations" in this chapter.

Additional supplementary fixture installations frequently are desirable. In industrial occupancies, the cost of time required for employees to travel from their usual work positions to fixture locations may be a most important consideration in determining the economic advantage of providing additional fixtures in convenient locations in close proximity to employee work positions. In all types of occupancies, the convenience of fixtures is of great importance in promoting improved sanitary habits on the part of users and in establishing a sanitary and healthful environment for the occupants.

FACILITIES FOR THE PHYSICALLY HANDICAPPED

It is shocking to most people to realize that nearly one in every seven persons in the United States has a permanent physical disability. For approximately 20 million Americans who are physically handicapped, many of the common features of existing public buildings pose severe architectural barriers preventing or impeding their use. Such barriers have lessened the social and economic gains made in rehabilitation of the handicapped and made it very difficult to project them into normal situations of education, recreation, and employment.

In 1961, a most significant change began in the design of buildings used by the public. Object of the change was to make all buildings and facilities, including plumbing, used by the public accessible to, and functional for, the physically handicapped, to, through, and within their doors, without loss of function, space, or facility where the general public is concerned.

Appropriate changes were set forth in the approved and published ANSI Standard A117.1-1961, Specifications for Making Buildings and

Facilities Accessible to, and Usable by, the Physically Handicapped. By 1971, many state and local laws and codes were enacted consistent therewith and mandating appropriate changes including various provisions related to plumbing systems in buildings. In 1980, the earlier standard was superseded by a new one, A117.1-1980, Specifications for Making Buildings and Facilities Accessible to and Usable by Physically Handicapped People.

Plumbing facility provisions are to a great extent contingent upon the specifications of the standard model wheelchair used by the physically handicapped, and upon its functioning. They are as follows:

1. Length: 42 in (1065 mm)
2. Width, when open: 26 in (660 mm)
3. Height of seat from floor: 19 in (485 mm)
4. Height of armrest from floor: 30 in (760 mm)
5. Height of pusher handles (rear) from floor: 36 in (915 mm)
6. Width, when collapsed: 11 in (280 mm)

Data relative to wheelchair functioning are as follows:

1. Fixed turning radius, wheel to wheel: 18 in (455 mm)
2. Fixed turning radius, front to rear (overall): 31½ in (800 mm)
3. Average turning space required (180 and 360°): 60 × 60 in (1525 × 1525 mm)
4. Minimum width of area with open ends, such as a corridor, to permit a 360° turn: 54 in (1370 mm)
5. Minimum width for two wheelchairs to pass each other: 60 in (1525 mm)

Recommended specific provisions dealing with toilet rooms may be summarized. It is essential that an appropriate number of toilet rooms, in accordance with the nature and use of a specific building or facility, be made accessible to, and usable by, the physically handicapped. Toilet rooms shall have space to allow traffic of individuals in wheelchairs, in accordance with the data cited in the foregoing relative to wheelchair specifications and functioning.

Toilet rooms shall have at least one toilet stall that:

1. Is 3 ft (915 mm) wide.

2. Is at least 4 ft, 8 in (1420 mm) deep.

3. Has a door (where doors are used) that is 32 in (815 mm) wide and swings out.

4. Has handrails on each side, 33 to 36 in (840 to 915 mm) high and parallel to the floor, 1¼ to 1½ in (31.3 to 37.5 mm) in outside diameter, with 1½ in (37.5 mm) clearance between rail and wall, and fastened securely at ends and center.

5. Has a water closet with the seat 17 to 19 in (430 to 485 mm) from the floor.

The design and mounting of the water closet are of considerable importance. A wall-mounted water closet with a narrow understructure that recedes sharply is most desirable. If a floor-mounted water closet must be used, it should not have a front that is wide and perpendicular to the floor at the front of the seat. The bowl should be shallow at the front of the seat and turn backward more than downward to allow the individual in a wheelchair to get close to the water closet with the seat of the wheelchair.

Toilet rooms shall have lavatories with narrow aprons, which when mounted at standard height are usable by individuals in wheelchairs; or shall have lavatories mounted higher, when particular designs demand, so that they are usable by individuals in wheelchairs. It is important that drain pipes and hot water pipes under a lavatory be covered or insulated so that a wheelchair-bound individual without sensation will not be burned.

Toilet rooms for men shall have wall-mounted urinals with the opening of the basin 17 in (430 mm) from the floor, or shall have floor-mounted urinals that are level with the main floor of the toilet room.

Doors, including doors and doorways of toilet rooms, shall have a clear opening of no less than 32 in (815 mm) when open and shall be operable by a single effort. The floor on the inside and outside of each doorway shall be level for a distance of at least 5 ft (1525 mm) from the door in the direction the door swings and shall extend 1 ft (305 mm) beyond each side of the door. Sharp inclines and abrupt changes in level shall be avoided at doorsills. As much as possible, thresholds shall be flush with the floor. Floors shall have a surface that is nonslip.

An appropriate number of water fountains or drinking water coolers and fountains shall be accessible to, and usable by, the physically handicapped. Water fountains or coolers shall be of a type having upfront spouts and controls. They shall be hand-operated, or hand-and-foot-operated. Conventional floor-mounted water coolers can be serviceable to individuals in wheelchairs if a small fountain is mounted on the side of the cooler 30 in (760 mm) above the floor. Wall-mounted, hand-operated coolers can serve the able-bodied and the physically handicapped equally well when the cooler is mounted with the outlet of the spout no higher than 36 in (915 mm) from the floor. Fully recessed water fountains are not recommended. Water fountains should not be set into an alcove unless the alcove is wider than a wheelchair.

STANDARD CODE REGULATIONS

Required Systems Buildings and portions thereof shall be provided with plumbing systems designed to dispose of sewage from all fixtures and to furnish cold water to every water closet and urinal, and hot and cold water to every sink, laundry tray, automatic laundry washing machine, lavatory, bathtub, and shower required therein, except as otherwise provided herein.

Performance Requirements for Materials Plumbing equipment and systems shall be made of approved materials, shall be free from defective workmanship, and shall be designed and installed so as to be durable, without need for frequent repairs or major replacements.

Before proceeding with an installation, the installer shall consult with the local authority having jurisdiction to determine the durability of materials and joints used under local conditions.

Installers shall observe the manufacturer's good practice recommendations regarding care, handling, installation, and adjustment of their products in order that the performance of such products will not be impaired by defects or damage during installation, or by bad installation practices.

Required Location, Number, and Type of Fixtures Fixtures shall be provided for the occupancies and under the conditions as set forth herein. The fixtures required for the stated number of dwelling units, rooms, or persons shall be required for any fraction thereof.

One- and Two-family Dwellings Where public water supply is available to a one- or two-family dwelling, provide within each dwelling unit:

One kitchen sink

One water closet

One bathtub or shower

One lavatory

Hot water need not be furnished unless required by law.

Multiple Dwellings Within each dwelling unit, provide:

One kitchen sink

One water closet

One bathtub or shower

One lavatory

Within each dwelling unit, not designed for use primarily by transients, provide one laundry tray or automatic laundry washing machine, or in a general laundry room, readily accessible from within the building, provide one two-compartment laundry tray for each 10 dwelling units or one automatic laundry washing machine for each 20 dwelling units.

Where multiple dwellings contain sleeping accommodations arranged as individual rooms or suites, provide for each six sleeping rooms:

One water closet

One bathtub or shower

One lavatory

Where multiple dwellings contain sleeping accommodations arranged as a dormitory, for each 15 persons so accommodated and located adjacent thereto provide:

One water closet

One bathtub or shower

One lavatory

Urinals may be substituted in men's toilet rooms for not more than one third of the required number of water closets.

Business Occupancy In business occupancies, water closets and lavatories shall be provided for employees in accordance with Table 2-1.

Table 2-1
WATER CLOSETS AND LAVATORIES FOR EMPLOYEES IN BUSINESS OCCUPANCIES

Number of water closets	Number of employees	Number of lavatories	Number of employees
1	1–15	1	1–20
2	16–35	2	21–40
3	36–55	3	41–60
4	56–80	4	61–80
5	81–110	5	81–100
6	111–150	6	101–125
7	151–190	7	126–150
		8	151–175
One additional water closet for each 40 employees in excess of 190		One additional lavatory for each 30 employees in excess of 175	

In business occupancies, one drinking fountain or equivalent fixture shall be provided for each 75 employees.

Urinals may be substituted in men's toilet rooms for not more than one third of the required number of water closets when more than 35 males are employed.

Toilet facilities shall be in separate rooms for each sex, where there are five or more employees, and shall be readily accessible and convenient to their regular working places.

Mercantile Occupancy In mercantile occupancies, plumbing fixtures shall be provided for employees in conformity with the same number, type, and location as required in business occupancies.

Industrial Occupancy In industrial occupancies, plumbing fixtures shall be provided for employees in conformity with the same number, type, and location as required in business occupancies; except that, in foundries, water closets, lavatories, and urinals shall be provided for employees in accordance with Table 2-2.

Storage Occupancy In storage occupancies, plumbing fixtures shall be provided for employees in conformity with the same number, type, and location as required in business occupancies, except that such fixtures may be in adjacent buildings, under the same ownership or control, where the maximum distance of travel from the employee's usual working place to the fixtures does not exceed 500 ft (152 m) horizontally.

Assembly Occupancy In assembly occupancies, other than places of worship and schools, plumbing fixtures shall be provided for occupants, based upon capacity, in accordance with Table 2-3.

Table 2-2
WATER CLOSETS, LAVATORIES, AND URINALS FOR EMPLOYEES IN FOUNDRIES

Number of water closets	Number of employees	Number of lavatories	Number of employees	Number of urinals	Number of male employees
1	1–10	1	1–8	1	11–29
2	11–25	2	9–16	2	30–79
3	26–50	3	17–30		
4	51–80	4	31–45		
5	81–125	5	46–65		
One additional water closet for each 45 employees in excess of 125		One additional lavatory for each 25 employees in excess of 65		One additional urinal for each 80 male employees in excess of 79	

Table 2-3
WATER CLOSETS, LAVATORIES, AND URINALS FOR OCCUPANTS IN ASSEMBLY OCCUPANCIES, OTHER THAN PLACES OF WORSHIP AND SCHOOLS

Number of water closets	Number of occupants	Number of lavatories	Number of occupants	Number of urinals	Number of male occupants
1	1–100	1	1–100	1	1–100
2	101–200	2	101–200	2	101–200
3	201–400	3	201–400	3	201–400
4	401–700	4	401–700	4	401–700
5	701–1100	5	701–1100	5	701–1100
One additional water closet for each 600 occupants in excess of 1100		One additional lavatory for each 1500 occupants in excess of 1100. Such lavatories need not be supplied with hot water		One additional urinal for each 300 occupants in excess of 1100	

Where motion pictures projection booths contain more than one projection machine, there shall be provided at least one water closet and one lavatory on the same level, within 20 ft (6 m) of the booth.

One drinking fountain or equivalent fixture shall be provided for each 1000 occupants, except that there shall be at least one such fixture at each assembly floor level and tier.

Fixtures provided for occupants shall be available for the use of employees. However, such fixtures shall consist of at least the same number and type as required for employees in business occupancies.

Toilet facilities shall be in separate rooms for each sex, and shall be readily accessible.

In places of worship, there shall be provided at least one water closet and one lavatory. Such fixtures may be in adjacent buildings under the same ownership or control, and shall be accessible during periods when the assembly space is occupied.

In schools, plumbing fixtures shall be provided for occupants, based on capacity, in accordance with the following:

For pupils' use:

One water closet for each 100 male pupils and one for each 35 females in elementary schools

One water closet for each 100 male pupils and one for each 45 females in secondary schools

One lavatory for each 50 pupils

One urinal for each 30 male pupils

One drinking fountain or equivalent fixture for each 150 pupils, but at least one such fixture at each floor having classrooms

Where more than 5 persons are employed, fixtures shall be provided for employees and shall consist of at least the same number and type as required in business occupancies. Such fixtures shall be located in rooms separate from those in which fixtures are provided for pupils.

Toilet facilities shall be in separate rooms for each sex, and shall be readily accessible and convenient.

Institutional Occupancy In institutional occupancies, supervised—plumbing fixtures shall be provided in accordance with the following:

Within each dwelling unit:

One kitchen sink

One water closet

One bathtub or shower

One lavatory

Where sleeping accommodations are arranged as individual rooms or suites there shall be located adjacent thereto for each six sleeping rooms:

One water closet

One bathtub or shower

One lavatory

Where sleeping accommodations are arranged as a dormitory, for each 15 persons so accommodated there shall be located adjacent thereto:

One water closet

One bathtub or shower

One lavatory

Toilet facilities, except those within dwelling units, shall be in separate rooms for each sex.

In institutional occupancies, restricted, other than hospitals—plumbing fixtures shall be provided in accordance with the following:

On each story so occupied:

One water closet for each 25 male occupants and one for each 20 females

One urinal for each 50 male occupants

One lavatory for each 10 occupants

One shower for each 10 occupants

One drinking fountain or equivalent fixture for each 50 occupants

Fixtures for employees shall consist of at least the same number and type as required in business occupancies.

Toilet facilities shall be in separate rooms for each sex.

In institutional occupancies, restricted, hospitals only—plumbing fixtures shall be provided in accordance with the following:

For patients' use:

One water closet and one lavatory for each 10 patients

One shower or bathtub for each 20 patients

One drinking fountain or equivalent fixture for each 100 patients

Fixtures for employees shall consist of at least the same number and type as required in business occupancies.

Toilet fixtures for employees shall be located in separate rooms from those in which fixtures for the use of patients are located.

Toilet facilities shall be in separate rooms for each sex, and shall be readily accessible and convenient.

In institutional occupancies, detained or confined, mental hospitals only—plumbing fixtures shall be provided in accordance with the following:

For patients' use:

One water closet, one lavatory, and one shower or bathtub, for each 8 patients

One drinking fountain or equivalent fixture for each 50 patients

Fixtures for employees shall consist of at least the same number and type as required in business occupancies.

Toilet fixtures for employees shall be located in separate rooms from those in which fixtures for the use of patients are located.

In institutional occupancies, detained or confined, penal institutions only—plumbing fixtures shall be provided in accordance with the following:

For inmate use:

One water closet and one lavatory in each cell

One shower at each floor on which cells are located

One water closet and one lavatory available at each exercise area

Lavatories for inmate use need not be supplied with hot water.

Fixtures for employees shall consist of at least the same number and type as required in business occupancies.

Toilet fixtures for employees shall be located in separate rooms from those in which fixtures for the use of inmates are located.

Miscellaneous Occupancy Temporary toilet facilities shall be provided for employees engaged in the construction, alteration, repair, or demolition of buildings on the basis of 1 unit for each 30 persons. Such units shall consist of water closets, chemical toilets, or privies, readily accessible to employees, shall be located not more than four stories above or below the place of work, and shall be sheltered from view and protected from any hazard of falling objects.

Temporary toilet facilities shall be maintained in a sanitary and serviceable condition. Upon completion of building work, such facilities and the sewage remaining therefrom shall be removed, the area shall be cleaned and disinfected, and privy pits shall be filled with clean earth.

Public Bathing Occupancy Facilities for bathers at swimming pools and other public bathing occupancies shall be in separate rooms for each sex, shall be accessible to bathers at all times, and shall be located so that bathers can use the facilities before entering the bathing area. The number and type of fixtures shall consist of at least the following:

One water closet for each 60 males and one water closet for each 40 females

One urinal for each 60 males

One lavatory for each 60 males and one lavatory for each 60 females

One shower for each 40 males and one shower for each 40 females, except that in schools such required showers shall be at least equal in number to one third the number of pupils in the largest class using the swimming pool at any one time

Public or Employee Dining Places Where food or drink is served, and the dishes, glasses, or cutlery for such service are to be reused, there shall be at least one machine or three-compartment sink of suitable type for the effective washing and sanitizing of such articles before reuse. Cold water need not be supplied to such machines and sinks.

Kitchens Serving Public or Employee Dining Places Every kitchen serving public or employee dining places shall have installed therein at least one lavatory for the personal use of kitchen employees.

Exposure to Harmful Materials or Excessive Heat Where there is exposure to skin contamination from poisonous, infectious, or irritating materials, there shall be provided for each five employees so exposed at least one lavatory.

Where there is exposure to excessive heat or to skin contamination from poisonous, infectious, or irritating materials, there shall be provided for each 15 persons so exposed at least one shower accessibly located. Where severely irritating materials are used, showers for emergency use shall be located within 30 ft of the work positions of such exposed persons, shall not be supplied with hot water, and need not have drainage provisions.

Wet Method of Dust Control Where the wet method of dust control is used, the floor of such space shall be provided with at least one floor drain.

Location and Usability of Fixtures for the Physically Handicapped Adequate clear floor space shall be provided for access to, and safe use of, plumbing fixtures for the physically handicapped. A clear floor space of at least 5 × 5 ft (1525 × 1525 mm) shall be available for turning a wheelchair in general toilet rooms, general bathrooms, and general laundry rooms equipped with fixtures for the physically handicapped.

Faucet and operating levers at plumbing fixtures for the physically handicapped shall be of wrist-blade or other suitable type not requiring the use of the hand for operation. Self-closing lavatory faucets shall be prohibited.

Where water closets for the physically handicapped are installed in general toilet rooms, they shall be located in stalls at least 36 in (915 mm) wide and 56 in (1420 mm) deep. Stall doors, where provided, shall swing out and shall afford a clear opening at least 32 in (815 mm) in width. Where stall doors are provided, the stalls shall be at least 66 in (1675 mm) deep. Where water closets for the physically handicapped in multiple dwelling and institutional occupancies are provided in dwelling units or in general toilet rooms, a horizontal handrail or grab bar of adequate length and 1¼ to 1½ in (31.3 to 37.5 mm) in diameter shall be securely mounted 33 to 36 in (840 to 915 mm) above the floor and conveniently located at one side of the water closet with at least 1½ in (37.5 mm) of clearance from walls. Where water closets are provided for physically handicapped persons whose movements are limited because of illness, or physical or mental handicap, in multiple dwelling or institutional occupancies water closets shall be equipped with seats located 17 to 19 in (430 to 485 mm) above the floor, and horizontal handrails of adequate length and at least 1¼ to 1½ in (31.3 to 37.5 mm) in diameter shall be securely mounted approximately 33 to 36 in (840 to 915 mm) above the floor on each side of the water closet, with at least 1½ in of clearance between handrails and adjacent walls.

Where provided for the physically handicapped, wall-mounted and pedestal urinals shall be installed so that rim levels are no more than 17 in (430 mm) above floor level; floor-set stall urinals shall be installed so that rim level is no higher than the toilet room floor level.

Lavatories and kitchen sinks for the physically handicapped shall have clearance of at least 29 in (735 mm) for lavatories and 27 in (685 mm) for sinks between fixture bottom and floor for a depth of 8 in so as to provide sufficient legroom, and the piping beneath the fixture shall be located or insulated so as to prevent injury to persons in wheelchairs.

Where bathtubs are provided for the physically handicapped, grab bars 1¼ to 1½ in (31.3 to 37.5 mm) in diameter shall be securely mounted on two adjacent walls of the tub or its enclosure, one bar horizontally 33 to 36 in (840 to 915 mm) above the floor and the second bar installed vertically with its lower end at such height.

Shower stalls for the physically handicapped shall have nonslip-type floors, a shower spray unit with a base at least 60 in (1.5 m) long that can be used as a fixed shower head or as a hand-held shower, and single-lever-type mixing valves designed to maintain constant water temperature under variable water pressure and temperature conditions. Mixing valves shall be installed approximately 38 to 48 in (965 to 1220 mm) above the stall floor and shall be operable from inside and outside the stall by a user at the threshold. Where such shower stalls are provided in dwelling units of multiple dwellings or institutional occupancies, grab bars 1¼ to 1½ in (31.3 to 37.5 mm) in diameter shall be securely mounted on two adjacent walls of the stall, one bar installed horizontally 33 to 36 in (840 to 915 mm) above the floor and the second bar installed vertically with its lower end 33 to 36 in (840 to 915 mm) above the floor. Where central bath facilities are provided for multiple dwellings or institutional occupancies, a shower stall for the physically handicapped shall have at least 9 ft^2 (0.84 m^2) of floor area, with minimum horizontal dimension of 3 ft (0.9 m), and a curb not exceeding 4 in (100 mm) in height at the stall entrance. Where central bath facilities are provided for physically handicapped persons whose movements are limited because of illness, or physical or mental handicap, in multiple dwelling or institutional occupancies a shower stall for the physically handicapped shall have at least 16 ft^2 (1.49 m^2) of floor area, with minimum horizontal dimension of 4 ft (1.22 m), and no threshold at the stall entrance. Where such shower stalls are provided in central bath facilities in multiple dwellings or institutional occupancies, they shall be equipped with a folding seat securely attached to a sidewall of the stall, opposite the mixing valve, and installed so that the top of the seat when in the lowered position shall be 18 in (455 mm) above the stall floor. Horizontal handrails 1¼ to 1½ in (31.3 to 37.5 mm) in diameter shall be securely mounted 33 to 36 in (840 to 915 mm) above the floor on the rear wall and on the wall opposite the seat. Handrails and grab bars shall provide at least 1½ in (37.5 mm) of clearance from walls.

Drinking fountains for the physically handicapped shall have spouts located no higher than 36 in (915 mm) above the floor and shall have basins projecting 17 to 19 in (430 to 485 mm) from the front of the wall or cabinet on which they are mounted. Such drinking fountains shall have upfront jets and controls and shall be designed for hand operation or hand-and-foot operation.

Where laundry facilities for multiple dwellings occupied by the physically handicapped are provided in a general laundry room, at least one laundry machine therein shall be of the front-loading type equipped with a side-hinged front door and front-mounted controls.

3

FIXTURES, FIXTURE TRAPS, AND INTERCEPTING DEVICES

STANDARDS FOR QUALITY OF FIXTURES

Sanitary quality should be built into plumbing fixtures. They should be suitably designed for their intended uses and for convenience in cleaning. They should be made of nonoxidizing, nonabsorbent material with smooth, abrasion-resistant, impervious surfaces; they should be free from defects and concealed fouling surfaces and reasonably durable for their intended services.

Over a period of about 120 years, experience with many fixture designs and kinds of materials has led to development of generally accepted standards with respect to strength, durability, corrosion resistance, abrasion resistance, performance under suitable tests, and other important qualities necessary in fixtures before they may be considered to be of satisfactory quality.

Standards for various kinds of plumbing fixtures and fittings have been developed most recently under procedures of the American National Standards Institute by the American National Standards Committee A112, a balanced committee having the assigned scope as follows:

> The recommendation of suitable standards, or the development of standards where none exist, for composition, dimensions, and/or mechanical and physical properties of materials, fixtures, devices, and equipment used or installed in plumbing systems.

American National Standards Committee A112 has recommended many existing standards as suitable for various materials, fixtures, devices, and equipment used or installed in plumbing systems. It has also developed many new and updated standards, and these have been given the A112 designation in the standard's reference number.

As a guide for architects, engineers, contractors, and suppliers, some of the recently developed standards applicable to plumbing fixtures and fittings are listed below:

A112.19.1-1979	Enameled cast iron
A112.19.2-1973	Vitreous china
A112.19.3-1976	Stainless steel, residential use
A112.19.4-1977	Porcelain enameled formed steel
A112.18.1M-1979	Finished and rough brass plumbing fixture fittings

WATER CLOSETS AND URINALS

Water closets of unsanitary design should be prohibited. Included in this category are all water closets which have one or more of the following unsanitary design features:

1. An invisible seal
2. Unventilated spaces
3. Walls which are not thoroughly washed at each flushing
4. A design which might permit siphonage of the contents of the water closet bowl back into the flush tank

Numerous obsolete types of water closets which had one or more such unsanitary features are often named in codes as prohibited types, such as the valve or Bramah water closet, introduced in 1788; the pan water closet, introduced in 1833; the plunger water closet, introduced in 1870; and the washout water closet, introduced in 1880. These are but a few of the obsolete types. If they were placed on display, few persons would know them by name, but their unsanitary features would be readily recognized.

All these early water closet models were rendered obsolete and relegated to the category of unsanitary water closets by the development in 1890 of the washdown water closet. Its design was the first to be considered sanitary. Since then, many improvements have been made in water closet design for quieter action, quicker function, and stronger siphonic action, as well as for uniting water closet and flush tank design as a unit and for installation of water closets on walls.

Many models of sanitary water closets, conforming to quality standards established by industry, are currently available. An appropriate selection may be made for any type of building occupancy. For public use, it is recommended that water closets be of the elongated bowl type so as to avoid unsanitary and objectionable conditions which might otherwise arise in such service. In kindergartens, in nurseries, and similar occupancies where fixtures are provided for the use and training of children under 6 years of age, water closets should be of suitable type, such as primary or junior models, 10 or 13 in (250 or 330 mm) in height, so as to avoid any hazard in use.

Water closets should be equipped with seats of smooth nonabsorbent material. Integral water closet seats should be of the same material as the fixture. However, it is preferable if the seats are made of a material having relatively low thermal conductivity so as to avoid excessive thermal shock when bare skin contacts material that is very cold. Seats should be of suitable shape for the type of water closet. Open-front-type seats should be provided for water closets intended for public use.

Trough-type urinals have been deemed by many health and code authorities to be of unsanitary design for use within buildings. The walls of such urinals are not thoroughly washed at each flushing. This results in objectionable odorous conditions in the vicinity of the fixture. Consequently, trough-type urinals are prohibited for use in buildings in many areas. However, they may be permitted by the authority having jurisdiction as a temporary outdoor facility for workmen during building construction or demolition, or for public use at temporary carnival sites.

FLUSH TANKS AND FLUSH VALVES (FLUSHOMETERS)

Each water closet and urinal should have a flushing device designed and installed so as to supply water at adequate rate and volume for satisfactory flushing of the interior of the fixture. A flush tank or direct supply flush valve (Flushometer) should be provided for this purpose.

Flush tanks should have sufficient water capacity for flushing thoroughly the water closets or urinals they are intended to serve. No flush tank should supply more than one water closet or urinal. However, where specially approved by the authority having jurisdiction, more than one urinal may be permitted to be flushed by a single flush tank, provided it is automatic in operation and has sufficient water capacity for flushing thoroughly and simultaneously all the urinals served thereby. Flush valves in tanks should be designed for manual operation, except where specially required to be automatic in operation. Flush-valve seats in water closet flush tanks should be at least 1 in above the rim level of the water closet connected thereto, except in the case of an approved water closet and flush tank combination specially designed so that when the water closet is clogged and the tank is flushed, the flush valve closes tightly and prevents water from spilling continuously over the rim of the water closet. Flush tanks should be provided with overflows of adequate size to prevent tank flooding at the maximum rate at which they are supplied with water. The overflow should be arranged so as to discharge into the water closet or urinal connected thereto.

Ball cocks in flush tanks should be designed to operate automatically, refilling the tank after each discharge and shutting off completely when the tank is filled to operational capacity. In lowdown flush tanks, ball cocks should be designed with means for bypassing an adequate amount of water to refill the trap seal while the tank refills after each flushing. Where flush tanks are supplied through ball cocks connected directly to the potable water supply system, such ball cocks should be equipped with approved means for protecting the system against contamination from back siphonage of tank water.

Flush pipes and fittings between flush tanks and water closets or urinals should be of adequate size to provide sufficient rate of flow for proper flushing of the fixture.

Flush valves through which water is supplied directly from the potable water supply system to water closets, urinals, or other fixtures should be equipped with means for protecting the system against contamination from back siphonage of water from the fixtures. Direct supply flush valves should be readily accessible for repairs, and they should be provided with convenient means for adjustment of the rate and volume of water discharged. They should be designed so that, when manually activated, they complete their normal cycle of operation, opening fully and closing positively under service pressure and delivering water in sufficient rate and volume to flush thoroughly the fixture supplied and refill the fixture trap seal.

LAVATORIES, BATHTUBS, AND SHOWERS

Lavatories should be provided with waste outlets not less than 1¼ in (31.3 mm) in diameter. Multiple-type lavatories, such as circular or straight-line multiple wash sinks or washfountains, should be considered equivalent to ordinary lavatories on the basis of one lavatory for each 18-in (455-mm) unit length of usable space at which hot and cold water are available along the perimeter of the fixture.

Bathtubs should be provided with waste outlets and overflows at least 1½ in (37.5 mm) in diameter and should be equipped with suitable stoppers. Where shower heads are installed above built-in bathtubs, waterproof joints should be provided between the tub and walls, and the walls should be constructed of smooth, noncorrosive and nonabsorbent waterproof materials to a height of not less than 6 ft (1.8 m) above the floor.

Waste outlets of shower floors and receptors should be at least 2 in (50 mm) in diameter so as to permit waste to drain off without puddling when the shower head discharges at normal maximum rate. A removable strainer should be provided in the waste outlet so that the shower trap may be cleaned from the waste outlet. Such provisions need not be made where shower heads are installed above bathtubs, or where emergency showers are provided for dousing persons exposed to contact with severely irritating chemicals.

Individual shower compartments should be of adequate floor area for satisfactory use by adults. For such service, the minimum floor area deemed necessary is 900 in^2 (58 cm^2), and the minimum span between walls for any compartment shape should be 30 in (760 mm). Floors under shower compartments should be smooth and structurally sound. Floors of public or institutional shower rooms should be drained in such manner that water from any shower head will not drain across areas occupied by other bathers.

Shower compartment floors, other than those installed directly on the ground or those having watertight receptors, should be constructed in a watertight shower pan of durable material. All sides of the pan should be turned up and extended at least 2 in (50 mm) above the finished floor level. The pan should be securely fastened to the shower waste outlet pipe at the seepage entrance so as to make a watertight joint between the pan and the outlet pipe. Where shower compartments are installed directly on the ground, they should have floors or receptors made of smooth, noncorrosive and nonabsorbent waterproof materials and be fastened securely to the fixture waste-outlet pipe so as to make a watertight joint therewith. Walls of shower compartments should be constructed of smooth, noncorrosive and nonabsorbent waterproof materials to a height of not less than 6 ft (1.8 m) above the floor.

SINKS AND LAUNDRY TRAYS

Sinks and laundry trays should be provided with waste outlets at least 1½ in (37.5 mm) in diameter to provide fixture drainage at a satisfactorily rapid rate. Each waste outlet of laundry trays should be provided with a suitable stopper so that water may be retained in the fixture for normal use.

A sink and one or two laundry trays, two or three sinks, or two or three laundry trays, grouped immediately adjacent to each other in the same room, may have their waste-outlet piping branched together into a single waste-outlet pipe for the group of fixtures and may be connected to the drainage system as a combination fixture. A one-piece combination fixture in which such sink or laundry tray compartments are grouped together should be treated and connected in the same manner.

Where a food waste grinder unit is to be installed in a sink, the waste opening in the fixture should be at least 3½ in (88 mm) in diameter. Food waste grinder units should be equipped with either automatic or hand-operated water supply controls so that the unit will operate only when water flows. This feature has been deemed necessary in order to minimize the incidence of stoppage development in the fixture drain. Such units should be installed only in sinks which are separately connected to the drainage system in order to avoid pumping ground food waste into other fixture compartments in the event of stoppage of the fixture drain.

Before installing a food waste grinder unit as a part of the plumbing system of a building in any given area, it is advisable to consult with the authority having jurisdiction regarding whether such units are permitted to be installed there. Some authorities do not permit such installations and strictly forbid the discharge of ground food waste into the public sewer system for reasons singular to their particular jurisdictions. In other areas, authorities may deliberately encourage such installations for their own reasons.

DRINKING FOUNTAINS

Nozzles of drinking fountains should be of nonoxidizing, impervious material. They should be located so that the lower edge of the nozzle orifice is at least ¾ in (19 mm) above the flood-level rim of the fixture and should be set at an angle so that no water can drip back onto the nozzle when the fixture is used. A nonoxidizing guard should be provided above and around the nozzle to prevent the mouth and nose of users from coming into contact with the nozzle. Guards should be designed to minimize the possibility of transmitting infection. Both nozzle and guard should be positioned so that the water jet does not strike the guard and cause spattering. Means must be provided for regulating and adjusting the supply of water to the drinking fountain nozzle so that a suitable jet of water occurs whenever the fountain is turned on by users.

Drinking fountain bowls should be designed free of corners to facilitate cleaning and to prevent any unnecessary splashing when the jet falls into the bowl. The waste outlet of the bowl should be at least 1 in (25 mm) in diameter and should be provided with a durable strainer. The height of the bowl and nozzle of a drinking fountain should be suitable for convenient use by the persons for whom it is intended.

Drinking fountains equipped with more than one nozzle above the receptacle should be deemed equivalent to the number of nozzles provided at reasonable spacing and accessible for users. Similarly, where properly installed drinking fountain nozzles are provided and approved for use at sinks or lavatories, such nozzles should be deemed equivalent drinking fountain fixtures.

DISHWASHING MACHINES AND FIXTURES

Domestic dishwashing machines may be provided in dwelling units to serve as convenient, laborsaving devices for washing and cleaning dishes, glasses, and cutlery. Many well-designed and highly efficient models are available. However, they should not be considered equivalent to, or suitable substitutes for, kitchen sinks which are required to be provided in dwelling units. Domestic machines may be designed to drain by gravity, or may be equipped with integral drainage pumps.

A separate trap should be provided for each domestic machine drained by gravity. Machines equipped with drainage pumps may discharge through a direct connection to the waste-outlet piping of an adjacent kitchen sink by means of a Y-branch fitting on the inlet side of the sink trap, provided that the pump discharge line rises to an elevation at least as high as the underside of the sink rim or counter before connecting to the Y-branch fitting. This latter provision has been found to be a satisfactory method of preventing backup of waste into the machine when the sink drain is clogged. Where the machines are indirectly connected to the drainage system, they should discharge through an air break into a fixture approved for such use.

Commercial dishwashing machines or three-compartment sinks of suitable type are required to be provided where food or drink is served for human consumption in public or employee dining establishments. In such establishments, all dishes, glasses, and cutlery which are to be reused must be washed and sanitized in conformity with standards established by the health authority having jurisdiction. Hot water supply for commercial dishwashing machines and dishwashing fixtures should be provided at 140 to 160°F (60 to 71.1°C) for washing, and at 180 to 190°F (82.2 to 87.8°C) for sanitizing. Commercial dishwashing machines process articles that have to meet sanitary standards. Consequently, such machines are required to be indirectly connected to the sanitary drainage system.

A minimum flow pressure of 15 psi [103 kilopascals (kPa)] should be available in the 180°F (82.2°C) hot water supply line in each commercial dishwashing machine during final sanitizing rinse spraying. Where the flow pressure exceeds 30 psi (206 kPa) either a pressure reducing valve or a flow control valve should be installed in the hot water supply line in order to keep the flow pressure between 15 and 30 psi (103 and 206 kPa).

FLOOR DRAINS

Floor drains should be of adequate size to serve the purpose for which they are intended without causing puddling at the drain inlet. The minimum size recommended for floor drains is 3 in (75 mm). Floor-drain inlets should be located so as to be readily accessible at all times and should be provided with removable strainers. The clear open area of such strainers should be at least two-thirds of the cross-sectional area of the drain pipe to which the floor drain connects.

Objections to the unnecessary installation of floor drains have been raised by authorities in many areas. Their objections have been based on the fact that the water seals of floor-drain traps evaporate over extended periods when they receive no wastes, and objectionable odors and gases escape from the drainage system into the building space where the floor drain is located.

When floor drains are installed, they should be provided with traps of the deep seal type. They should also be provided with convenient water supply by means of a faucet located not more than 3 ft (0.9 m) above the floor area drained, unless other suitable means for maintaining the trap seal are permitted.

FIXTURE OVERFLOWS AND STRAINERS

Integral overflow passageways and separate overflow pipes, provided as a means of permitting water to overflow from fixtures at a level below the flood-level rim, should be connected to the fixture waste-outlet fitting or pipe on the outlet side of the outlet stopper, plug or control valve, but on the inlet side of the fixture trap. In this way, water may overflow from the fixture and be conveyed into the drainage system, and the fixture trap serves to prevent drainage system gases and odors from escaping through the overflow.

The use of *standing-waste-and-overflow* fittings on fixtures has been condemned as promoting unsanitary conditions, and such fittings are generally prohibited. Where overflows are provided on fixtures, the waste fittings should be arranged so that water cannot rise in the overflow when the fixture is being filled, or remain in the overflow when the fixture is empty.

Durable strainers should be provided in the waste outlets of all plumbing fixtures except those which are designed with integral fixture traps. Fixed strainers should be provided except where specially required to be of removable type. The clear waterway area of strainers should be not less than that required of the fixture waste outlet so as not to impair satisfactorily rapid fixture drainage.

SPECIAL USE FIXTURES

Fixtures intended to receive and discharge objectionable quantities of detrimental wastes, such as substances which would clog the pipes, destroy the pipes or their joints, interfere unduly with the sewage disposal process, or produce explosive mixtures, should not be connected to the building sanitary drainage system unless such fixtures are provided with efficient and suitable means for the satisfactory treatment and handling of such wastes to render them unobjectionable and harmless. Intercepting strainers or intercepting fixture traps may be used where suitable. Where such means are unsuitable, fixtures receiving detrimental wastes should be connected to an independent drainage system specially designed to dispose of the wastes in an acceptable manner.

Fixtures which receive the discharge of indirect waste pipes should be of suitable shape and capacity to prevent splashing and flooding. No fixtures provided for domestic or culinary purposes should be used to receive the discharge of an indirect waste pipe, except that a sink or laundry tray in a dwelling unit may serve to receive indirect wastes from domestic appliances. However, in no case should a water closet, urinal, bathtub, or shower be used to receive indirect wastes, for such service would impede normal use of such fixtures and promote unsanitary conditions.

INTERCEPTING STRAINERS

An intercepting strainer, basket, or equivalent acceptable device should be provided at the waste outlet of every fixture or receptacle which receives wastes containing large, objectionable solid substances to prevent them from entering the drainage system where they could clog the piping. The device should be designed to intercept solids ½ in and larger in size. It should be easily removable for cleaning purposes.

Such a device should be provided at the waste outlet of a fixture or receptacle receiving the wash from garbage cans, or wastes containing strings, rags, buttons, paper, broken glass, bottle tops, feathers, entrails, or similar solid substances. Fixtures which receive sewage containing large, objectionable solids from indirect waste pipes should be equipped with a readily removable metal basket or beehive strainer, not less than 4 in (100 mm) in height, installed at the waste outlet of the fixture.

SWIMMING POOLS

There are three different types of swimming pools: (1) fill-and-draw, (2) flow-through, and (3) recirculating. The fill-and-draw type is the oldest. This type of pool is filled with clear water of sanitary quality, which is retained in the pool until it becomes turbid or unfit for use. Then the pool is drained, cleaned, and refilled with clear water of sanitary quality. Such pools have become relatively obsolete and are no longer installed. They are recognized as being hazardous to health because of their unreliability in maintaining the sanitary quality of the pool water.

The flow-through type of pool is one into which clear water of sanitary quality flows continuously, thereby displacing pool water at the same rate through overflows. This type of pool is little used in practice. In most cases, the cost of supplying the large amount of clear water needed for maintaining the sanitary quality of the pool water is relatively prohibitive. In addition, unsatisfactory pool water quality can occur in sections of the pool where clear water does not circulate adequately.

Recirculating-type pools use the same pool water over and over again, pumping water continuously from the pool, through a water treatment system, and back into the pool again. Only a small amount of clear water need be added to the pool from time to time to make up for the amount lost in operation and evaporation. Public pools are almost exclusively of this type because, through the use of suitable water treatment equipment and the recirculation of pool water, clear water of sanitary quality can be maintained in the pool at all times.

Design, installation, and maintenance of swimming pools should conform to regulations of the health authority having jurisdiction. Frequently, it is found that such regulations do not apply to pools located on the premises of a one-family dwelling and intended exclusively for private family use. However, the need for protecting the health of users of such pools has been clearly demonstrated and reported by authorities in the field.

Swimming pools should be of watertight construction. Surfaces within pools should be of smooth, nonabsorbent material with rounded corners and nonskid floors and should be constructed so as to be easily cleaned and disinfected. Preferably, pool surfaces should be light in color so that accumulations of dirt on the bottom and sidewalls will be visible and make evident the need for cleaning.

Public pools should be provided with overflow drains at the high waterline and should be designed so as to prevent the entrance of dirt, sand, foreign matter, or floor drainage from surrounding areas. Water inlets and outlets of pools should be located and spaced so as to secure satisfactory dispersion and complete circulation.

Each pool should be provided with at least one waste outlet so located that the entire pool can be emptied when necessary. The waste outlet should be equipped with a vortex and suction-reducing device consisting of an outlet strainer or grate having a total open area at least four times the cross-sectional area of the pool drain pipe. The drain from the pool should be equipped with a readily accessible gate valve so that water may be retained in the pool or emptied therefrom. The size of the waste outlet and drainage piping from a pool should be such that the pool may be completely drained in a period of 12 h for private family use pools, and in 4 h for all other pools, but the minimum size recommended in any case is 3 in (75 mm).

Filtering, sterilizing, and auxiliary equipment, where required by the health authority having jurisdiction, should be adequate to maintain the sanitary quality of pool water during each period that the pool is in use. Equipment containing gases or disinfectants capable of giving off irritating, toxic, or flammable fumes should be located in ventilated rooms or enclosures separated from the spaces used by bathers.

The recommended method for determining the maximum number of bathers permitted to use a pool at any one time is to allow 25 ft^2 (2.3 m^2) of pool area for each bather. However, the sanitary quality of pool water may be maintained for this maximum number of bathers only if the recirculation pump has sufficient capacity to recirculate chlorinated water with the rapidity necessary to maintain residual chlorine throughout the pool during such maximum use. Hence, the usable capacity of the pool is based upon the capacity of the recirculating pump and the water treating equipment. Experience has shown that the maximum capacity of any given pool is available when the recirculating pump is able to recirculate the entire contents of the pool in 8 h or less, and the freshly chlorinated water is introduced through a number of inlets located on all four sides of the pool.

RELIGIOUS, ORNAMENTAL, AND AQUARIUM FIXTURES

Baptistries, ornamental and lily pools, aquaria, ornamental fountain basins, and similar special fixtures need not be provided with connections to the water supply and sanitary drainage systems of buildings as required for ordinary plumbing fixtures. However, where water supply or sanitary drainage system connections are provided for such fixtures, the connections should be made in the same manner as would be required for ordinary plumbing fixtures.

FIXTURE INSTALLATION

Plumbing fixtures should be located in spaces which are adequately lighted and ventilated in conformity with applicable building regulations.

This is a basic sanitary requirement. Wherever fixtures are installed, moisture is added to the air when water flows, spattering is prone to occur, and fixture odors develop. Lack of adequate lighting or ventilation promotes poor maintenance and unsanitary conditions. Artificial light of suitable intensity should be provided in all rooms or spaces where fixtures are installed. Ventilation may be either natural or mechanical.

Water closets, urinals, bathtubs, and showers should be located only in rooms or compartments which are adequately ventilated directly to the outer air, or are provided with independent mechanical ventilating systems which exhaust air from such spaces to the outer air. For natural ventilation, the minimum openable area of windows should be at least 1½ ft^2 (0.14 m^2) for a private toilet room or bathroom and at least 1 ft^2 (0.09 m^2) per water closet or urinal, but no less than 3 ft^2 (0.28 m^2) for a toilet room or bathroom used by the public or employees. For mechanical ventilation, air should be exhausted at a rate of 25 ft^3/min (0.012 m^3/s) for private bathrooms or toilet rooms, and at a rate of 40 ft^3/min (0.019 m^3/s) per water closet or urinal for a toilet room or bathroom used by the public or employees.

Fixtures which receive the discharge of indirect waste pipes should not be located in any unventilated storeroom or closet, as unsanitary conditions are bound to result. They should be located in well-lighted and amply ventilated spaces where the use of such fixtures does not constitute a nuisance. Preferably, they should be located where any need for maintenance will be readily apparent to building occupants.

In one- and two-family dwellings, it is recommended that the floors of toilet rooms and bathrooms be of materials that are impervious to moisture so that they may be washed and cleaned thoroughly. In buildings other than one- and two-family dwellings, toilet rooms and bathrooms should be provided with waterproof floors and with waterproofing extending at least 6 in (150 mm) above the floors, except at doors. This is required so that the floors of such rooms may be quickly and easily maintained in sanitary condition by mopping or hosing. Walls and floors adjacent to urinals should be finished with noncorrosive and nonabsorbent materials extending at least 1 ft (0.3 m) in front of the urinal lip, 1 ft (0.3 m) on each side of the urinal, and 4 ft (1.2 m) above the floor. This too has been found necessary as a sanitary measure for normal service.

Water closets, urinals, bathtubs, and showers should not be located on the next floor directly above space used for the manufacture, preparation, packaging, storage, or display of food unless an additional watertight barrier is provided to intervene between the toilet room or bathroom floor and such space immediately below (see Fig. 3-1).

In general, it is recommended that no drinking fountain or equivalent

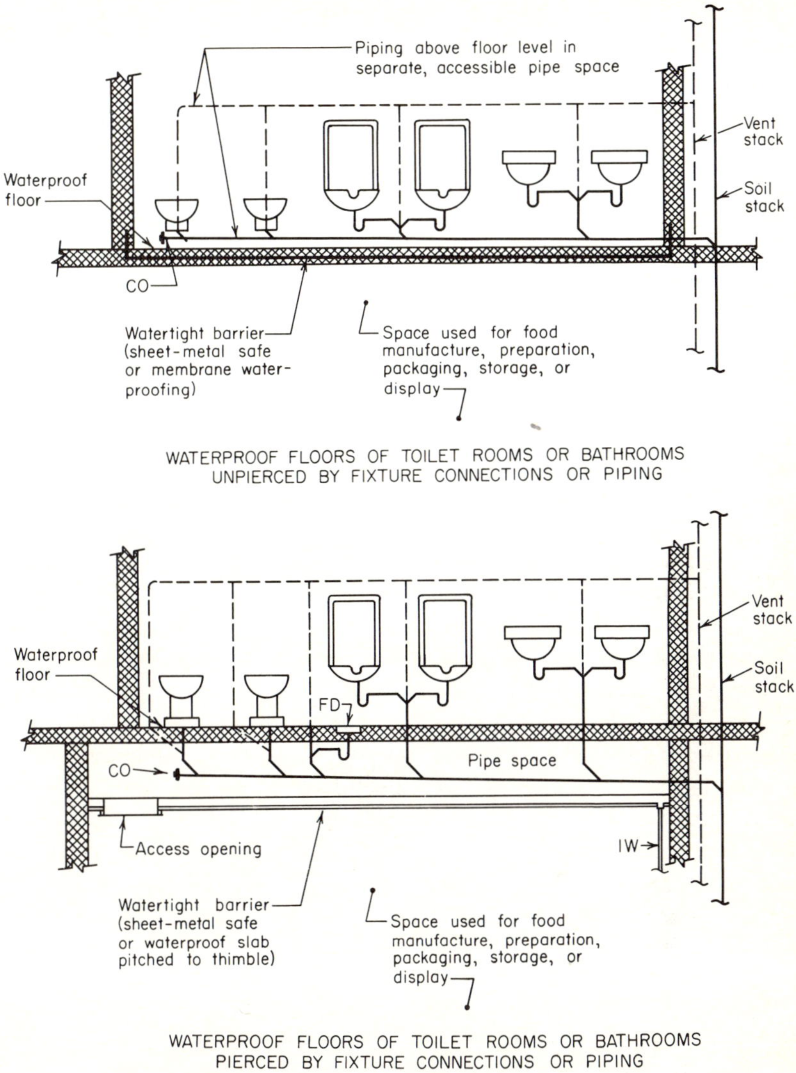

Fig. 3-1 Watertight floor or intervening watertight barrier.

drinking fountain fixture be located in bathrooms or toilet rooms. However, it has been deemed permissible to provide drinking fountains in rooms containing no more than one water closet or urinal, such as may be considered to be the condition existing in bathrooms or toilet rooms for private use where drinking water normally is obtained at lavatories.

Fixtures and equipment should be located so as not to interfere with the normal operation of windows, doors, or exit openings. Fixtures should be set level and in proper alignment with adjacent walls. They should be installed with regard to spacing so as to be reasonably accessible for their intended use, and for cleaning and repairs. Where fixtures are installed in contact with walls or floors, the space at the outer edge of fixture contact should be sealed against water seepage so that moisture cannot get between the fixture and the wall and promote development of vermin.

Wherever practicable, fixture supply and drain pipes should be run to piping connections in the nearest wall rather than through the floor. Similarly, wall-hung and built-in fixtures are preferred over floor-standing types because of greater ease in cleaning and maintaining floors. Fixtures having concealed packing or gasket-type slip-joint connections should be provided with an access panel or utility space to make the slip joint accessible for repair when necessary.

Wall-hung water closets and urinals should be rigidly secured against the wall and rigidly held in place by a durable concealed support so that no strain is transmitted to the piping connections. Fixtures having floor outlets should be attached securely to the floor. Fixture connections between drainage pipes and water closets, pedestal urinals, floor-outlet service sinks, and earthenware trap standards should be made by means of approved flanges. They should be soldered, screwed, or otherwise securely attached to the drainage pipes in a manner suitable for the type of drainage piping. The flange should be installed on a firm base made of materials impervious to moisture. An acceptable gasket, washer, or setting compound should be placed between the earthenware fixture outlet and the flange, and the fixture should be bolted tightly to the flange so as to form a gastight joint. Commercial putty and plaster of paris are unsuitable for use as fixture gasket material for they tend to wash away in service.

FIXTURE TRAPS

Each fixture directly connected to the sanitary drainage system should be equipped with a water seal trap to prevent the escape of drainage system odors and gases into the atmosphere at fixture locations. Similarly, fixtures which discharge wastes indirectly through drainage piping exceeding 4 ft (1.2 m) in developed length, measured from the fixture outlet, should be equipped with a water seal trap. In this case, any greater length of untrapped waste piping is considered to be productive of an excessive and objectionable amount of odor and to warrant installation of a fixture trap to prevent air circulation through the fixture waste outlet. However, no trap is deemed to be necessary in the drain from a swimming pool which discharges indirectly, or from a private swimming pool which discharges through an independent drainage and disposal system specifically provided for that purpose.

A single trap may be provided for a sink and one or two laundry trays, two or three sinks, or two or three laundry trays grouped immediately adjacent to each other in the same room. But in this combination or group of fixtures, no sink should be equipped with a food waste grinder unit because, in the event of a stoppage in the fixture trap or drain, ground food waste would be pumped into other fixtures of the group when the grinder was used.

Several designs of fixture traps used in the past and found to be unsatisfactory and objectionable are generally stated in codes as prohibited types, such as a trap which depends for its seal upon the action of movable parts; crown-vented traps; and bell traps, except where installed in refrigerator safes or receptors. Movable parts in traps soon become coated with waste matter, rendering their performance ineffective. Crown vents on traps have a similar record. They soon become clogged with deposits of waste matter centrifuged into the vent at the crown of the trap, at which point they are worth very little. Bell traps cannot perform their function when the bell is removed, displaced from normal position, or broken. Nevertheless, this type of trap is permitted for refrigerator safes and receptors which are set at floor level. In this location, the bell is visible, may be easily removed, and may serve as a convenient means of access for cleaning the indirect waste piping to which the trap connects.

In order to avoid stoppages in fixture traps, they should be self-cleaning, except for those which are specially designed to intercept grease and sediment. Traps which are integral with fixtures should have uniform interiors and smooth waterways. No movable parts should be provided in traps, nor should they be designed with interior partitions, except when they are integral with fixtures or are designed for grease or sediment interception. Bodies of drum traps should be either 3 or 4 in (75 or 100 mm) in diameter. In the design of intercepting traps, provision should be made to prevent them from becoming airbound.

All fixture traps should be designed with a water seal of not less than 2 in (50 mm) and not more than 4 in (100 mm) in depth, except where permitted for use in special conditions. Traps with less than 2 in (50 mm) of seal depth should be avoided, in view of the fact that the criterion for sizing the vent-piping system is based upon the use of traps with a minimum seal depth of 2 in (50 mm).

Convenient access to the interior of fixture traps is necessary so that they may be readily cleared whenever stoppages occur. An accessible brass screw cleanout or plug should be provided in the water seal of each common fixture trap, preferably at the bottom of the trap where it may serve as a drain plug. However, such cleanout provisions are unnecessary where the fixture trap is integral, or combined with the fixture, and the trap seal is readily accessible from the fixture, or where a portion of the trap is easily removable for cleaning purposes. Where a fixture serves as a receptor for indirect wastes and is set below floor level, a running trap with a cleanout extended to floor level should be provided adjacent to the fixture. Special fixture traps designed for the

Table 3-1
MINIMUM SIZES OF TRAPS FOR VARIOUS PLUMBING FIXTURES

Fixture	Size	
	in	mm
Bathtub (with or without overhead shower)	1½	37.5
Bidet	1½	37.5
Combination sink and wash tray	1½	37.5
Combination sink and wash tray with food waste grinder unit	1½*	37.5
Dental unit or cuspidor	1¼	31.3
Dental lavatory	1¼	31.3
Drinking fountain	1¼	31.3
Dishwasher, commercial	2	50.0
Dishwasher, domestic	1½	37.5
Floor drain	3	75.0
Kitchen sink, domestic	1½	37.5
Kitchen sink, domestic, with food waste grinder unit	1½	37.5
Lavatory, common	1¼	31.3
Lavatory (barber shop, beauty parlor, or surgeon's)	1½	37.5
Lavatory, multiple type (washfountain or wash sink)	1½	37.5
Laundry tray (1 or 2 compartments)	1½	37.5
Shower stall	2	50.0
Sink (surgeon's)	1½	37.5
Sink (flushing rim type, flush valve supplied)	3	75.0
Sink (service type with floor outlet trap standard)	3	75.0
Sink (service type with P trap)	2	50.0
Sink, commercial (pot, scullery, or similar type)	2	50.0
Sink, commercial (with food waste grinder unit)	2	50.0
Urinal (pedestal type, integral trap)	3 nominal	75.0
Urinal (all types with integral traps except pedestal type)	2 nominal	50.0
Urinal (stall washout type, separate trap)	2	50.0
Urinal (wall hung washout type, separate trap)	1½	37.5
Water closet	3 nominal	75.0

* Separate trap required for wash tray and separate trap required for sink compartment with food waste grinder unit.

interception of grease, hair, or similar substances may have gastight covers, handholes, or other cleanout provisions held in place by lugs or bolts.

Traps should be installed level with respect to their water seals so that the seal strength which they are designed to afford may be fully utilized to resist pneumatic pressure fluctuations in the drainage system. They should be located in areas where they are not subject to damage from freezing, or should be provided with adequate protection against frost conditions. It is recommended that fixture traps be installed as close as practicable to the fixture waste outlet in order to minimize loss of trap seal due to momentum effects when the fixture discharges and to minimize odors from the fouled interior of piping between the waste outlet and the trap. The maximum developed length from the fixture waste outlet to the trap usually is limited to 24 in, or 2 ft (0.6 m). However, this length may be permitted to be as much as 48 in (1.2 m) where fixtures are located remote from all walls, such as "island" sinks, washfountains, and lavatories, provided the fixture has a large interior flat bottom 120 (7.7 cm^2) or more in area, or is not equipped with a waste-outlet stopper or plug. Excessive trap seal losses do not occur with such fixtures when they discharge. The greater length of piping between the fixture trap and the waste outlet is no more than would be exposed to room atmosphere where a combination fixture was installed, and should be no more productive of objectionable odors.

The size of the trap for any given fixture should be sufficient to permit satisfactorily rapid fixture drainage. In no case should a fixture trap be larger than the fixture drain to which it connects, for then an obstruction or ledge would be formed at the trap outlet and promote stoppage of the trap. Recommended minimum sizes for fixture traps are given in Table 3-1.

USE OF INTERCEPTING FIXTURE TRAPS

Intercepting fixture traps should be provided only where they are deemed to constitute the most suitable and acceptable means of removing objectionable substances from detrimental wastes. Such traps should be installed strictly in accordance with their approvals in respect to type, size, rating, and location. No wastes should be discharged through intercepting traps other than those for which the traps are designed to handle. Each intercepting trap should be installed and located to provide ready accessibility to the cover and other means for maintaining the device in efficient operating condition. All intercepting traps require maintenance. Accumulated intercepted matter must be removed periodically in order for them to operate efficiently and perform satisfactorily.

Approved grease-intercepting traps should be installed in the drains of fixtures through which an objectionable amount of grease usually enters the building drainage system, such as in the following establishments: restaurants, hotel kitchens or bars, factory cafeterias or restaurants, clubs and other commercial kitchens. In such establishments, grease-intercepting traps should be installed in the fixture drains of pot, scullery, or food-scrap sinks, and in floor drains which receive waste or spillage from soup or stock kettles. However, no grease-intercepting trap should be installed in the fixture drain of a sink that is equipped with a food waste grinder unit, for the trap would intercept and be rapidly filled with ground food waste.

In commercial establishments, an approved sediment-intercepting trap for intercepting plaster, hair, silt, sand, or similar solid substances should be installed in the fixture drain of each fixture at which such substances are introduced into the drainage system in objectionable quantities. Approved sediment-intercepting traps of suitable types should be installed in the fixture drains of dental and orthopedic laboratory sinks as well as of sinks receiving wastes resulting from various activities such as hair removal processes and commercial car washing.

VAPOR RELIEF PIPES OR LOCAL VENTS FOR FIXTURES

Vapor relief pipes, sometimes termed local vents or local ventilation pipes, may be provided for fixtures. Many years ago, water closets and urinals were equipped with special ventilation piping connected directly to the fixtures and extended outdoors, so as to ventilate foul odors from the unsanitary types of fixtures used at the time. These pipes were called local vents or local ventilation pipes as they were intended to ventilate the odorous fixtures and remove air from adjacent areas, rather than to ventilate the interior of the sanitary drainage system. With the development of sanitary types of water closets and urinals and the requirement of adequate ventilation of toilet rooms and bathrooms, the need for installing local ventilation pipes for such fixtures disappeared.

However, numerous hospital fixtures are designed to use steam or boiling water for cleaning and for sterilizing. Steam is used in fixtures designed to sterilize surgical instruments, bandages, dressings, bottles, pans, bedpans, and numerous other items. To dissipate steam and water vapor from such fixtures before they are opened, vapor relief pipes may be connected to the fixtures and extended outdoors. Vapor relief pipes for bedpan washers or bedpan steamers should not be connected to vapor relief pipes serving other fixtures so as to prevent the escape of foul odors from the vapor relief connections at such other fixtures. Vapor relief pipes should be independent of other vent pipes, ventilating ducts, and flues.

Where vapor relief pipes are provided for fixtures on two or more floors and are connected as branches to a vapor relief stack, the stack should be extended independently through the building roof, or to an approved location in the open air, and terminated as required for stacks of the sanitary drainage system. Vapor relief pipes in which condensation can collect should be provided with drip pipes. The drip pipe for an individual vapor relief pipe may be connected to the waste-outlet piping on the inlet side of the fixture served by the vapor relief pipe. Other drip pipes should drain in the same manner as indirect waste pipes and discharge into a fixture or receptacle approved for such use.

The size of individual vapor relief pipes should be at least as large as the relief outlet of the fixture in each case. Vapor relief stacks and branch pipes serving two or more individual vapor relief pipes should be at least one standard pipe size larger than the largest individual pipe connecting thereto, but in no case less than 1¼ in (31.3 mm) in size. Vapor relief stacks should extend upward undiminished in size from the lowest vapor relief branch to the vent terminal in the open air. Drip pipes connected at the base of vapor relief stacks should be at least 1¼ in (31.3 mm) in size.

STANDARD CODE REGULATIONS

Quality of Fixtures Plumbing fixtures shall have smooth impervious surfaces, shall be durable for the uses intended, and shall be free from defects and concealed fouling surfaces.

Water Closets and Urinals Pan, valve, plunger, offset, washout, latrine, frost-proof, and other water closets having an invisible seal or unventilated spaces or having walls which are not thoroughly washed at each flushing, shall be prohibited. Water closets which might permit the contents of the water closet to be siphoned back into the flush tank shall be prohibited.

Trough urinals shall be prohibited, except where specially approved by the local authority having jurisdiction.

Water closets for public use shall be of the elongated bowl type.

In schools, nurseries, and other similar occupancies where fixtures are provided for the use of children under 6 years of age, water closets shall be of suitable type.

Water closets shall be equipped with seats of smooth nonabsorbent material. Integral water closet seats shall be of the same material as the fixture. Seats of water closets provided for public use shall be of the open-front type.

Walls and floors adjacent to urinals shall be finished with noncorrosive, nonabsorbent material extending at least 1 ft (0.3 m) in front of the urinal lip, 1 ft (0.3 m) on each side of the urinal, and 4 ft (1.2 m) above the floor.

Flushing devices shall be provided at each water closet and urinal, except as otherwise required and shall be designed and installed so as to supply water at adequate volume and rate to flush satisfactorily the water closets or urinals with which they are connected.

Flushing Devices Flush tanks shall have sufficient flow capacity to flush properly the water closets or urinals with which they are connected.

No flush tank shall supply more than one water closet or urinal, except that, where approved, a single flush tank may be used to flush more than one urinal provided the tank is automatic in operation and of sufficient capacity to provide the amount of water required to flush and cleanse properly and simultaneously all the urinals connected thereto.

Flush pipes and fittings between flush tanks and water closets or urinals shall be of adequate size to provide sufficient rate of flow for proper flushing.

Where flush tanks are supplied through ball cocks which are connected directly to the potable water supply system, such ball cocks shall be installed as described in "Protective Methods For Fixtures With Below-Rim Potable Water Supply Outlets" in the standard code regulations in Chap. 4.

Ball cocks in flush tanks shall be designed to operate automatically, refilling the tank after each discharge and shutting off completely when the tank is filled to operational capacity. In lowdown flush tanks, ball cocks shall also be designed with means for bypassing an adequate amount of water to the water closet seal so as to refill the trap seal while refilling the tank after each flushing.

Flush valves in tanks shall be designed for manual operation, except those in flush tanks which are required to be automatic in operation.

Seats of flush valves in tanks for flushing water closets shall be at least 1 in (25 mm) above the rim level of the water closets connected thereto, except in approved water closet and flush tank combinations designed so that, when the water closet is clogged and the tank is flushed, the flush valve will close tightly and water will not spill continuously over the rim of the water closet.

Flush tanks shall be provided with overflows of adequate size to prevent tank flooding at the maximum rate at which the tanks are supplied with water. The overflow of a flush tank shall discharge into the water closet or urinal connected thereto.

Flush valves connected directly to the potable water supply system shall be installed as described in "Protective Methods For Fixtures With Below-Rim Potable Water Supply Outlets."

Flush valves shall be readily accessible for repairs. Convenient means shall be provided for adjustment of the volume and rate of water discharged by flush valves.

Flush valves shall be designed so that when manually activated, they shall complete their cycle of operation, opening fully and closing positively under service pressure, and shall deliver water in sufficient volume and rate to flush thoroughly the fixture supplied and refill the fixture trap seal.

Lavatories Lavatories shall have waste outlets not less than 1¼ in (31.3 mm) in diameter.

Multiple type lavatories, such as circular or straight-line multiple wash sinks or washfountains, shall be considered equivalent to a number of ordinary lavatories on the basis of one lavatory for each 18-in (455-mm) unit length of usable space at which hot and cold water are available along the perimeter of the fixture.

Bathtubs Bathtubs shall be provided with waste outlets and overflows at least 1½ in (37.5 mm) in diameter, and shall be equipped with a suitable stopper.

Showers Waste outlets of showers shall have removable strainers and shall be at least 2 in (50 mm) in diameter, except for bathtubs serving as receptors for shower heads installed above, and except for emergency showers for which no drains are provided.

Shower compartments, except those installed directly on the ground or those having watertight enameled metal receptors or an approved equivalent, shall have floors constructed in a watertight shower pan of durable material. The pan shall turn up on all sides at least 2 in (50 mm) above the finished floor level, and shall be securely fastened to the fixture waste outlet pipe at the seepage entrance, making a watertight joint between the pan and the outlet pipe.

Shower compartments directly on the ground shall have floors or receptors constructed of smooth, noncorrosive, and nonabsorbent waterproof materials, and shall have such floors or receptors securely fastened to the fixture waste outlet pipe, making a watertight joint thereto.

Individual shower compartments shall have floor areas of at least 900 in^2 (58 cm^2) and if rectangular, square, or triangular in plan, shall be not less than 30 in (760 mm) in least dimension.

Floors under shower compartments shall be smooth and structurally sound.

Floors of public or institutional shower rooms shall be drained in such manner that water from any shower head will not drain across areas occupied by other bathers.

Shower compartments shall have walls constructed of smooth, noncorrosive and nonabsorbent waterproof materials to a height of not less than 6 ft (1.8 m) above the floor.

Built-in bathtubs above which shower heads are installed shall have waterproof joints between the tub and walls, and such walls shall be constructed of smooth, noncorrosive, and nonabsorbent waterproof materials to a height of not less than 6 ft (1.8 m) above the floor.

Laundry Trays Each compartment of a laundry tray shall be provided with a waste outlet at least 1½ in (37.5 mm) in diameter, and shall be equipped with a suitable stopper.

Sinks Sinks shall be provided with waste outlets at least 1½ in (37.5 mm) in diameter.

No food waste grinder unit shall be installed as part of the plumbing system unless the use of such units is specially approved.

A sink equipped with a food waste grinder shall have a waste opening at least 3½ in (88 mm) in diameter.

Food waste grinder units installed in sinks shall be equipped with either automatic or hand-operated water supply controls, so that the unit will operate only when water flows.

Drinking Fountains and Equivalent Fixtures Drinking fountains shall conform to American Standard Specifications for Drinking Fountains, ANSI A112.11.3-1973.

Drinking fountain nozzles shall be placed so that the lower edge of the nozzle orifice is at an elevation not less than ¾ in (19 mm) above the flood-level rim of the receptacle.

Drinking fountains equipped with more than one nozzle shall be deemed equivalent in number to the number of nozzles provided at reasonable spacing and accessible for users. Where properly installed drinking fountain nozzles are provided and approved for use at sinks or lavatories, such nozzles shall be deemed equivalent drinking fountain fixtures.

Dishwashing Machines and Fixtures A separate trap shall be provided for a dishwashing machine which drains by gravity and is directly connected to the drainage system. Machines having drainage pumps may discharge into the waste outlet piping of an adjacent kitchen sink by means of a Y-branch fitting on the inlet side of the sink trap provided that the pump discharge line rises to an elevation at least as high as the underside of the sink rim or counter. Where indirectly connected, dishwashing machines shall discharge through an air break into a fixture approved for such use.

Commercial dishwashing machines and dishwashing fixtures using hot water shall be provided with water at 140 to 160°F (60 to 71.1°C) for washing, and at 180 to 190°F (82.2 to 87.8°C) for sanitizing.

Floor Drains Floor drains shall be provided with removable strainers. The open area of the strainer shall be at least two thirds of the cross-sectional area of the drain pipe to which the floor drain connects.

The floor drain inlet shall be located so that it is readily accessible at all times.

Traps of floor drains shall be of the deep seal type. Floor drains shall be provided with water supply by means of a faucet located not more than 3 ft (0.9 m) above the floor area drained, except where other suitable means for maintaining the trap seal are approved.

Floor drains shall be of such size as to serve efficiently the purpose for which they are intended, but in no case less than 3 in (75 mm) in nominal diameter.

Fixture Overflows Fixtures provided with overflows shall have their wastes arranged so that standing water in the fixture cannot rise in the overflow when the stopper is closed or remain in the overflow when the fixture is empty.

The overflow pipe from a fixture shall be connected on the inlet side of the fixture trap, except that overflows for flush tanks may discharge into the water closets or urinals served by their flush tanks; but such overflows shall not be connected with any part of the drainage system.

Common Fixture Strainers Common plumbing fixtures, except water closets and siphon-action washdown or blowout urinals, shall be provided with durable strainers installed in the fixture waste outlets. Such strainers shall have waterway areas adequate for satisfactorily rapid fixture drainage.

Special Use Fixtures No fixture shall be provided or used to receive and discharge into the building drainage system any objectionable quantities of detrimental wastes, such as substances which will clog the pipes, destroy the pipes or their joints, interfere unduly with the sewage disposal process, or produce explosive mixtures, unless such fixtures are provided with efficient and approved means for the satisfactory treatment and handling of such wastes to render them unobjectionable and harmless.

The means provided for the treatment and handling of wastes at such fixtures shall conform to the provisions of this chapter, or such fixtures shall be connected to an independent drainage system conforming to the provisions for such systems in Chap. 8.

No fixture used for domestic or culinary purposes shall be used to receive the discharge of an indirect waste pipe, except that sinks and wash trays in dwelling units may receive the indirect wastes from domestic appliances. No water closet, urinal, bathtub, or shower shall be used to receive indirect wastes.

Fixtures which receive the discharge of indirect waste pipes shall be of such shape and capacity as to prevent splashing and flooding.

No garbage can washer shall discharge through a trap serving any other device or fixture.

Intercepting Strainers for Special Use Fixtures An approved intercepting strainer, basket, or equivalent device shall be installed in the waste outlet of every receptacle receiving wastes which contain large, objectionable solid substances to prevent their passage into the drainage system. The intercepting device shall be designed to intercept solids ½ in (13 mm) and larger in size, and to be easily removable for cleaning purposes.

Plumbing receptacles permitted to receive wastes containing strings, rags, buttons, or similar solid substances shall be equipped with an approved intercepting strainer, basket, or equivalent device.

Plumbing receptacles permitted to receive wastes containing feathers, entrails, or similar solid substances shall be equipped with an approved intercepting strainer, basket, or equivalent device.

A fixture which receives from indirect waste pipes discharges containing large, objectionable solid substances shall be equipped with a readily removable metal basket or beehive strainer, not less than 4 in (100 mm) in height, installed at the waste outlet in the fixture.

The receptacle receiving the wash from garbage cans shall be provided with a strainer, basket, or similar intercepting device to prevent the discharge of large particles into the building sanitary drainage system.

Swimming Pools Swimming pools, except those for private family use, shall be designed, installed, and maintained in accordance with applicable regulations of the health authority having jurisdiction.

Swimming pools shall be of watertight construction. Inside surfaces shall be of smooth, nonabsorbent material with rounded corners and nonskid floors, and shall be constructed so as to be easily cleaned.

Swimming pools shall be provided with at least one drain outlet so located that the entire pool can be emptied, and the drain therefrom shall be equipped with a readily accessible gate valve.

Each drain outlet shall be equipped with a vortex and suction reducing device consisting of an outlet strainer or grate having a total open area at least equal to four times the cross-sectional area of the pool drain pipe.

The drain outlet and the drain piping therefrom shall be of such size as to permit the pool to be completely drained in a period of 12 h for private family use pools, and of 4 h for other pools, but shall be not less than 3 in (75 mm) in size.

Filtering, sterilizing, and auxiliary equipment, where required by the health authority having jurisdiction, shall be adequate to maintain the sanitary quality of pool water during each period the pool is in use.

Equipment containing gases or disinfectants capable of giving off irritating, toxic, or flammable fumes shall be located in ventilated rooms.

The installation shall be designed to prevent dirt, sand, foreign matter, and water from surrounding areas from entering the pool.

Religious, Ornamental, and Aquarium Fixtures Baptistries, ornamental and lily pools, aquaria, ornamental fountain basins, and similar specialty fixtures, when provided with water supply or drainage connections to the plumbing system, shall have such connections in conformity with these regulations.

Fixture Installation Plumbing fixtures, except drinking fountains and single lavatories, shall be located only in rooms or compartments lighted and ventilated in conformity with applicable building regulations.

Water closets, urinals, bathtubs, and showers shall be located only in rooms or compartments which are ventilated directly to the outer air or provided with independent mechanical ventilating systems which exhaust air from such spaces to the outer air.

Fixtures which receive indirect wastes shall be located in properly lighted and ventilated spaces where the use of such fixtures does not constitute a nuisance, and shall not be located in any unventilated storeroom or closet.

In buildings other than one- and two-family dwellings, water closets, urinals, bathtubs, and showers shall be located only in toilet rooms or bathrooms provided with waterproof floors and with waterproofing extending at least 6 in (150 mm) above the floors, except at doors.

Water closets, urinals, bathtubs, and showers shall not be located on the next floor directly above space used for manufacture, preparation, packaging, storage, or display of food, except that they may be so located if an additional watertight barrier is provided to intervene between the toilet room or bathroom floor and such space immediately below.

Drinking fountains and equivalent fixtures provided as sources of drinking water shall not be located in rooms containing more than one water closet or urinal.

Fixtures and equipment shall be located so as not to interfere with normal operation of windows, doors, or exit openings.

In schools, nurseries, and other similar occupancies where fixtures are provided for the use of children under 6 years of age, such fixtures shall be located convenient to the spaces in which the children study, play, or sleep, and shall be installed so as to be fully and safely usable.

Plumbing fixtures shall be installed with regard to spacing so as to be reasonably accessible for their intended use, and for cleaning and repairs. Where practical, fixture supply and drain pipes shall be run to piping connections in the nearest wall rather than through the floor. Fixtures having concealed packing or gasket type slip-joint connections shall be provided with an access panel or utility space arranged so as to make the slip joint accessible for repair.

Fixtures shall be set level and in proper alignment with adjacent walls.

Where fixtures are installed in contact with walls or floors, the space at the outer edge of fixture contact shall be sealed against water seepage.

Wall-hung water closets and urinals shall be rigidly secured against the wall and shall be rigidly supported by a durable concealed support so that no strain is transmitted to the piping connections.

Fixtures having floor outlets shall be rigidly secured to the floor.

Fixture connections between drainage pipes and water closets, pedestal urinals, floor-outlet service sinks, and earthenware trap standards, shall be made by means of approved flanges which shall be soldered, screwed, or otherwise securely attached to the drainage pipes in a manner approved for the type of drainage piping. The flange shall be set on an approved firm base made of materials impervious to moisture. The connection shall be bolted, with an approved gasket, washer or setting compound between the earthenware and the flange. The use of commercial putty or plaster instead of an approved gasket, washer, or setting compound is prohibited.

Fixture Traps Plumbing fixtures, other than those having integral traps, shall be provided with separate water seal traps placed as close as possible to the fixture waste outlets, except that

> A combination plumbing fixture, not equipped with a food waste grinder unit, may be installed on one trap if one compartment is not more than 6 in (150 mm) deeper than the other.
>
> One trap may be installed for a group of not more than three single laundry trays, or three single sinks, or one sink and two laundry trays, immediately adjacent to each other in the same room, if the trap is centrally located when three such fixtures are installed.
>
> No trap need be provided for fixtures and equipment which discharge their wastes indirectly through drainage piping not exceeding 4 ft (1.2 m) in developed length, measured from the fixture outlet.
>
> No trap need be provided for swimming pools which discharge their wastes indirectly or for private swimming pools which discharge through an independent drainage and disposal system.

The following traps are prohibited: a trap which depends for its seal upon the action of movable parts; crown-vented traps; and bell traps, except where installed in refrigerator safes or receptors.

All fixture traps, except grease and sediment intercepting traps, shall be self-cleaning; traps which are integral with fixtures shall have uniform interiors and smooth waterways; traps shall have no movable parts; traps shall have no interior partitions except where such traps are integral with fixtures or are designed for grease or sediment interception; and the bodies of drum traps shall be either 3 or 4 in (75 or 100 mm) in diameter. Intercepting traps shall be designed so as not to become airbound.

Each fixture trap shall have a water seal of not less than 2 in (50 mm) and not more than 4 in (100 mm) in depth, except for traps having a deeper seal and approved for use in special conditions.

The maximum developed length from the fixture waste outlet to the trap shall be not more than 24 in (0.6 m), except that where a fixture is located remote from all walls such developed length shall be not more than 48 in (1.2 m), provided the fixture either has an interior flat bottom exceeding 120 in^2 (7.7 cm^2) in area or is not equipped with a waste outlet stopper or plug.

Traps shall be set level with respect to their water seals and, where necessary, shall be protected against damage from freezing.

Fixture traps, except those which are integral or combined with fixtures wherein the trap seal is readily accessible, and except those having a portion of the trap readily removable for cleaning purposes, shall have accessible brass screw cleanouts or plugs so installed as to be watertight; special approved traps used for the interception of grease, plaster, hair, or similar substances may have covers, handholes, or other cleanout provisions held in place by lugs or bolts.

Where a receptor for indirect wastes is set below floor level, it shall be provided with a running trap set adjacent thereto with the trap cleanout extended to the floor level.

The size of trap for any given fixture shall be sufficient to drain the fixture rapidly but not smaller than as stated in Table 3-1. No trap shall be larger than the fixture drain to which it is connected.

An approved grease intercepting trap shall be installed in the fixture drain of sinks, floor drains, and other fixtures through which grease usually is introduced into the drainage system in objectionable quantities in the following establishments: restaurants, hotel kitchens or bars, factory cafeterias or restaurants, clubs and other commercial kitchens. In such establishments, grease traps shall be installed in the fixture drains of pot, scullery, or food scrap sinks, and of floor drains receiving waste or spillage from soup or stock kettles.

No grease intercepting trap shall be installed in the fixture drain of a sink equipped with a food waste grinder unit.

In commercial establishments, an approved sediment intercepting trap for intercepting plaster, hair, silt, sand or similar solid substances shall be installed in the fixture drain of each fixture through which sediment usually is introduced into the drainage system in objectionable quantities. Sediment intercepting traps shall be installed in the fixture drains of dental laboratory sinks, orthopedic laboratory sinks, sinks receiving wastes resulting from hair removal processes, and commercial car washing equipment.

Intercepting traps shall be installed in accordance with approvals as to type, size, rating, and location, and so that no wastes shall be discharged through the traps other than those which the traps are designed to handle. Each intercepting trap shall be installed so as to provide ready accessibility to the cover and other means for maintaining the device in efficient operating condition.

Intercepting traps shall be maintained in efficient operating condition by periodic removal of accumulated intercepted matter.

Vapor Relief Pipes for Fixtures Where fixtures are provided with vapor relief pipes, such pipes shall be independent of other pipes, ventilating ducts, and flues.

Vapor relief pipes in which condensation can collect shall be provided with drip pipes. Such drip pipes shall be either connected to the waste outlet piping on the inlet side of the trap of the fixture served by the relief pipe, or shall drain into a fixture or receptacle approved for such use.

Vapor relief pipes provided for bedpan washers and bedpan steamers shall be independent of vapor relief pipes serving other types of fixtures.

Where vapor relief pipes are provided for fixtures on two or more floors and are connected as branches to a vapor relief stack, the stack shall be extended independently through the roof, or to an approved location in the open air and shall be terminated as described in "Vent Extensions through Roofs," in Chap. 8.

Individual vapor relief pipes for fixtures shall be at least as large as the fixture relief outlets. Vapor relief stacks and branch pipes serving two or more individual vapor relief pipes shall be at least one standard pipe size larger than the largest individual pipe connecting thereto, but shall be at least 1¼ in (31.3 mm) in size. Vapor relief stacks shall extend upward undiminished in size from lowest vapor relief branch to the vent terminal in the open air. Drip pipes connected at the base of vapor relief stacks shall be at least 1¼ in (31.3 mm) in size.

4

WATER SUPPLY SYSTEMS

GENERAL DESIGN CONSIDERATIONS

This chapter deals with general design considerations for building water supply systems. They are essential elements in the planning of systems to provide a safe, adequate, and reliable supply of potable water to all fixtures and equipment.

Of primary consideration is the source of water supply, its availability, quality, adequacy, reliability, chemical and bacteriological characteristics, and need for treatment for corrosion control.

Protection of the potable water supply is essential. Potable supply systems must not be connected with unsafe water sources nor be subject to the hazards of contamination by backflow or back siphonage.

Plumbing fixtures, devices, and appurtenances should be supplied with water in sufficient volume and at pressures adequate to enable them to function properly and without undue noise under normal conditions of use. Pressure at fixture water outlets should be limited to a safe level, an essential consideration in systems for tower buildings and in high water pressure districts.

Piping systems require adequate provision for support and attachment to the building structure. Provision also is necessary for accommodation of normal expansion and contraction of piping, and for vibration effects. These are important considerations in the layout and installation of piping.

A most important consideration before any potable water supply system is put into service is that it should be subjected to effective testing and proved tight. A final consideration is that the system should be subjected to the prescribed method of disinfection for potable water service.

SOURCE OF WATER SUPPLY

Pure and wholesome potable water from a source approved by the health authority having jurisdiction should be available at all times on the premises of every building in which plumbing fixtures are installed. The potable water supply system of the building should be connected to a public potable water supply system where available. However, where a public system is not available, an approved private source of potable water must be provided, conforming to regulations of the health authority having jurisdiction, and the building system must be supplied therefrom.

It is mandatory that building systems be connected to available public systems as a positive measure to protect the health of building occupants. Public systems supply water which has been subjected to adequate and effective treatment so as to assure that it is of satisfactory quality for drinking purposes. Additional treatment generally is applied to eliminate unpleasant taste and odor and excessive turbidity. The use of available public systems for building water supply is also recommended because they afford convenience, reliability, capacity, and trouble-free service for the life of any given building, features which rarely can be matched by water obtained from private sources.

In areas where public water supply systems have been installed, regulations specify when such systems shall be deemed available for supplying building systems. Availability may be considered to be just 100 ft (30.5 m) in the case of a one- or two-family dwelling, and as much as 500 ft (152 m) for other types of building occupancies, the distance being measurable along the street from the public system to the premises on which the building is situated. But distance should not be the only criterion for determining when one should supply buildings from a public system. All applicable factors should be considered in each individual case.

QUALITY OF WATER SUPPLY FOR FIXTURES

Only potable water should be supplied to fixtures and equipment at which water is provided or used for drinking; cooking; food preparation; bottling, canning, or packing; washing of dishes, glasses, cutlery, or kitchen utensils; or for similar domestic purposes. Potable water should conform to the chemical and bacteriological quality standards established by the health authority having jurisdiction.

As a health protection measure, it is recommended that all plumbing fixtures and water-using equipment in buildings be provided with water exclusively from the potable water supply system of the building wherever circumstances permit. The presence of a second system, of nonpotable or uncertain quality, in a building poses a hazard which should be avoided if possible. Experience with such installations in the past has amply demonstrated that cross-connections between the potable and unsafe water supply systems are made sometimes unknowingly, but generally because of emergencies, and that such cross-connections remain in existence for extended periods and can cause contamination of the potable water supply system. This should be prevented.

Nonpotable water should not be supplied to any fixtures unless specially approved by the health authority having jurisdiction. The use of nonpotable water should be limited to those things which do not require potable water, such as water closets, urinals, car-washing equipment, cooling jackets, condensers, and similar apparatus. Wherever a nonpotable water supply system is installed in a building, all nonpotable water faucets and outlets and all piping conveying nonpotable water should be adequately and properly identified by conspicuous markings as hazards, using appropriate signs, colors, and other standard symbols as required by the authority having jurisdiction.

Saltwater supply systems have been installed aboard ships and in some buildings at offshore installations where potable water was in critically short supply. Such systems have been used to provide the water requirements of fixtures at which potable water was not necessary. But experience has indicated that ordinary plumbing materials and equipment, which perform with reasonable durability in potable water supply service, generally are unsuitable for handling seawater because of its much higher corrosivity. Special piping materials and equipment designed for durability in saltwater supply service should be used for such systems.

The use of salt water for washing clothes has been proposed on numerous occasions, but this has many disadvantages and undesirable features. Fresh or potable water is much better for this purpose and should be utilized if possible. When salt water is pumped from the ocean, it may contain pollution, sand, and suspended organic matter. Treatment to remove these from the water is necessary, or the clothes may become more unsanitary and objectionable after laundering than they were before. The high salt content of seawater makes ordinary laundry soaps and detergents less effective in cleansing action and results in the production of a considerable amount of curd or scum. Generally, it is necessary to use detergents designed specifically for salt water. A considerable residue of salt remains in any cloth after it has been washed in salt water and allowed to dry thoroughly. As some of the salt is hygroscopic, it causes the cloth to feel slightly damp or clammy when touched. To remove this salt residue, it is necessary to rinse the cloth thoroughly in fresh water.

WATER SERVICES AND WATER METERS

Water services connecting to public water supply systems or extending under public ways should be designed and installed in accordance with regulations of the appropriate authority having jurisdiction in the particular area. Regulations in different areas vary greatly because of local conditions, such as capacity limitations of the public system, size of the public mains available for water service connections, maximum and minimum pressures normally prevailing in the public mains, chemical properties of the water supply, maintenance and operating procedures established by the public water authorities in accordance with local conditions, soil conditions below ground, public roadway construction and maintenance standards, roadway traffic and loads, underground structures and utility piping systems located beneath public roadways, and numerous other conditions singular to given areas in which water services may be installed.

There may be several regulatory agencies which exercise jurisdiction over water service installations in a given area, such as a public water supply system authority, public highway authority, local board of health, public safety department, and other governmental agencies appropriately designated by law and charged with such duties and responsibilities. Local health agencies usually exercise jurisdiction over water services which convey water to buildings from private well systems located on the premises. In such cases, the health agency may serve appropriately as the single agency exercising jurisdiction over private well water supply systems and plumbing systems for buildings. Each building generally is required to be provided with a separate water service to serve its own water supply requirements. However, where several buildings on the same premises are designed to remain permanently under a single ownership, they may be permitted to be supplied by a common water service equipped with appropriate control valves on the branches to buildings so that each building may be separately shut off when necessary.

Water meters may be required to be installed so as to register the amount of water supplied to a building from a public water supply system. Such meters should be designed, located, and installed in accordance with regulations of the public water authority having jurisdiction. Meters should be located and installed so that they are readily accessible for reading and inspection and adequately protected against damage by frost. Where the supply of water to a premises is to be fully metered, the meter should be installed immediately adjacent to the point of entry of the water service. However, where a building is located remote from the property line or conditions exist in a building that prevent installation of the meter at the point of entry of the water service, the meter may be permitted to be installed outside of the building in a watertight and frost-proof pit or meter box equipped with a suitable cover.

All water meters should be of types approved by the public water authority for the water supply conditions under which they are to perform. The type and size of meter to be installed for any given condition should be such as will ensure accurate registration of flow through the meter. Generally, the size of meters is restricted to those which will ensure accuracy on the basis of consumption and occupancy of the premises, and in no case should it be more than one standard size larger than the tap or connection to the public water main.

Each water meter should be installed so that it may be read and tested conveniently. The dial should face upward, and be located not more than about 3 ft (1.0 m) above floor level. Meters should not be within pits in buildings unless specially permitted by the public water authority, in which case the pits should be of approved type and dimensions and provided with suitable covers. Convenient means for testing meters in place for accuracy of registration should be provided. A control valve should be installed immediately adjacent to and on the inlet side of the meter. On the outlet side of the meter, there should be a T branch equipped with a faucet or valve for testing purposes. Meters 3 in (75 mm) and larger in size are generally required to be equipped with 2-in (50-mm) gate valves for test purposes. No water supply connections are permitted to be made to test faucets or test valves. For meters 1 in (25 mm) and larger in size, a gate valve should also be installed on the outlet side of the meter, adjacent to, and on the outlet side of, the test branch connection.

WATER TREATMENT FOR CORROSION CONTROL

In many water areas, the addition of certain chemicals to the potable water supply in buildings is permitted under strict control and regulation by public water supply and health authorities. The purpose of this chemical treatment of water is to control corrosion and scaling in domestic water supply systems without adversely affecting the potability of the water.

Such treatment does not guarantee 100 percent control of corrosion. Nor is it claimed to be a means of stopping even slight leaks in defective piping. However, where properly applied, it should result in slowing down the rate of corrosion of the piping system and thereby increase its normal period of satisfactory service. Further, it can relieve to a great extent rusty water conditions due to corrosion of ferrous parts of the system.

The particular chemicals which should be added to the water supply in order to control corrosion are dependent upon the chemical characteristics of the water supplied to the building system and the type of piping material installed. To determine the type and quantity of chemicals needed in each case is not a task for an amateur. It requires a person with a satisfactory knowledge of water chemistry and corrosion control methods. Such a person should be in responsible charge and supervise the maintenance of this type of water treatment installation.

For work of this kind, it is necessary to maintain a suitable laboratory equipped to analyze water samples in accordance with the standard method of water analysis. In many large cities, competent water treatment firms are engaged in business and are available to provide this service to building operators. Such firms are recognized and qualified by public water supply and health authorities and are subject to their regulatory jurisdiction.

Chemicals permitted to be added to the potable water supply system of a building are as follows: sodium silicates, sodium carbonate, sodium bicarbonate, sodium hydroxide, sodium phosphates, calcium carbonate, calcium bicarbonate, and calcium hydroxide. None of these should increase the total alkalinity, total hardness, or total silica content of the water by more than 50 parts per million (ppm). The addition of excessive amounts of such chemicals could be harmful.

Special mechanical devices and apparatus have been designed and constructed to regulate the chemical dosage of the water. The devices are installed on the water supply distribution system in such a manner that the chemicals are added to the water in accordance with the velocity of water flow. Thus, the prescribed and limited dosage is carefully adjusted to the volume of water flowing in the system.

The treatment device is generally connected to the water supply system near the water service control valve. Such work must comply with the requirements of the public water supply and health authorities having jurisdiction and with the specifications for installation of the water treatment device. The installation of all necessary pipe, fittings, valves, and vacuum breakers for connection of the treatment device to the system is plumbing work.

This process of chemically treating the potable water supply is designed specifically to control corrosion and scaling in the various parts of the domestic system. It is not recommended for building heating systems. Other processes are available and are better adapted for such systems because of the different conditions.

PROTECTION OF POTABLE WATER SUPPLY

Building potable water supply systems should not be subject to contamination from any source. Chemicals that could produce toxic conditions in the system should not be introduced into any part of the piping, except as may be specially approved by the health authority having jurisdiction. Similarly, piping materials that could produce toxic conditions and any piping that has been used for conveying fluids other than potable water should be prohibited from use in the potable water supply system.

State sanitary code provisions enable health officials to supervise public water supplies in matters pertaining to protection of sources of supply, operation of water treatment facilities, and general control of the supply to ensure the delivery of water of safe, sanitary quality. For protection and control to be adequate, they must extend all the way to the water outlets of building systems.

There should be no cross-connection or physical connection between any private water supply system and a public potable water supply system, except as may be specially approved in conformity with existing regulations by the health authority having jurisdiction. Under certain specific conditions, cross-connections may be permitted by health authorities subject to such examination, inspection, and maintenance procedures as may be necessary to ensure against contamination of the public potable water supply system.

There should be no interconnection between any part of a potable water supply system and any drainage or vent-piping system. Included as part of the potable water supply system in this regard are emptying and overflow pipes of potable water supply tanks or storage tanks, discharge pipes of relief valves, and discharge pipes of condensers and cooling jackets supplied directly from the potable water system.

Potable water supply piping, water outlets, vacuum breakers, and similar equipment should be arranged and located so as to prevent them from being submerged in any contaminated or polluted liquid or substance, and thereby to avoid any potential contamination from such sources. However, where this may be impractical to do, as in the case of certain specific fixtures and equipment, adequate protection must be provided by some other means acceptable for approval by the authority having jurisdiction.

Existing regulations prohibit direct connection of the potable water supply system to certain fixtures and equipment because of the severity of the hazard deemed to exist. Sterilizers are not permitted to be directly connected because of the possibility that potable water may enter a sterilizer without being noticed when the supply valve does not shut off completely, and thereby contaminate the sterilized contents of the equipment. Such equipment should be supplied through an air gap at which any flow of water may be readily seen. Operating, dissection, embalming, and mortuary tables, or similar equipment, are not permitted to be directly connected to the potable water supply system. Hoses used in conjunction with such tables should terminate at a sufficient distance from any part of the table or attachments so that they may not be entered into bodies on the table. A reasonable distance in this case is deemed to be at least 1 ft (300 mm) from the table. Bidets are not permitted to be directly connected because they may be equipped with bottom-supply douche outlets, and the water piping to such outlets can readily be contaminated by particles of human waste from the fixture. The hazard in this case has been deemed sufficiently severe as to warrant prohibiting a direct connection.

Aspirators, ejectors, water siphons, and similar equipment, or any chemical solution tank or apparatus are not permitted to be directly connected to the potable water supply system because of the severity of hazard which can occur with such equipment and connections, as evidenced by the records of water supply contamination cases and past epidemic conditions. Under certain conditions, permission may be granted for direct connection to such equipment for a particular use deemed to be relatively nonhazardous when adequate protection is otherwise provided by means acceptable to the authority having jurisdiction.

Priming connections for pumps, used for pumping nonpotable water or liquids, are not permitted to be made by means of a direct connection to the potable water supply system. Contamination of potable water supply systems and large sections of public water distribution system mains have occurred in the past where such direct connections were made to the discharge or high-pressure side of the pumps. The recommended method for priming pumps is to provide a fixed air gap of suitable size in a vertical section of the potable water supply branch pipe and connect the outlet side of the air gap to the inlet or suction side of the pump.

Direct connections of potable water supply piping to condensers or cooling jackets of refrigeration units should be equipped at least with a check valve so as to prevent contamination by leakage of refrigerant into the cooling water, except in cases where the water supply piping is entirely outside of the piping or tank containing the refrigerant and two independent wall thicknesses of metal separate the refrigerant from the potable water supply. Where refrigeration units contain more than 20 lb (9.1 kg) of refrigerant, additional protection should be provided in the form of a pressure relief valve installed in the water supply piping between the check valve and the unit. The relief valve should be set so as to relieve at 5 psi (34.5 kPa) higher than the maximum water pressure in the water supply system at the point of installation.

When water is used for cooling, heating, processing, or similar purposes, its temperature, taste, palatability, potability, clarity, and other qualities may be adversely affected. This has been amply evidenced by numerous investigations of water supply complaints from building occupants. Consequently, after water has been used for such purposes, it may not be permitted to be reintroduced into the potable water supply system nor be distributed to plumbing fixtures requiring potable water supply. This water should be discharged into a fixture or receptacle approved to receive such liquids, and the discharge pipe should terminate at a suitable height above the rim of the receiving fixture so as to provide an effective air gap.

All potable water supply outlets should, wherever practicable, terminate at sufficient height above the flood-level rim of the fixture or receptacle so as to provide the minimum air gap required by standard regulations. The minimum required air gap measured vertically from the outlet end of a faucet or other water supply outlet to the flood-level rim of the receiving fixture or receptacle shall be at least twice the diameter of an outlet of equivalent circular cross section when remote from all walls, and at least three times such diameter when the outside edge of the outlet is less than three diameters distant from a single wall or less than four diameters distant from each of two intersecting walls.

Below-rim potable water supply outlets should be prohibited in every instance, except where such outlets are absolutely essential for proper function of fixtures. For such fixtures, approved protective methods must be applied to ensure against any hazard of potential contamination of the potable water supply system by means of backflow or back siphonage of liquids from the fixtures through their below-rim water supply outlets.

PROTECTION AGAINST CONTAMINATION BY BACKFLOW AND BACK SIPHONAGE

A severe outbreak of amoebic dysentery occurred in Chicago in 1933 at the time when a World's Fair was being held there. The outbreak was of epidemic proportions and was definitely traced to the contamination of water supply systems in a number of buildings. Investigations and tests by the Chicago Board of Health conclusively proved that water supply contamination could occur in almost every existing potable water supply system in buildings. It was definitely established that this could occur because of the design of many plumbing fixtures and the widespread use of below-rim water supply outlets, features which permitted the water supply to become contaminated by backflow or back siphonage of the liquid contents of fixtures.

For many years, plumbing contractor associations, on national, state, and local levels, had contended that water supply system contamination occurred frequently in buildings because of below-rim water supply outlets at fixtures. They had appealed repeatedly to regulatory agencies for more adequate regulations to be invoked to protect the health of building occupants. But the lack of conclusive proof, affirmed and attested by recognized health agencies and officials, resulted in the failure of plumbing regulatory agencies to take effective action for correcting the condition.

However, the spotlight of publicity was focused on the Chicago amoebic dysentery epidemic, and worldwide interest was aroused regarding the severe health hazards involved in plumbing systems that were defective or deficient in design. As a result, health officials, water supply officials, plumbing contractors, and plumbing equipment manufacturers broke all precedents and earnestly collaborated to devise ways and means of eliminating all potential sources of contamination. To Dr. Herman N. Bundesen, Commissioner of the Chicago Board of Health, his associates, and the Chicago Plumbing Contractors Association, much credit is due for initiating such collaboration. Many important health benefits have resulted therefrom, and it is evident that continued collaboration of all parties of interest is both advisable and necessary for the future.

The most important single recommendation which has yet been devised for preventing contamination by backflow or back siphonage is that, wherever possible, potable water supply outlets should terminate at a sufficient height above the flood-level rim of each fixture so as to provide at least the minimum air gap required to eliminate any possibility of draining or siphoning liquids from fixtures up into a water outlet. The minimum required air gap was discussed in the preceding section of this chapter.

There are several kinds of plumbing fixtures and various items of plumbing equipment that require below-rim or submerged water supply connections for satisfactory operation or function. Where such connections are necessary, it is recommended that the water supply piping to each fixture or equipment be equipped with an approved backflow preventer or vacuum breaker specifically designed to prevent back siphonage. Vacuum breakers did not exist at the time of the Chicago epidemic. But, by 1945, an adequate number of types of vacuum breakers had been developed and approved for all necessary applications.

There are five general types of vacuum breakers: atmospheric type, pressure type, flush-valve type, flush-tank–ball cock type, and hose-outlet type. The atmospheric type is designed for installation in the supply piping on the outlet side of the last control valve. The pressure type is designed for installation in the supply piping on the pressure side of the outlet valve or faucet. Inlet and outlet of this type are 90° apart. Where back pressure or water hammer can develop on the outlet side of this type, a check valve should be installed between the vacuum breaker and the fixture in order to prevent unseating of the air-port valve and resultant leakage or seepage therefrom. The flush-valve type is designed to be installed on the flush-valve outlet above its connection to the fixture. The flush-tank–ball cock type is designed as an integral part located on the discharge side of the ball cock seat in the top section of the device. The hose-outlet type is designed to be screwed onto threaded hose bib outlets. The outlet end of this type of vacuum breaker may be equipped with either a serrated tip for hose attachment or a threaded outlet onto which laboratory aspirators or hose couplings may be screwed.

The basic principle involved in the design of all vacuum breakers is simply this: when a vacuum occurs in a fixture supply pipe, the vacuum breaker must permit air from the atmosphere to enter the fixture supply pipe at a sufficiently rapid rate and at a satisfactorily high elevation above the flood-level rim of the fixture so as to break the vacuum as it occurs, and thereby prevent back siphonage of liquid from a fixture through its below-rim water supply outlet and fixture supply pipe. In addition, the design of vacuum breakers should be such that they perform their function without leakage or spitting from the air ports under normal internal pressure conditions and with varying flows, and without permitting any rise of liquid from the fixture to occur in the supply piping for the complete range of vacuum intensities, from fractional to maximum vacuum conditions.

Each fixture having below-rim potable water supply outlets should be individually equipped with approved vacuum breakers of the same nominal size as the fixture supply pipes and of suitable type for the use intended. The use of a single vacuum breaker on a branch water pipe supplying a number of outlets at a group or battery of fixtures has been proved to be unacceptable. It has been conclusively established by tests at public water supply department laboratories that back siphonage can occur readily at individual fixtures during simultaneous flow and cause contamination of the branch supply piping.

Vacuum breakers should be installed in strict conformity with the terms of their approvals. They must be installed in proper position for satisfactory operation and should be located at least 6 in (150 mm) above the flood-level rim of the fixture or receptacle being supplied, except where they may be required by regulations to be located at a higher elevation for certain specific items of equipment. For lawn sprinkler or irrigation systems, the water supply pipe should be equipped with a vacuum breaker located at least 12 in (0.3 m) above the highest sprinkler or discharge outlet of the system.

Potable water supply piping to water preheating equipment designed to absorb heat from waste water, such as preheating coils submerged in laundry waste water sumps, should be equipped with a vacuum breaker and also a check valve located between the vacuum breaker and the preheating equipment. In the event that a hot water storage tank is supplied through the preheating equipment and also by an independent cold water supply pipe, additional protection should be provided for the independent supply. It should be equipped with a vacuum breaker located at least 6 in (150 mm) above the highest elevation of the tank and with a check valve located between the vacuum breaker and the tank.

Flush valves which are directly connected to the potable water supply system should be equipped with vacuum breakers of the flush-valve type, located on the discharge side of the flush valve and at least 6 in (150 mm) above the top of the fixture supplied.

Flush tanks which are directly connected to the potable water supply system should be equipped with approved ball cocks. Where the ball cock is in contact with tank water, it should be equipped with a vacuum breaker located at least 1 in (25 mm) above the overflow outlet of the flush tank. Where the ball cock is not in contact with tank water, the ball cock outlet should be located at least 1 in (25 mm) above the overflow outlet of the flush tank, or a vacuum breaker should be installed as previously stated.

Hose-coupling water outlets and serrated-tip water outlets should be individually equipped with a vacuum breaker installed in the individual supply pipe to the water outlet, or installed on the water outlet with provision for hose attachment to the outlet side of the vacuum breaker. For such outlets, the vacuum breaker should be located at least 6 in (150 mm) above the highest point of hose usage. Hose-coupling water outlets for bedpan washing or flushing should also be equipped with a check valve located between the vacuum breaker and the hose coupling. Vacuum-breaker protection need not be provided at hose outlets where no hazard is deemed to exist, such as standpipe system hose connections, and drain cocks for building heating systems.

To ensure that vacuum breakers perform satisfactorily under service conditions, they must be maintained in good working condition. Consequently, they should be subjected to periodic inspections and tests, in place. They should be located so as to be accessible for maintenance and inspection and so that leakage at vacuum breakers is readily evident. It is recommended that they be installed in the same room as the fixtures they serve, and exposed to view, except for those which are integral parts of approved ball cocks in flush tanks. Lack of proper maintenance or inspection of vacuum-breaker function eventually will result in leakage, failure of the device to perform satisfactorily, and potential contamination of the potable water supply system.

BACKFLOW PREVENTION DEVICES FOR SPECIAL FIXTURES AND EQUIPMENT

In view of the special hazards involved with the following plumbing fixtures and equipment, connections to the potable water supply system should be protected against backflow with any one or more of the devices as indicated:

LOW INLET TO RECEPTACLES CONTAINING TOXIC SUBSTANCES (VATS, STORAGE CONTAINERS, PLUMBING FIXTURES):

1. An approved air-gap fitting
2. Reduced-pressure-principle back-pressure backflow preventer
3. Pressure vacuum-breaker unit
4. Atmospheric vacuum-breaker unit

LOW INLET TO RECEPTACLES CONTAINING NONTOXIC SUBSTANCES (STEAM, AIR, FOOD, BEVERAGES, ETC.):

1. An approved air-gap fitting
2. Reduced-pressure-principle back-pressure backflow preventer
3. Pressure vacuum-breaker unit
4. Atmospheric vacuum-breaker unit
5. Approved double-check-valve assembly

OUTLETS WITH HOSE ATTACHMENTS OR REFRIGERATION SYSTEMS WHICH MAY CONSTITUTE A CROSS CONNECTION:

1. An approved air-gap fitting
2. Reduced-pressure-principle back-pressure backflow preventer
3. Pressure vacuum-breaker unit
4. Atmospheric vacuum-breaker unit
5. Double-check-valve assembly (where approved)

COILS OR JACKETS USED AS HEAT EXCHANGERS IN COMPRESSORS, DEGREASERS, AND SIMILAR EQUIPMENT INVOLVING TOXIC SUBSTANCES:

1. An approved air-gap fitting
2. Reduced-pressure-principle back-pressure backflow preventer
3. Pressure vacuum-breaker unit

DIRECT CONNECTIONS SUBJECT TO BACK PRESSURE INVOLVING NONTOXIC SUBSTANCES:

1. An approved air-gap fitting
2. Reduced-pressure-principle back-pressure backflow preventer
3. Approved double-check-valve assembly

WATER SUPPLY REQUIREMENTS

Adequacy of Supply Building water supply systems should be designed and installed so as to provide at all times a supply of water to plumbing fixtures, devices, and appurtenances in sufficient volume and at pressures adequate to enable them to function satisfactorily and without excessive noise under all normal conditions of use. Where standpipe or automatic sprinkler systems are supplied by potable water services from a public water supply system, the services should be connected to water mains which are reliable and of adequate capacity. Water services for standpipe or automatic sprinkler systems should be designed and installed to provide at all times a supply of water in sufficient volume to enable them to function satisfactorily.

Adjustment and Maintenance for Conservation The water supply distributing system of the building should be designed and adjusted to supply fixtures and equipment with the minimum quantity of water consistent with proper performance and cleaning. The system should be maintained to prevent leakage and excessive waste of water.

Excessive Pressures Where water pressures are over 80 psi (550 kPa), or where self-closing faucets or valves are installed, approved air chambers or mechanical devices should be provided to prevent pressure hazard, water hammer, and objectionable line noises. It is recommended that pressure in fixture supply pipes be limited to no more than 80 psi (550 kPa) by the appropriate installation of pressure-reducing valves, such as in the branch water supply piping to fixtures on a single floor or in a particular zone of the building system.

Minimum Available Pressure The minimum pressure available at all times at water outlets should be at least 8 psi (55 kPa) for fixtures generally. For flush valves directly supplied from the water distributing system, the minimum pressure available should be at least 15 psi (103 kPa) for all flush-valve supplied fixtures, except blowout-type water closets and urinals at which the minimum pressure should be at least 25 psi (172 kPa).

Where the water pressure available at the public water main or other source of supply is insufficient to maintain the minimum pressures required at all water outlets at all times, a pressure booster pump system, approved as to capacity and reliability, or an automatically controlled water supply tank of either the pressure or gravity type should be provided. The pumping equipment and tanks should be designed and installed so as to provide the required minimum available pressures at all times. Where a pressure booster pump system is provided, a low-pressure cutoff switch should be installed within 5 ft (1.5 m) of the pump inlet. The switch should be set for operation to prevent the creation of a vacuum or negative pressure on the suction side of the pump, thus cutting off water supply to other outlets.

MULTIZONE WATER SUPPLY SYSTEMS FOR TOWER BUILDINGS

What is the maximum static pressure to which any plumbing fixture in a building should be limited under no-flow conditions? Up until 1968, generally accepted standard codes did not provide such criteria for design. Before that, designers posed limits for themselves based upon their individual judgments.

In 1968, the current criterion for design was established and recommended generally. It is as follows: *"Pressure at any fixture shall be limited to no more than 80 psi (550 kPa) pressure under no-flow conditions"* (italics added).

The above criterion is presented also in the 1978 edition of the National Standard Plumbing Code, sponsored jointly by the National Association of Plumbing-Heating-Cooling Contractors and the American Society of Plumbing Engineers.

In accordance with this standard, appropriate recommendations may be developed for the design of multizone water supply systems for tower-type buildings. In the construction of systems for such buildings, designers may apply various satisfactory methods to achieve performance objectives.

However, as a guide to designers, general recommendations may be presented for a preferred design method. They are as follows:

1. Provide each water supply zone with its own separate system of hot and cold water supply, its own hot water heater, and its own circulation pump for hot water recirculation.
2. Provide a static pressure or 25 psi (172 kPa) at the top of the downfeed risers in each zone.

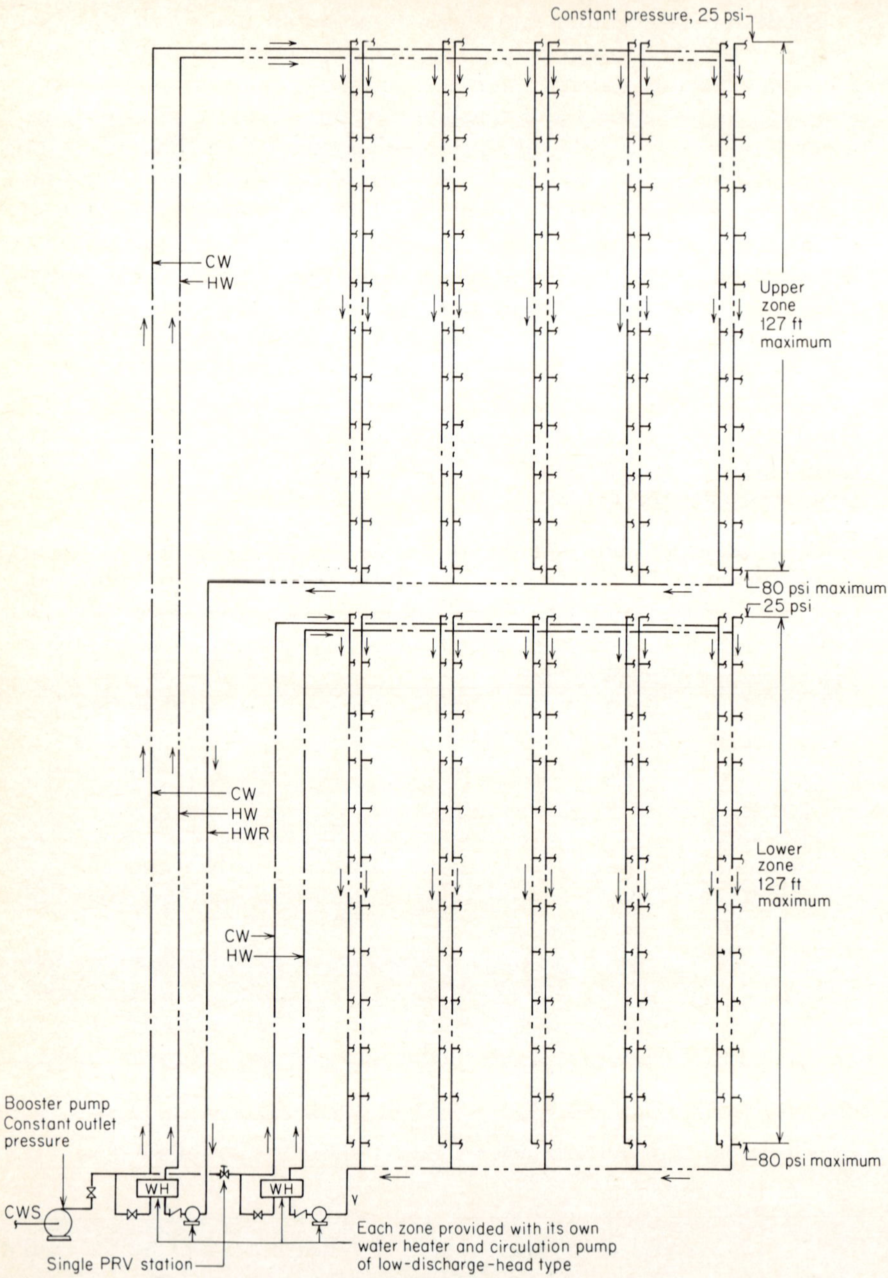

Fig. 4-1 Preferred method—separate system for each zone.

3. Limit each water supply zone to a maximum height of 127 ft (38.7 m), corresponding to a maximum static pressure differential of 55 psi (378 kPa) between the top and bottom of the zone, thereby limiting pressure at the lowest fixture outlets to 80 psi (550 kPa).

These recommendations are shown in Fig. 4-1.

The number of stories which may be included in a single 127-ft (38.7-m) zone depends upon the height of stories in the building. For example, a 127-ft (38.7-m)-high zone may be applied to include ten 12.5-ft (3.8-m) stories, twelve 10-ft (3.0-m) stories, or fourteen 9-ft (2.7-m) stories.

Figure 4-1 shows a single pressure-reducing valve in the main line supplying both hot and cold water distributing systems in the lower zone. This pressure-reducing valve should be adjusted to provide a static pressure of 25 psi (172 kPa) at the top of the downfeed risers in the zone. This zone has its own hot water heater, recirculation piping, and circulation pump. The pump may be of the low-discharge-head type, as it need only provide sufficient discharge head to overcome friction head in the recirculation piping at the recirculation design rate of flow.

Similarly, the upper zone has its own hot water heater, recirculation piping, and low-head-type circulation pump. Thus, there can be no interference between zones to affect their satisfactory performance. For the upper zone to be provided with a static pressure of 25 psi (172 kPa) at the top of the downfeed risers of this zone, the control and operation of the booster pump must be set to provide and maintain such static pressure at all times.

In the selection of pressure-reducing valves, booster pumps, and hot water circulation pumps, the manufacturers' recommendations should be carefully observed.

These general recommendations are offered as a guide in the design of multizone water supply systems in tower-type high-rise buildings of any height. They may be applied to achieve satisfactory performance in service and avoidance of excessively high pressure at fixture outlets in accordance with current design standards.

WATER SUPPLY TANKS

In the interest of economy and speed in delivery, it is recommended that standard sizes of water supply tanks be used wherever possible. A wide range of standard sizes is available in steel pressure tanks for hydropneumatic water supply systems. These may be most advantageously used where the peak water demand rate is relatively low, such as in small buildings. For large buildings and high peak water demand rates, elevated gravity water supply tanks of either wood or steel are generally preferred. They are available in standard sizes of 5,000; 10,000; 15,000; 20,000; 25,000; 30,000; 40,000; 50,000; 60,000; 75,000; and 100,000 gal (18,900; 37,800; 56,700; 75,600; 94,500; 113,400; 151,200; 189,000; 226,800; 283,500; 378,000 L) net capacity. Standard steel tanks are also available with capacities of 150,000; 200,000; 300,000; and 500,000 gal (567,000; 756,000; 1,130,000, and 1,890,000 L).

Tank storage capacity required for domestic water supply should be based upon the peak demand load on the water supply system and should be adequate to satisfy that demand for at least ½ h. The peak demand rate, in gallons per minute (or liters per second) may be estimated by the standard method described in Chap. 6. The ½-h limit is purely an arbitrary one that is deemed reasonable in the light of experience. In this way, the tank storage capacity required for domestic water supply for any type of building occupancy may be determined in the same manner. For multiple dwellings and large residential apartment developments, this method provides for the design population normally assumed, approximately 15 gal (56.3 L) of storage capacity per person for buildings with 100 dwelling units, 9 gal (33.8 L) per person for buildings with 500 dwelling units, and 8 gal (30 L) per person for buildings with 1000 dwelling units.

It is recommended that the capacity of any single tank in or on a building roof should not exceed 30,000 gal (113,000 L) as a matter of safety in the event of emergencies. Where tanks are located on flat roofs and the total capacity exceeds 30,000 gal (113,000 L), drain pipes from the tanks should discharge so as to distribute water over separate drainage areas of the roof.

Water supply tanks should be designed and constructed so as to be watertight, vermin-proof, rodent-proof, resistant to corrosion, and capable of withstanding the pressures under which they are to operate. Safe and easy means of access should be provided for inspection and maintenance. Tank supports should be of noncombustible construction, and neither the tank nor its supports should be used to support equipment or structures other than for tank use unless specially designed to carry the additional load imposed. Tanks should not be located over openings in floor or roof construction. Floor or roof openings through which piping for tanks must pass should be made watertight.

Potable water supply tanks for domestic supply and standpipe or automatic sprinkler systems should be designed and installed so as to furnish water in sufficient quantity and pressure for such systems. Tanks used to supply water to a domestic system and a standpipe or automatic sprinkler system should be provided with an outlet for the domestic supply located a sufficient distance above the bottom of the tank to maintain the minimum reserve required for fire protection service. Potable water supply tanks which provide water for domestic supply and also for standpipe and automatic sprinkler systems should have an outlet for the standpipe system located a sufficient distance above the bottom of the tank to maintain the minimum reserve required for the sprinkler system.

The elevation at which a gravity water supply tank should be installed is dependent upon the minimum pressure requirements at the topmost outlets of the systems being supplied from the tank. For domestic water supply, the minimum level in the tank should be at an elevation sufficient to provide the minimum available pressure required at the topmost fixture outlets on the system, and to compensate for friction losses in flow from the tank to such outlets during peak demand on the system. For standpipe system water supply, the elevation of the supply connection from the tank should be sufficient to provide at least 65 psi (448 kPa) available at the highest standpipe hose rack on the system. For sprinkler system water supply, the elevation of the bottom of the tank should be sufficient to provide the minimum pressure required at the highest line of sprinklers on the system.

Ball cocks or other suitable automatic valves should be provided to control the supply of water to gravity water supply tanks which are furnished directly by pressure from the public water supply system, or which may be so supplied at times when pressure in the public system is maximum. Such ball cocks or valves are necessary to prevent overflowing of the tank under the circumstances.

Water supply inlets to gravity potable water supply tanks should terminate at a sufficient height above the overflow pipe of the tank so as to provide an effective air gap. It is recommended that the supply inlet terminate at least 4 in (100 mm) above the top of the overflow pipe.

Overflow pipes should be provided for all gravity water supply tanks, and they should be arranged so as to discharge within 6 in (150 mm) of the roof, or into a suitable catch basin, or into an open, water-supplied fixture approved for such use. Overflow pipes should be at least one size larger than the tank supply pipes and should be adequate in overflow capacity to match the rate of tank filling when the water level rises to the top of the overflow pipe connection in the tank. Where several gravity potable water supply tanks are installed and connected to supply a single water supply system for one or more buildings, all gravity tank overflow pipe connections on the system should be at the same elevation.

Emptying pipes should be provided for all water supply tanks. Such pipes should be located and arranged so as to prevent damage from water discharged when the tanks must be drained during an emergency.

Emptying pipes should also discharge in the same manner as required for overflow pipes. Each emptying pipe should be equipped with a suitable type of approved valve of the same diameter as the pipe. Each tank should be provided with an emptying pipe not smaller than 4 in (100 mm), for tanks over 10,000 gal (37,800 L) in capacity; 3 in (75 mm) for tanks over 5000 gal (18,900 liters) but not over 10,000 gal (37,800 L) in capacity; and 2½ in (63.5 mm) for tanks of smaller capacity.

To protect potable water supply tanks against contamination, certain additional precautions should be observed. On gravity tanks, all overflow and air vent pipes connected to the tanks should be provided with durable screens of not less than 100 mesh. No gravity potable water supply tanks, nor service holes of pressure potable water supply tanks, should be located directly beneath any soil or waste piping so as to prevent any sewage leakage from such piping gaining entrance into the tank water. When water supply tanks are to be repaired, painted, or lined, they should be shut off from the system, and the materials used for the work should not be of a type that would affect either the taste or potability of the water supply from the tanks when they are restored to service.

PIPING INSTALLATION

Water supply piping should be installed in practical alignment and should be pitched so that the entire system, or parts thereof, can be drained by means of a drain cock, faucet, or valve at the lowest point, or points. Traps, sags, and vertically bowed arrangements in the piping should be avoided so as to permit the entire system to be drained as well as to prevent pocketing of sediment in low sections and pocketing of air in vertically bowed piping.

There are certain locations in buildings where water supply piping should not be installed, such as in stairways, in a hoistway or under an elevator or counterweight, or where such piping would interfere with the normal operation of windows, doors, or other building openings. In addition, underground water supply piping which is to be installed parallel to building walls should be located at least 3 ft (0.915 m) distant from footings or bearing walls so as to avoid any hazard to structural safety such as might otherwise occur as a result of the washing away of bearing soil beneath footings in the event of severe leakage from the parallel underground water supply piping.

Unless adequate provision is made to protect water supply piping against damage from frost conditions, such piping should not be installed outside of buildings, or concealed in exterior walls in climate zones where freezing temperatures may occur. Similarly, such piping should not be installed in rooms or spaces of buildings where freezing temperatures may occur normally, such as in food freezer rooms, lockers, refrigerators, cold storage rooms, etc.

Protection against damage from external corrosion should be provided for water supply piping which must be installed in or beneath cinders or other corrosive material. Although adequate protection may be afforded in many instances by application of one or more coats of suitable paints and wrapping of joints, it is recommended that piping installations in highly corrosive beds be avoided wherever possible or that the piping be encased in a special bed of chemically neutralized, noncorrosive material.

Water supply piping should be installed in such a manner as to avoid damage and breakage due to strain accompanying normal expansion and contraction of the piping and due to building settlement. Where piping passes through foundation or bearing walls, protection should be provided by means of iron or steel pipe sleeves two sizes larger than the pipe passing through the wall, or by means of masonry relieving arches directly above the top of the piping. Flexible sealing material should be caulked into the annular space between the pipes and the sleeves or arches.

Underground water supply piping should be laid on a firm natural bed of earth for its entire length or on an equally firm means of continuous support. Tunneling is not recommended as a satisfactory method for installing such piping because of the resulting misalignment of the piping when the soil above it settles down upon it in time and a rut forms on the ground surface at grade level. Open trenchwork is generally recommended for such piping installations. Proper compactness of backfill should be assured without damage to piping. Clean earth, sand, or screened gravel should be placed under, around, and above the piping, to at least 1 ft (0.3 m) above it, and compacted carefully. Thereafter, backfilling should be completed to grade, compacting the backfill at least every 2 ft (0.6 m). Heavy boulders and corrosive cinder fill should not be allowed in the trench as backfill material.

Aboveground water supply piping should be securely supported and attached to the building construction. Where it is deemed necessary to prevent movement of the piping, it should be attached securely to an anchor rigidly affixed to the building construction. Hangers, piers, and pipe anchors should be of durable materials having adequate strength to perform their respective functions for the anticipated life of the building.

The maximum distances between supports for aboveground water supply piping of the various types of materials commonly used are as follows:

Vertical piping

Screwed pipe (standard pipe size): every other story

Copper and red brass pipe (type TP): every other story

Copper water tube (types K and L): every story, but not more than 10-ft (3.0-m) intervals

Plastic pipe (schedule 40, rigid): not more than 4-ft (1.2-m) intervals

Horizontal piping

Screwed pipe (standard pipe size): 10-ft (3.0-m) intervals for piping 1¼ in (31.8 mm) and larger; 8-ft (2.4-m) intervals for smaller piping

Copper and red brass pipe (type TP): 10-ft (3.0-m) intervals for piping 1¼ in (31.8 mm) and larger; 8-ft (2.4-m) intervals for smaller piping

Copper water tube (types K and L): 10-ft (3.0-m) intervals for piping 1½ in (38.1 mm) and larger; 6-ft (1.8-m) intervals for smaller piping

Plastic pipe (schedule 40, rigid): not more than 4-ft (1.2-m) intervals

EXPANSION AND CONTRACTION OF PIPING

Changes in pipe temperature cause corresponding changes in pipe length. Piping increases in length or expands with temperature rise and decreases in length or contracts with temperature drop. The proportional change in length corresponding to a 1°F change in temperature is known as the *coefficient of linear expansion.* This varies for different kinds of piping materials as shown in Table 4-1.

The total change in pipe length may be calculated for a given type of material, length of pipe, and temperature change by means of the following formula:

$$L_2 - L_1 = C_e L_1 (T_2 - T_1) \qquad [4\text{-}1]$$

where L_1 = pipe length at temperature T_1, ft
L_2 = pipe length at temperature T_2, ft
C_e = coefficient of linear expansion corresponding to the kind of piping material
T_1 = pipe temperature when length is L_1, °F
T_2 = pipe temperature when length is L_2, °F

Table 4-1
PIPING EXPANSION WITH TEMPERATURE CHANGE

Piping material	Coefficient of linear expansion, C_e	Total expansion in 100-ft (30.5-m) pipe length for 100°F (55.5°C) temperature change			
		ft	(m)	in	(mm)
Cast iron	0.00000595	0.0595	0.018	0.714	18.1
Steel	0.0000065	0.065	0.020	0.780	19.8
Wrought iron	0.0000068	0.068	0.021	0.816	20.7
Copper	0.0000095	0.095	0.029	1,140	28.9
Red brass	0.0000104	0.104	0.032	1.248	31.7
ABS, type I	0.000056	0.560	0.171	6.720	171.0
PVC, type I	0.000028	0.280	0.085	3.360	85.3
PVC, type II	0.000055	0.550	0.168	6.600	168.0

Total linear expansion or contraction occurring in a 100-ft (30.5-m) length of piping, when subjected to a 100°F (55.5°C) change in temperature, has been calculated for several kinds of standard water supply piping materials and is shown in Table 4-1. As domestic hot water supply piping is generally subjected to a 100°F (55.5°C) variation in temperature, the amount of expansion per 100 ft (30.5 m) of piping shown in the table may be applied directly in designing such piping. Provision should be made in the arrangement and support of the piping so as to permit the necessary amount of expansion. If the piping is restrained from moving, then it will be subjected to internal compressive stress during temperature rise and to internal tensile stress during temperature drop. Although the pipe itself may be able to withstand such stress, failure may and frequently does occur at pipe joints and fittings when the piping cannot expand and contract freely.

Allowance for expansion and contraction should be made in all hot water supply mains, risers, and branches therefrom. It is recommended that suitable expansion loops and swing joints be provided at appropriate locations in the piping runs. It is also recommended as a reasonable basis for designing hot water piping that the total amount of expansion to be absorbed at any loop or swing arrangement be limited to no more than 1½ in (37.5 mm).

In horizontal mains, a horizontal U bend or four-elbow expansion loop of adequate developed length, as shown in Fig. 4-2, generally is the most convenient and practical means of absorbing expansion in long straight runs of piping. The expansion loop should be of sufficient developed length so as to absorb the amount of expansion by flexure or spring without causing excessive stress in the piping and joints. Similarly, any 90° offset in a horizontal main may absorb expansion by flexure

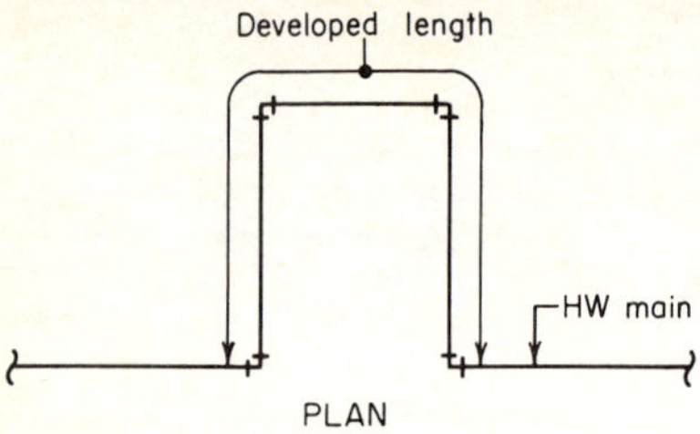

Fig. 4-2 Four-elbow horizontal U bend.

of the pipe if the offset is of sufficient length. Branch connections to horizontal mains should not restrict movement of the main, but instead should move with it. If the branch is of sufficient length, it may flex without excessive strain. But if the branch is relatively short, it should be provided with a four-elbow swing joint of adequate developed length, as shown in Fig. 4-3.

In risers, it is recommended that six-elbow swing joints or loops, as shown in Fig. 4-4, be provided midway between riser anchors. Such swing joints should be of adequate developed length to absorb the amount of expansion by flexure of the piping in the swing arrangement, rather than by tightening or loosening of joints at fittings. At the base of each riser, it is recommended that the downward expansion from the anchor above should be absorbed by means of a suitable four-elbow swing joint or loop, rather than permitting such expansion to be transmitted directly into the horizontal supply piping to the riser. Branch connections to risers, like those to horizontal mains, should not restrict movement of the riser, but rather should move with it. As riser branches usually are relatively short and of insufficient length to absorb riser expansion by flexure, it is recommended that each branch be provided with a suitable four-elbow swing joint of adequate developed length as shown in Fig. 4-5. All parts of swing joints and expansion joints should be free to move as the riser expands and contracts; consequently, they should not be built into wall construction in a manner that would restrict necessary movement.

Sliding- or slip-type expansion joints are not recommended except where limited space conditions cause practical difficulties. In such cases,

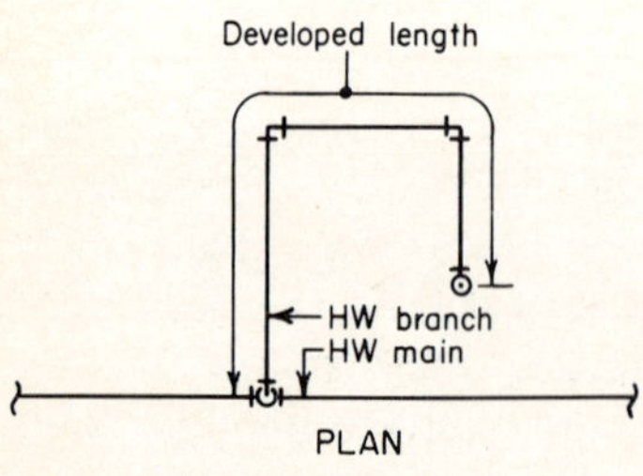

Fig. 4-3 Four-elbow horizontal swing joint.

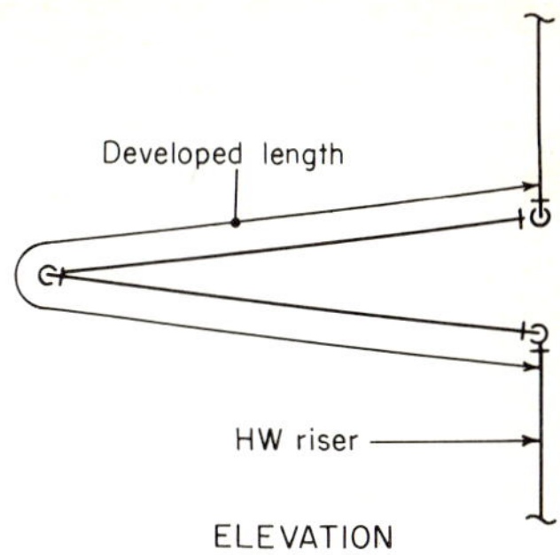

Fig. 4-4 Six-elbow riser swing joint.

adequate provision should be made for convenient access to the slip-type expansion joints so that they can be maintained periodically.

The developed length of piping required for absorbing a given amount of expansion by means of flexure or spring in an expansion loop or swing joint, without causing excessive strain in the piping, can be calculated. This may be done by means of the following formulas:

For steel or wrought-iron piping:

$$L = 6.16\sqrt{de} \qquad [4\text{-}2]$$

For red brass piping:

$$L = 6.83\sqrt{de} \qquad [4\text{-}3]$$

For copper piping:

$$L = 7.4\sqrt{de} \qquad [4\text{-}4]$$

For ABS, type I piping:

$$L = 1.71\sqrt{de} \qquad [4\text{-}5]$$

For PVC, type II piping:

$$L = 1.89\sqrt{de} \qquad [4\text{-}6]$$

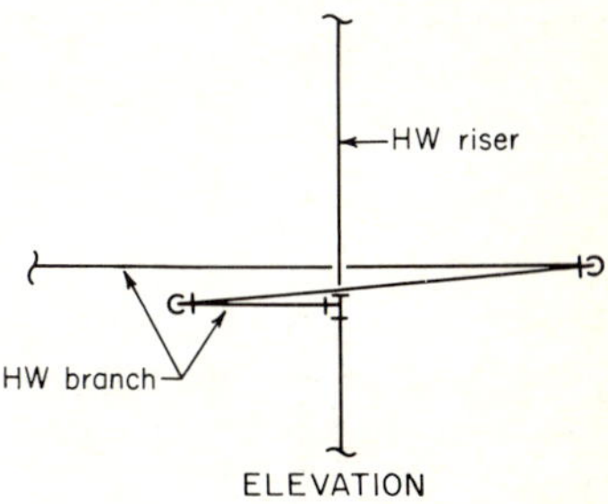

Fig. 4-5 Four-elbow swing joint on riser branch.

Table 4-2
DEVELOPED LENGTH OF EXPANSION JOINTS AND SWING JOINTS

For Absorbing 1½ in (37.5 mm) of Expansion

Nominal pipe size		Standard steel pipe		Standard red brass pipe		Standard copper pipe		Schedule 40 ABS pipe		Schedule 40 PVC pipe	
in	(mm)	ft	(m)	ft	(m)	ft	(m)	ft	(m)	ft	(m)
½	12.7	6.92	2.11	7.66	2.34	8.30	2.53	1.92	0.586	2.12	0.647
¾	19.0	7.70	2.35	8.55	2.61	9.25	2.82	2.14	0.653	2.36	0.720
1	25.4	8.68	2.65	9.60	2.93	10.4	3.18	2.40	0.732	2.65	0.808
1¼	31.8	9.75	2.97	10.8	3.29	11.7	3.57	2.70	0.824	2.98	0.909
1½	38.1	10.4	3.18	11.5	3.51	12.5	3.81	2.89	0.881	3.19	0.973
2	50.8	11.5	3.51	12.7	3.87	13.8	4.21	3.19	0.973	3.52	1.07
1½	63.5	12.8	3.90	14.2	4.33	15.4	4.70	3.56	1.09	3.93	1.20
3	76.2	14.2	4.33	15.7	4.79	17.0	5.19	3.93	1.20	4.34	1.32
4	102.0	16.0	4.88	17.7	5.40	19.2	5.86	4.44	1.35	4.90	1.49

where L = developed length of piping in expansion loop or swing joint, ft
d = outside diameter of piping, in
e = amount of expansion to be absorbed, in

These formulas may be applied to determine the developed length of piping required for absorbing 1½ in (37.5 mm) of expansion, the limit previously recommended as a reasonable basis for design. This has been done, and the values are given in Table 4-2.

STANDARD PIPE THREAD

Standard screwed or threaded pipes, valves, and fittings are provided with a standard thread conforming to the American Standard taper pipe thread, ANSI B2.1-1968. This type of thread was originally known as the Briggs standard thread for wrought-iron pipe. Its definite detail dimensions were formulated about 1862 by Robert Briggs, superintendent of the Pascal Iron Works of Morris, Tasker & Company, Philadel-

Table 4-3
STANDARD PIPE THREADS

Nominal pipe size, in	Die size classification	Number of threads per inch	Nominal length of thread on pipe, in	Tight-fit makeup length of thread, in
⅛	Tiny	27	⅜	¼
¼	Very small	18	9/16	⅜
⅜	Very small	18	9/16	⅜
½	Small	14	¾	½
¾	Small	14	¾	9/16
1	Medium	11½	15/16	11/16
1¼	Medium	11½	15/16	11/16
1½	Medium	11½	1	11/16
2	Medium	11½	1⅛	¾
2½	Large	8	1½	15/16
3	Large	8	1 9/16	1
4	Large	8	1 11/16	1⅛
5	Large	8	1 13/16	1¼
6	Large	8	1⅞	1 5/16
8	Large	8	2 1/16	1 7/16
10	Large	8	2 5/16	1⅝
12	Large	8	2½	1¾
15	Large	8	2¾	1¾

phia, Pa. By 1886, most American manufacturers threaded pipe in accordance with the Briggs standard. In December of 1886, the American Society of Mechanical Engineers adopted it as standard pipe thread practice, and master plug and ring gages were developed.

In December of 1919, the American Standards Association adopted a revised standard for taper pipe threads prepared by a committee on the standardization of pipe threads, organized so as to re-edit and expand the Briggs standard. Further revisions of the pipe thread standard have been made since then, the latest being dated 1968.

The standard taper pipe thread has an angle of 60° between its sides, is slightly rounded at top and bottom, and has a taper of ¾ in/ft. Several important features of this pipe thread are tabulated in Table 4-3. Of special interest to those involved in pipe layout and dimensioning is the last column headed, "Tight-fit makeup length of thread," which provides the normal allowance recommended for thread engagement with fittings for tight joints.

OFFSET CALCULATIONS

In the layout and dimensioning of piping arrangements, calculations must be made frequently to determine the exact distances between fitting centers of offsets in the run of piping. Where 90° elbows are used, this is no problem. However, where elbows of less offset degree are used, special calculations must be made.

Figure 4-6 illustrates a common pipe offset between parallel runs of piping. Angle β represents the degree of offset of each of the elbows. The only known dimension in most instances is the offset distance O between centers of the parallel runs of the piping. To be determined are P, the length of piping along the offset between fitting centers, and A, the distance at which one fitting center is in advance of the other fitting center.

The cosecant of angle β is equal to P/O, and therefore P = offset × cosecant of angle β. Similarly, the cotangent of angle β is equal to A/O, and A = offset × cotangent of angle β. Values of the cosecant and cotangent for each of the offset angles provided by various kinds

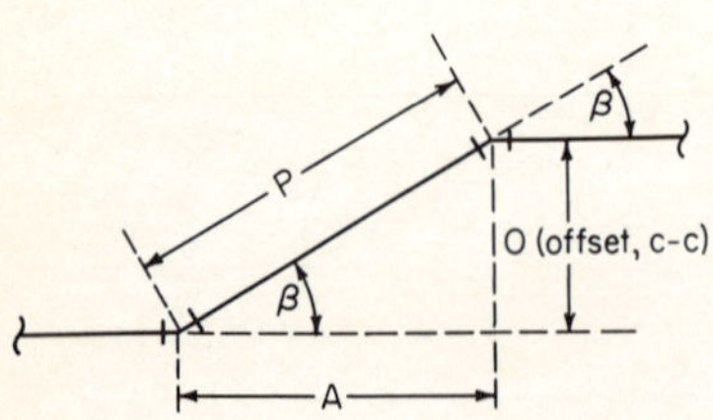

Fig. 4-6 Common pipe offset.

Table 4-4
COMMON PIPE OFFSETS

β (elbow), deg	P (piping length c-c)	A (advance of fitting centers c-c)
90	1.00 × offset	0.00 × offset
60	1.15 × offset	0.58 × offset
45	1.41 × offset	1.00 × offset
30	2.00 × offset	1.73 × offset
22½	2.61 × offset	2.41 × offset
11¼	5.12 × offset	5.02 × offset
5⅝	10.20 × offset	10.15 × offset

of standard elbows have been determined and applied to the above equations. The results are tabulated in Table 4-4 for convenient use in calculations.

Figure 4-7 illustrates a rolled pipe offset between parallel runs of piping which are not in the same horizontal or vertical plane. One run is a vertical distance V above and a horizontal distance H from the other run. A rolled pipe offset is a common pipe offset that may be considered to have been rotated from a vertical plane to a plane between the vertical and horizontal. The known dimensions of rolled offsets are the distances V and H. They may be considered to be the lengths of the two short sides of a right triangle, the hypotenuse of which is the offset distance O between parallel runs of the piping. Hence, the offset distance O is equal to $(V^2 + H^2)^{1/2}$. Lengths of P and A for the rolled pipe offset may be determined in the same manner as for a common pipe offset, as in Table 4-5.

Figure 4-8 illustrates offsets in parallel pipes made in such a manner that they maintain uniform span between centers of the pipe runs. The

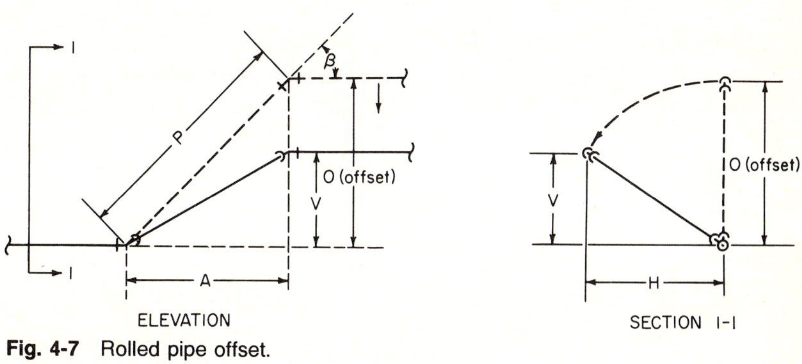

Fig. 4-7 Rolled pipe offset.

Table 4-5
ROLLED PIPE OFFSETS

β (elbow), deg	P (piping length c-c)	A (advance of fitting centers c-c)
90	$1.00(V^2 + H^2)^{1/2}$	$0.00(V^2 + H^2)^{1/2}$
60	$1.15(V^2 + H^2)^{1/2}$	$0.58(V^2 + H^2)^{1/2}$
45	$1.41(V^2 + H^2)^{1/2}$	$1.00(V^2 + H^2)^{1/2}$
30	$2.00(V^2 + H^2)^{1/2}$	$1.73(V^2 + H^2)^{1/2}$
22½	$2.61(V^2 + H^2)^{1/2}$	$2.41(V^2 + H^2)^{1/2}$
11¼	$5.12(V^2 + H^2)^{1/2}$	$5.02(V^2 + H^2)^{1/2}$
5⅝	$10.20(V^2 + H^2)^{1/2}$	$10.15(V^2 + H^2)^{1/2}$

only known dimension in this instance is the span S. To be determined is the dimension Z, the distance by which the outer elbow center leads the inner elbow center. In the illustration, the offset angle of the fitting is labeled β. It can be proved that the tangent of ½ angle β is equal to Z/S. Hence, $Z =$ span $\times$ tangent of ½ angle β. For various kinds of standard elbows, values of the tangents of one-half their offset angles have been determined and applied to the above equation. The results are tabulated in Table 4-6 for use in making convenient calculations for such parallel offset piping arrangements.

VALVING

All control valves in the water supply distributing system, except those which control the supply to single fixtures, should be designed to provide, when fully opened, no less cross-sectional area throughout the water passage than that of the piping in which the valve is installed. Standard gate, ball, and butterfly valves and ground-key stopcocks are recognized as satisfying this requirement. Globe and angle-type compression valves may be used to control or throttle the supply to single fixtures.

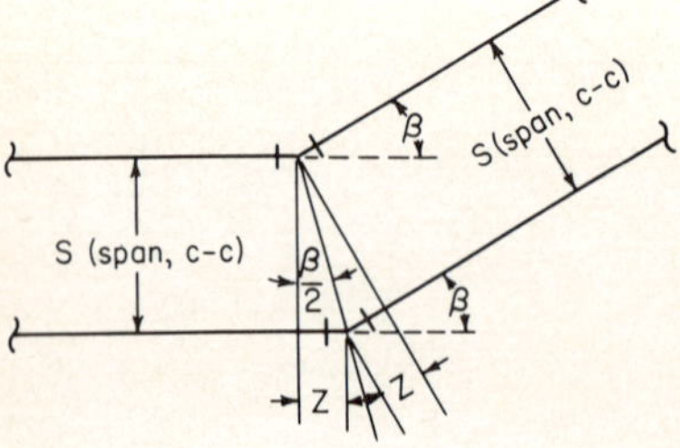

Fig. 4-8 Parallel pipe offsets with uniform span.

Table 4-6
PARALLEL PIPE OFFSETS WITH UNIFORM SPAN

β (elbow), deg	Z (outer elbow lead over inner elbow c-c)
90	1.00 × span
60	0.577 × span
45	0.414 × span
30	0.268 × span
22½	0.199 × span
11¼	0.098 × span
5⅝	0.049 × span

No combination stop-and-waste valve or cock should be installed underground in water supply piping. This prohibition has been deemed necessary in order to prevent contamination of the water supply piping due to entrance of underground liquids when the valve is shut off and the waste outlet is open.

Water service pipes should be equipped with an approved-type gate, ball, or butterfly valve or ground-key stopcock near the curbline, between the property line and the curbline. Water service curb valves should have an approved curb valve box extending to the grade and provided with an approved-type cover for easy access in order to be able to operate the valve when necessary. Curb valves and valve boxes should not be located under a driveway where they may be subjected to concentrated loads which could result in damage to the water service pipe. Water service pipes should also be equipped with an approved gate, ball, or butterfly valve or ground-key stopcock inside the building near the point of entry. Such valves are needed when it is necessary to shut off all water furnished to the water supply distributing system of a building.

All water supply control valves should be installed in accessible locations so that they can be readily shut off whenever the need arises. At water supply tanks, a valve should be installed in the water supply piping so as to control the supply of water to the water supply distributing system. At the base of each water supply riser, except in one-family dwellings, a valve should be installed to control the supply to the riser. In the cold water supply branch piping to water heating equipment, a valve should be provided to control the supply to such equipment. In two-family or multiple dwellings, one or more valves should be provided in the water supply piping so as to shut off the water supply to fixtures and equipment in each dwelling unit without interfering with the water supply to other dwelling units or portions of the building. In occupancies other than dwelling units, valves should be provided in the water supply piping so as to shut off the water supply to individual fixtures and equipment or to the rooms in which they are located. All control valves in the water supply distributing system should be at least the same size as the piping in which they are installed.

WATER SUPPLY PIPING TESTS

All piping of the potable water supply system should be proved watertight by the application of a water pressure test, using potable water, so as to disclose leaks and defects. No part of the piping system should be covered or concealed until it has been thoroughly tested and approved. The test pressure in any case should be at least equal to the maximum water pressure at which the piping system is to operate.

It is recommended that all potable water supply piping inside buildings should be tested, prior to covering or concealment and before fixtures, faucets, and trim are installed. As a test of the rough piping installation, it should be subjected to a hydrostatic test at 125 psig (862 kPa) minimum for a period of 3 h and should be proved watertight without any loss of pressure. When the system has been completed and all fixtures, faucets, hose connections, and trim have been installed, the entire system should be retested by subjecting it to a hydrostatic test at 75 psig (517 kPa) for a period of 3 h, and it should be proved watertight without any loss of pressure.

DISINFECTION METHODS

Disinfection of newly installed potable water supply systems, including water supply tanks, is frequently required by authorities having jurisdiction before the systems may be placed in service. This may also be required by official orders when inspection of existing water supply systems reveals that a system has become contaminated, or that it is necessary to remove scum, foreign matter, or other objectionable matter or surface encrustation from a potable water supply tank so as to correct what is deemed an unsanitary condition.

During disinfection of a system or tank, its water supply connections should be disconnected, plugged, or effectively shut off so as to prevent any foreign matter or contamination from entering the water supply thereto. For disinfection, one of the following methods may be applied.

1. The system, or part thereof, should be filled with a hypochlorite solution containing 50 ppm of available chlorine and allowed to stand for a period of at least 24 h, after which the system

should be flushed with potable water before being returned to service.

2. The system, or part thereof, should be filled with a hypochlorite solution containing 200 ppm of available chlorine and allowed to stand for a period of at least 3 h, after which the system should be flushed with potable water before being returned to service.

3. Where it is not practicable to disinfect a potable water supply tank by either of the previous methods, the entire interior of the tank should be swabbed with a hypochlorite solution containing 200 ppm of available chlorine and allowed to stand for a period of at least 3 h, after which the tank should be flushed with potable water before being returned to service.

4. For potable water filters and similar devices, the hypochlorite solution dosage and duration of application should be as specially prescribed by the health authority having jurisdiction under the circumstances prevailing for the particular installation.

STANDARD CODE REGULATIONS

Source and Quality of Water Supply Buildings in which plumbing fixtures are installed shall be provided with ample supply of potable water by connection to a public water supply system. Where a public system is not available, an approved source of private water supply shall be provided.

For a one- or two-family dwelling, a public water supply system shall be deemed available when such system is within 100 ft (30.5 m) of the premises on which the dwelling is located, measured along a street, and a connection may be made lawfully thereto.

For buildings of any other occupancies, a public water supply system shall be deemed available when such system is within 500 ft (152 m) of the premises on which the building is located, measured along a street, and a connection may be made lawfully thereto.

Only potable water shall be supplied to fixtures and equipment at which water is provided for drinking, cooking, food preparation, bottling, canning or packing, washing of dishes, glasses, cutlery, or kitchen utensils, or for similar domestic purposes. Such water shall conform to the chemical and bacteriological quality standards for potable water established by the health authority having jurisdiction.

Nonpotable water shall not be supplied to fixtures or equipment unless specially approved by the health authority having jurisdiction, and the supply of nonpotable water shall be limited to water closets, urinals, and other fixtures and equipment which do not require potable water supply. All nonpotable water faucets and outlets, and piping conveying nonpotale water shall be adequately and properly identified by conspicuous markings as hazards.

Protection of Potable Water Supply No piping materials that could produce toxic conditions in the potable water supply system shall be used in such system.

No chemicals that could produce toxic conditions in the potable water system shall be introduced into such system except as specially approved by the health authority having jurisdiction.

No piping that has been used for other than a potable water supply system shall be used for conveying potable water.

No private water supply system shall be cross-connected with any public potable water supply system unless specially approved in conformity with existing regulations by the health authority having jurisdiction.

No part of a potable water supply system, including discharge pipes from relief valves and other extensions of the water supply system, shall be connected to any drainage or vent piping.

Potable water supply piping, water discharge outlets, vacuum breakers, and similar equipment shall be located so as to prevent them from being submerged in any contaminated or polluted liquid or substance, except as otherwise provided by regulations or as may be specially approved for swimming pool system connections conforming to regulations of the health authority having jurisdiction.

No direct connection of the potable water supply system shall be made to the following fixtures and equipment:

1. Bidets
2. Operating, dissection, embalming, and mortuary tables or similar equipment—hose used for water supply shall terminate at least 12 in (300 mm) away from every point of the table or attachments
3. Pumps used for nonpotable water—potable water supply connections for priming purposes may be made to the inlet side of such pumps, provided a fixed air gap of approved type is installed in a vertical section of the water supply piping connections
4. Sterilizers, aspirators, water siphons, or similar equipment, or any chemical solution tank or apparatus, except where specially approved

Where a direct connection of the potable water supply system is made to condensers or cooling jackets of refrigeration units, the water supply pipe thereto shall be equipped with an approved check valve, except in such installations where the water supply piping is entirely outside of the piping or tank containing the refrigerant, and two independent wall thicknesses of metal separate the refrigerant from the potable water supply. Refrigeration units containing more than 20 lb (9.1 kg) of refrigerant shall be provided also with an approved pressure relief valve installed adjacent to and at the outlet side of the check valve. Such relief valve shall be set to relieve 5 psi (34.5 kPa) above the maximum water pressure at the point of installation.

Water used for cooling, heating, processing, or similar purposes shall not be reintroduced into the potable water supply system nor be distributed to plumbing fixtures requiring potable water supply. When discharged to the building drainage system, such water shall be discharged into a fixture or receptacle approved for such use through an air gap conforming to regulations.

Where practicable, potable water supply outlets shall have the outlet end spaced a distance above the flood level rim of the fixture or receptacle, to provide an air gap conforming to regulations, except as otherwise provided herein.

The minimum required air gap, measured vertically downward from the outlet end of a faucet or other water outlet to the flood level rim of the receiving fixture or receptacle, shall be at least twice the diameter of an outlet of equivalent circular cross section; but the minimum required air gap shall be at least three times such diameter where the outside edge of the outlet is less than three diameters away from the wall or, where walls intersect, less than four diameters away from each wall.

Below-rim potable water supply outlets to fixtures shall be prohibited except where such outlets are absolutely essential for proper functioning of the fixture, and protective methods have been applied in conformity with the regulations given below.

Protective Methods for Fixtures with Below-Rim Potable Water Supply Outlets Fixtures with below-rim potable water supply outlets shall be individually equipped with approved vacuum breakers, of the same nominal size as the fixture supply pipes. Vacuum breakers shall be of suitable type for the use intended.

Vacuum breakers shall be installed in proper position for satisfactory operation. They shall be located at least 6 in (150 mm) above the flood level rim of the fixture or receptacle supplied, except as may be otherwise required by regulations.

Vacuum breakers installed in a potable water supply system shall be maintained in good working condition. Vacuum breakers, except those on ball cocks, shall be exposed in the same room as the fixtures they serve, and shall be readily accessible for inspection.

Potable water supply lines to water preheating equipment utilizing waste water as a source of heat shall be equipped with a vacuum breaker and with a check valve located between the vacuum breaker and the preheating equipment. Where a hot water storage tank is supplied through such preheating equipment and has an independent cold water supply line, such cold water supply line shall be equipped with a vacuum breaker located at least 6 (150 mm) above the highest elevation of the tank, and with a check valve located between the vacuum breaker and the tank.

Each hose coupling water outlet and serrated tip water outlet shall be individually equipped with a vacuum breaker installed on the individual supply line to the water outlet, or installed on the water outlet with provision for hose attachment to the outlet side of the vacuum breaker. However, this requirement shall not apply to drain cocks for building heating systems. Such vacuum breaker shall be at least 6 in (150 mm) above the highest point of hose usage. Hose coupling water outlets for bedpan washing shall also be equipped with a check valve located between the vacuum breaker and the hose coupling.

Flush valves which are directly connected to the potable water supply system shall be equipped with vacuum breakers. Such vacuum breakers shall be located on the discharge side of the flush valves and at least 6 in (150 mm) above the top of the fixture supplied.

Flush tanks shall be equipped with approved ball cocks. Where ball cocks are in contact with tank water, they shall be equipped with vacuum breakers at least 1 in (25 mm) above the overflow outlet of the flush tank. Where ball cocks are not in contact with tank water, the ball cock outlet shall be installed at least 1 in (25 mm) above the overflow outlet of the flush tank or a vacuum breaker shall be installed as in the preceding provision.

The supply pipe to a lawn sprinkler or irrigation system shall be equipped with a vacuum breaker at least 12 in (0.3 m) above the highest sprinkler or discharge outlet of the system.

Backflow Prevention Devices for Special Fixtures and Equipment In view of the significant hazards involved with the following special plumbing fixtures and equipment, connections to the potable water supply system shall be protected against backflow with any one or more of the devices as indicated:

LOW INLET TO RECEPTACLES CONTAINING TOXIC SUBSTANCES (VATS, STORAGE CONTAINERS, PLUMBING FIXTURES)

1. An approved air-gap fitting
2. Reduced-pressure-principle back-pressure backflow preventer
3. Pressure vacuum-breaker unit
4. Atmospheric vacuum-breaker unit

LOW INLET TO RECEPTACLES CONTAINING NONTOXIC SUBSTANCES (STEAM, AIR, FOOD, BEVERAGES, ETC.)

1. An approved air-gap fitting
2. Reduced-pressure-principle back-pressure backflow preventer
3. Pressure vacuum-breaker unit
4. Atmospheric vacuum-breaker unit
5. Approved double-check-valve assembly

OUTLETS WITH HOSE ATTACHMENTS OR REFRIGERATION SYSTEMS WHICH MAY CONSTITUTE A CROSS CONNECTION

1. An approved air-gap fitting
2. Reduced-pressure-principle back-pressure backflow preventer
3. Pressure vacuum-breaker unit
4. Atmospheric vaccum-breaker unit
5. Double-check-valve assembly (where approved)

COILS OR JACKETS USED AS HEAT EXCHANGERS IN COMPRESSORS, DEGREASERS, AND SIMILAR EQUIPMENT INVOLVING TOXIC SUBSTANCES

1. An approved air-gap fitting
2. Reduced-pressure-principle back-pressure backflow preventer
3. Pressure vacuum-breaker unit

DIRECT CONNECTIONS SUBJECT TO BACK PRESSURE INVOLVING NONTOXIC SUBSTANCES

1. An approved air-gap fitting
2. Reduced-pressure-principle back-pressure backflow preventer
3. Approved double-check-valve assembly

Water Supply Requirements Plumbing fixtures and equipment shall be provided with water in sufficient volume and at pressures adequate to enable them to function satisfactorily and without excessive noise under normal conditions of use.

The water supply distributing system shall be designed and adjusted to supply fixtures and equipment with the minimum quantity of water consistent with proper performance and cleaning. The system shall be maintained to prevent leakage and excessive waste of water.

The minimum pressure available at all times at water outlets shall be 8 psi (55 kPa), except at direct supply flush valves at which the minimum shall be 15 psi (103 kPa), and except at other equipment requiring higher pressure at which the minimum shall be that required for satisfactory performance.

Where the water pressure available at the public water main is insufficient to maintain the minimum pressures required at all water outlets, an automatically controlled water supply tank of either the pressure or gravity type shall be provided. Such tank shall be designed and installed to provide the required minimum pressures.

Where water pressures are over 80 psig (550 kPa), or where self-closing faucets or valves are installed, approved air chambers or mechanical devices shall be provided to prevent pressure hazard, water hammer, and objectionable line noises.

Where street main pressure exceeds 80 psi (550 kPa), an approved pressure-reducing valve and an approved relief device shall be installed in the water service pipe near its entrance to the building to reduce the water pressure to 80 psi (550 kPa) or lower except where the water service pipe supplies water directly to a water pressure booster system, an elevated gravity water tank, or to pumps provided in connection with a hydropneumatic or elevated gravity water supply tank system. Pressure at any fixture shall be limited to no more than 80 psi (550 kPa) under no-flow conditions.

Where standpipe or automatic sprinkler systems are supplied by potable water services, the services shall be connected to water mains which are reliable and of adequate capacity. Water services for standpipe or automatic sprinkler systems shall be designed and installed to provide at all times a supply of water in sufficient volume to enable them to function satisfactorily.

Water Supply Tanks Water supply tanks shall be designed and constructed so as to be watertight, vermin-proof and rodent-proof, resistant to corrosion, and capable of withstanding the pressures under which they are to operate. Tanks shall be provided with safe and easy means of access for inspection.

The capacity of any single tank in or on a building shall not exceed 30,000 gal (113,000 L). Where tanks are located on flat roofs and the total capacity exceeds 30,000 gal, drain pipes from the tanks shall discharge so as to distribute water over separate drainage areas of the roof.

Supports for tanks shall be of noncombustible construction. Tanks and their supports shall not be used to support equipment or structures other than for tank use, except where specially designed for such other use.

Tanks shall not be located over openings in floor or roof construction. Openings in floor or roof for piping are permitted provided they are made watertight.

Potable water supply tanks for domestic supply and standpipe or automatic sprinkler systems shall be designed and installed to furnish water in sufficient quantity and pressure for such systems. A tank used to supply water both to a domestic system and a standpipe or automatic sprinkler system shall have the outlet for the domestic supply located a sufficient distance above the bottom of the tank to maintain the minimum reserve required for fire protection service. Potable water supply tanks which supply water for domestic supply and also for standpipe and automatic sprinkler systems shall have the outlet for the standpipe system located a sufficient distance above the bottom of the tank to maintain the minimum reserve required for the sprinkler system.

Where water supply tanks of the gravity type are supplied directly by pressure of the public water main, or may be so supplied at times, ball cocks or other automatic valves shall be installed to control the supply of water to such tanks and prevent their overflowing.

Table 4-7
SIZES OF OVERFLOW PIPES

Tank capacity		Overflow pipe size	
gal	**L**	**in**	**mm**
0–750	0–2,842	1	25.0
751–1,500	2,843–5,684	1½	37.5
1,501–3,000	5,685–11,369	2	50.0
3,001–5,000	11,370–18,948	2½	62.5
5,001–7,500	18,949–28,421	3	75.0
More than 7,500	Over 28,421	4	100.0

Potable water supply inlets within tanks of the gravity type shall terminate at an elevation above the overflow pipe not less than is required for an air gap; the elevation shall be not less than 4 in (100 mm) above the top of the overflow pipe.

Overflow pipes shall be provided for water supply tanks of the gravity type. Such pipes shall discharge above and within 6 in (150 mm) of a roof or catch basin, or they shall discharge into an open water-supplied fixture approved for such use. Overflow pipes shall be at least one pipe size larger than tank fill pipes, and not less than the sizes given in Table 4-7.

Emptying pipes shall be provided for all water supply tanks, located and arranged so as to prevent damage from water discharged, and shall discharge as required for overflow pipes. Each tank shall be provided with an emptying pipe not smaller in size than the sizes given in Table 4-8.

Each emptying pipe shall be provided with an approved valve of the same diameter as the pipe.

No potable water supply tanks of the gravity type, nor service holes of potable water supply tanks of the pressure type, shall be located directly beneath any soil or waste piping.

Overflow pipes and air vent pipes of potable water tanks of the gravity type shall be provided with durable screens of not less than 100 mesh.

When water supply tanks are lined, painted, or repaired they shall be disconnected from the system. No material shall be used during such operations which may affect the taste or potability of the water supply.

Water Supply Piping Installation No water supply piping shall be located in stairways, nor so as to interfere with normal operation of windows, doors, or other openings. No water supply piping shall be installed in a hoistway or under an elevator or counterweight.

Water service pipes connected to public water mains or extending under public ways shall be installed in accordance with regulations of the authority having jurisdiction.

No water supply piping shall be installed outside of buildings, or concealed in exterior walls, or located where it may be subjected to freezing temperatures unless adequate provision is made to protect such piping against damage from freezing.

Table 4-8
SIZES OF EMPTYING PIPES

Tank capacity		Emptying pipe size	
gal	**L**	**in**	**mm**
0–5,000	0–18,948	2½	62.5
5,001–10,000	18,949–36,895	3	75.0
More than 10,000	Over 36,895	4	100.0

Water supply piping passing through or under cinders or other corrosive material shall be provided with approved coating, wrapping, or other means of protecting against damage from external corrosion.

Water supply piping shall be installed so as not to be subject to undue strain. Provision shall be made to protect the piping against damage from strain due to normal expansion and contraction, and to building settlement.

Water supply piping passing through foundation or bearing walls shall be protected against breakage by means of sleeves or arches, or approved equivalent protection shall be provided. The space between sleeves or arches and the pipes passing through the wall shall be filled with approved sealing material. Sleeves shall be of iron or steel pipe two standard sizes larger than the pipe passing through.

Excavation for the installation of underground water supply piping shall be open trenchwork. Such piping shall be supported on a firm bed for its entire length. Precautions shall be taken to assure proper compactness of backfill without damage to the piping. Trenches shall be backfilled and compacted to at least 12 in (0.3 m) above the top of piping with clean earth, sand, or screened gravel, which shall not contain boulders, cinders, or other substances which may damage or break the piping or cause corrosive action. Thereafter, backfilling shall be completed up to grade and be properly compacted.

Water supply piping aboveground shall be securely attached to the building construction at no greater distances between supports than given in the following:

Vertical piping

Screwed pipe (standard pipe size): every other story

Copper and red brass pipe (type TP): every other story

Copper water tube (types K and L): every story, but not more than 10-ft (3.0-m) intervals

Plastic pipe (schedule 40, rigid): not more than 4-ft (1.2-m) intervals

Horizontal piping

Screwed pipe (standard pipe size): 10-ft (3.0-m) intervals for piping 1¼ in (31.8 mm) and larger; 8-ft (2.4-m) intervals for smaller piping

Copper and red brass pipe (type TP): 10-ft (3.0-m) intervals for piping 1¼ in (31.8-mm) and larger; 8-ft (2.4-m) intervals for smaller piping

Copper water tube (types K and L): 10-ft (3.0-m) intervals for piping 1½ in (38.1-mm) and larger; 6-ft (1.8-m) intervals for smaller piping

Plastic pipe (schedule 40, rigid): not more than 4-ft (1.2-m) intervals

Hangers, anchors, and piers for the support and attachment of piping shall be of approved material and have sufficient strength to support the piping and its contents.

Water supply piping shall be pitched so that the entire system or parts thereof can be drained and a drain cock, faucet or valve shall be provided at the lowest point. The formation of traps or sags shall be avoided wherever possible.

Water Supply Control Valves No combination stop-and-waste valve or cock shall be installed underground in water supply piping.

An approved type gate, ball, or butterfly valve or ground key stopcock shall be installed in the water service pipe near the curbline, between the property line and the curbline. Such control valve shall be provided with an approved curb valve box and shall not be located under a driveway.

A gate, ball, or butterfly valve or ground key stopcock shall be installed in the water service pipe inside the building near the point of entry.

A gate valve shall be installed in water piping supplied from a water supply tank and shall be located at or near such tank.

A control valve shall be installed at the foot of each water supply riser, except in one-family dwellings.

In two-family or multiple dwellings, the water supply piping shall be equipped with one or more valves to shut off the water supply to fixtures and equipment in each dwelling unit without interference with the water supply to other dwelling units or other portions of the building. Such dwelling unit control valves shall be located within the dwelling unit so controlled.

A control valve shall be installed in the cold water supply branch to water heating equipment.

Water supply piping to fixtures and equipment, in occupancies other than dwelling units, shall be equipped with valves to shut off the water supply to individual fixtures and equipment or to the room where they are located.

Control valves installed at the outlets of water meters shall be gate valves or ground key stopcocks, and at least as large as the size of the supply piping at the inlets of the water meters.

Water supply control valves shall be installed in accessible locations so that they can be readily shut off.

Water supply control valves in the distributing system, except those which control the supply to single fixtures, shall be designed to provide, when fully open, cross-sectional area throughout the water passage not less than that of the pipe in which the valve is installed.

Methods for Disinfecting Potable Water Supply Systems Whenever disinfection of the potable water supply system, or part thereof, is officially ordered, one of the following methods shall be used before the system, or part thereof, is placed in operation or returned to service:

> The system, or part thereof, shall be filled with a water solution containing 50 parts per million of available chlorine and allowed to stand for 24 h before flushing and returning to service
>
> The system, or part thereof, shall be filled with a water solution containing 200 parts per million of available chlorine and allowed to stand for 3 h before flushing and returning to service

For a potable water storage tank, where it is not practicable to disinfect by the foregoing methods, the entire interior of the tank shall be swabbed with a water solution containing 200 parts per million of available chlorine and allowed to stand for 3 h before flushing and returning to service

For potable water filters and similar devices, the dosage used shall be specially approved under the circumstances prevailing

Test Methods Water supply systems shall be proved watertight by applying a water pressure test, using potable water in piping of the potable water supply system. Test pressure shall be equal to at least the maximum pressure at which the piping is to serve. Such tests shall be applied to water service pipes and rough piping installations prior to covering or concealment.

5

FLOW IN WATER SUPPLY PIPING

PRINCIPLES OF HYDRAULICS

This chapter deals with principles of hydraulics that are relevant to flow in water supply piping for buildings. Such principles concern physical and mechanical properties of water and their application in engineering design of water supply piping systems.

The physical properties of water include its density and viscosity corresponding to temperature, boiling point corresponding to pressure, streamline and turbulent conditions of flow at low and high velocities, and quantity rates of flow.

Energy principles include potential and kinetic energy, static and velocity heads, and pressurized flow from nozzles, outlets, and orifices. Friction head loss is an essential energy consideration. Significant factors are pressure loss for flow through different kinds of pipe under various service conditions, and losses through meters, equipment, valves, and fittings. The dynamic force of flowing water may produce objectionable water hammer, shock, noise, and cavitation or erosion of metal surfaces.

Appropriate application of the relevant principles of hydraulics can achieve satisfactory engineering design and adequacy of performance of water supply piping systems for buildings.

PHYSICAL PROPERTIES OF WATER

In any consideration of flow in water supply piping, several of the physical properties of water are of special concern. Those most pertinent to the subject are the density, viscosity, compressibility, and boiling point of water. All these properties are involved in the flow of water.

The density of water, that is, its weight per unit of volume, varies with change in temperature. At 39.3°F (4°C), water is at maximum density, weighing 62.424 lb/ft³ (1000 kg/m³). Both above and below this temperature, water expands with temperature change and becomes less in unit weight. Weight per cubic foot of water corresponding to various temperatures is given in Table 5-1.

When a fluid flows, its natural characteristics of adhesion and cohesion result in the development of internal resistance to flow. This resistance is called the viscosity of the fluid. It is a measurable property which varies greatly from one fluid to another, and in liquids decreases with rise in temperature. The absolute and kinematic viscosities of water corresponding to various temperatures are given in Table 5-2.

Water is perfectly elastic, compressing when pressure is imposed, and returning to original volume when the pressure is removed. One pound per square inch (6.895 kPa) of pressure compresses water about 1/300,000 of its volume, and 1 psf (992.88 kPa) produces a volumetric change of 1/43,200,000. The compressibility of water may be written as $1/K$, in which K is the coefficient of compressibility having a value of 43,200,000 lb/ft². This value is so high that water is considered to be relatively incompressible.

Table 5-1
DENSITY OF WATER

Temperature		Density		Temperature		Density	
°F	°C	lb/ft³	kg/m³	°F	°C	lb/ft³	kg/m³
32	0.00	62.416	999.84	100	37.78	61.998	993.15
35	1.67	62.421	999.92	110	43.33	61.865	991.02
39.3	4.00	62.424	1000.00	120	48.89	61.719	988.68
40	4.44	62.423	999.96	130	54.44	61.555	986.05
45	7.22	62.419	999.89	140	60.00	61.386	983.34
50	10.00	62.408	999.71	150	65.56	61.203	980.41
55	12.78	62.390	999.43	160	71.11	61.006	977.26
60	15.56	62.366	999.04	170	76.67	60.799	973.94
65	18.33	62.336	998.56	180	82.22	60.586	970.53
70	21.11	62.300	997.98	190	87.78	60.351	966.76
75	23.89	62.261	997.36	200	93.33	60.135	963.30
80	26.67	62.217	996.65	210	98.89	59.881	959.23
85	29.44	62.169	995.89	212	100.00	59.843	958.63
90	32.22	62.118	995.07				
95	35.00	62.061	994.16				

Table 5-2
VISCOSITY OF WATER

Temperature		Absolute viscosity		
°F	°C	cP	pdl·s/ft²	Kinematic viscosity, ft²/s
32	0.00	1.79	0.001203	0.0000193
50	10.00	1.31	0.000880	0.0000141
60	15.56	1.12	0.000753	0.0000121
70	21.11	0.98	0.000668	0.0000107
80	26.67	0.86	0.000578	0.00000929
100	37.78	0.68	0.000457	0.00000737
120	48.89	0.56	0.000376	0.00000609
140	60.00	0.47	0.000316	0.00000515
160	71.11	0.40	0.000269	0.00000441
180	82.22	0.35	0.000235	0.00000388

Note: One centipoise (cP) = 0.000672 poundal-second per square foot (pdl·s/ft²).
Absolute viscosity of water at 68.4°F is equal to 1 cP.
Kinematic viscosity (square foot per second) is equal to absolute viscosity (poundal-second per square foot) divided by density (pound per cubic foot).

The temperature at which water boils varies with the pressure to which it is subjected. At sea level, an open pot of water boils at 212°F (100°C). This is the boiling point of water at 14.7 psi (101.354 kPa) absolute pressure, or 0 psig. At an elevation of 5200 ft (1585 m) above sea level, where the atmospheric pressure is normally about 12.1 psia (83.43 kPa), an open pot of water boils at 202°F (94.4°C). In a closed water supply system, operated at a service pressure of 50 psi (344.75 kPa) above atmospheric pressure, the water may not boil until its temperature reaches almost 300°F (148.9°C). Boiling points corresponding to various pressures are given in Tables 5-3 and 5-4.

STREAMLINE AND TURBULENT FLOW

Two distinctly different conditions of flow may occur as water passes through the supply system under pressure. This was first demonstrated experimentally in 1883 by Osborne Reynolds. He showed in his critical velocity color band experiments that at relatively low velocities of flow, all particles of the fluid move in paths which are parallel and in line with the stream. This condition is termed *streamline flow.*

Sometimes this type of flow is referred to as *laminar* or *viscous* flow because the resistance developed during flow is caused by sliding friction between adjacent layers or particles of the fluid. In this condition, the frictional resistance is directly proportional to the viscosity of the fluid, while the roughness of the pipe surface is an insignificant factor.

Streamline flow becomes unstable and changes abruptly to what is termed *turbulent* flow when the velocity is increased beyond certain limits. In turbulent flow, fluid particles move along in such irregular paths that the condition is one of a turbulent mass moving through the piping and rubbing in contact with its interior surfaces. Under this condition, resistance occurs principally because of friction between the fluid and the pipe surface, and is directly proportional to the roughness of the pipe surface, while the viscosity of the fluid is a relatively significant factor.

In his series of experiments using parallel glass tubes of various diameters, Reynolds determined the critical velocities at which the change from streamline to turbulent flow occurred and at which the change from turbulent to streamline flow occurred. The latter change takes place at a lower velocity; hence, this velocity is regarded as the true critical velocity of flow transition. It was found to vary directly with the absolute viscosity of the fluid and inversely with the density of the fluid and diameter of the pipe.

Table 5-3
BOILING POINTS OF WATER AT SUBATMOSPHERIC PRESSURES

Absolute		Boiling point	
psi	kPa	°F	°C
1	6.895	101.8	38.8
2	13.790	126.1	52.3
3	20.684	141.5	60.8
4	27.579	153.0	67.2
5	34.474	162.3	72.4
6	41.369	170.1	76.7
7	48.263	176.9	80.5
8	55.158	182.9	83.8
9	62.053	188.3	86.8
10	68.948	193.3	89.6
11	75.842	197.8	92.1
12	82.737	202.0	94.4
13	89.632	205.9	96.6
14	96.527	209.6	98.7
14.7	101.352	212.0	100.0

Note: Atmospheric pressure at sea level is 14.7 psia or 0 psig.

Table 5-4
BOILING POINTS OF WATER AT ABOVE-ATMOSPHERIC PRESSURES

Gauge		Boiling point	
psi	kPa	°F	°C
0	6.895	212.0	100.0
10	68.948	239.3	115.2
20	137.895	258.6	125.9
30	206.843	274.0	134.4
40	275.790	286.7	141.5
50	344.738	297.6	147.6
60	413.685	307.3	152.9
70	482.633	316.0	157.8
80	551.581	323.9	162.2
90	620.528	331.2	166.2
100	689.476	337.9	169.9
110	758.428	344.1	173.4
120	827.376	350.0	176.7
130	896.324	355.6	179.8
140	965.272	360.8	182.7
150	1034.220	365.9	185.5

An equation for determining the true critical velocity for any given fluid and size of pipe is as follows:

$$V_c = \frac{R_c \mu}{Dw} = \frac{2000\mu}{Dw} = \frac{2000\nu}{D} \qquad [5\text{-}1]$$

where V_c = true critical velocity, fps
R_c = Reynolds number, 2000, corresponding to the lower critical velocity
μ = absolute viscosity of the fluid, pdl·s/ft²
ν = kinematic viscosity of the fluid, ft²/s
D = diameter of the pipe, ft
w = density of the fluid, lb/ft³

Applying the above equation to water at 50°F (10°C), the true critical velocities of flow in pipes of ½-, 1-, and 2-in (12.5-, 25.-, and 50-mm) diameters are respectively 0.676, 0.338, and 0.169 fps (0.206, 0.103, and 0.052 m/s). For water at 140°F (60°C), the velocities for the same pipe sizes are respectively 0.247, 0.124, and 0.0617 fps (0.075, 0.038, and 0.019 m/s).

The critical velocity at which the change from streamline to turbulent flow occurs is recognized as being about twice that of the lower or true critical velocity. A Reynolds number of 4000 therefore may be applied in the preceding equation so as to determine the upper critical velocity for each size of pipe and water temperature condition.

Such very low velocities at which streamline flow may exist in the water supply system of a building may be found to occur when demand is considerably below the normal maximum. This should be recognized from the fact that the economical sizing of systems is based upon maximum flow velocities ranging from a low of 4 fps to a high of 8 fps. Consequently, at normal design velocities the prevailing condition is turbulent flow.

VELOCITY

Pitot tube measurements of flow at any given point in piping show that the velocity is not uniform throughout the cross section of flow. It varies from a minimum at the pipe wall to a maximum at the center of the pipe. The average velocity at the cross section is less than that which prevails at the center. For different conditions of flow and roughness of pipe surface, the average velocity may be stated as a percentage of the maximum.

With streamline flow, average velocity is approximately one-half the maximum. For turbulent flow, average velocity is about 80 percent of the maximum when the pipe surface is rough, and about 86 percent of the maximum when the pipe surface is smooth. In the case of a jet of water, flowing from an orifice into the atmosphere, the average and maximum velocities are approximately equal at the *vena contracta,* where the jet is narrowest, for at this point the velocity is relatively uniform across the cross section of flow.

Maximum and minimum velocities for any given flow condition are not applicable in flow calculations. Only the average velocity may be applied in this instance. Hence, wherever the term *velocity* is used elsewhere in this text, what is meant is the average velocity through a cross section of flow at a given point.

QUANTITY RATE OF FLOW THROUGH PIPING

When water flows through a channel, the quantity rate of flow past any given point is related to the cross-sectional area of flow and the velocity of flow at that point. This relationship is expressed in a rational equation, commonly termed the *basic flow formula,* which is as follows:

$$Q = AV \tag{5-2}$$

where Q = quantity rate of flow, ft³/s
A = cross-sectional area of flow, ft²
V = velocity of flow, fps

In a pressurized water supply system, the piping is filled with water and is of circular cross section. The quantity, in this case, is the volume contained in a cylinder of the same cross-sectional area and of length equal to the velocity. Thus, the quantity rate of flow is in direct proportion to the velocity of flow and to the cross-sectional area of flow or square of the pipe diameter.

For plumbing calculations, the basic flow formula is inconvenient because the types of plumbing piping available are of circular cross section. Pipe sizes are ordinarily stated in terms of inches of diameter, and quantity rates of flow are customarily referred to in terms of gallons per minute of flow.

To adapt the basic flow formula so that convenient and direct plumbing calculations may be made, appropriate factors must be applied to convert the terms to those commonly used in the plumbing industry. When this is done, the result is an adapted flow formula which is as follows:

$$q = 2.448 d^2 V \qquad [5\text{-}3]$$

where q = quantity rate of flow, gpm
d = diameter of the pipe, in
V = velocity of flow, fps

POTENTIAL AND KINETIC ENERGY

Energy is defined as the capacity for performing work. A body may possess mechanical energy because of its elevation above some reference level or its motion with respect to some reference point.

The energy of a body due to elevation is termed *potential energy*. The amount of energy gained by a body in being raised from one level to a higher level is equal to the work required to accomplish the lift. As the work done is the product of the weight and the height through which it is lifted, the following equation may be written.

$$E_p = Wh \qquad [5\text{-}4]$$

where E_p = potential energy gain, ft·lb
W = weight of the body, lb
h = height body is raised, ft

From the above equation, it can be noted that the potential energy gain is equal to the height when the weight is 1 lb. Hence, the potential energy gain per pound of weight can be expressed as follows:

$$e_p = h \qquad [5\text{-}5]$$

where e_p = potential energy gain per pound weight, ft·lb
h = height one pound body is raised, ft

Energy which a body possesses because of its motion is called *kinetic energy*. The amount of energy gained by a body in being put into motion with respect to some reference point is the product of one-half its mass times the square of its velocity. As the mass of a body is equal to its weight W divided by the gravitational acceleration g the gain in kinetic energy may be expressed as follows:

$$E_k = \frac{1}{2}\frac{W}{g}V^2 \qquad [5\text{-}6]$$

where E_k = kinetic energy gain, ft·lb
W = weight of the body, lb
g = gravitational acceleration, 32.2 ft/s²
V = velocity, fps

It may be seen from the above equation that the kinetic energy gain is equal to the term $V^2/2g$ when the weight of the body is 1 lb. Hence, the kinetic energy gain per pound of weight may be expressed as follows:

$$e_k = \frac{V^2}{2g} \qquad [5\text{-}7]$$

where e_k = kinetic energy gain per pound weight, ft·lb
V = velocity, fps
g = gravitational acceleration, 32.2 ft/s²

STATIC HEAD

At any point below the surface of a body of water, at rest and exposed to the atmosphere, pressure is produced by the weight of water lying above the point. The pressure is equal and effective in all directions and is directly proportional to the depth of the point below the surface. Thus, the pressure may be expressed in terms of still water depth below the free surface, or *static head* of water above the point. This is often referred to as *hydrostatic head*. It is the measure of potential energy due to elevation of the water above a reference point.

As the pressure is due to the weight of water, static head may be converted from feet of head to pounds per square inch of pressure.

This may be done by determining the weight of a column of water equal in height to the head and 1 in^2 in cross-sectional area, as expressed in the following equation:

$$p = \frac{w}{144} h \qquad [5\text{-}8]$$

where p = pressure, psi
w = density or weight of fluid, lb/ft^3
h = static head of fluid, ft

For water at 50°F (10°C), the pressure due to a 1 ft (0.3 m) head of water is equal to 62.408/144 or 0.433 psi (2.99 kPa). Conversely, the static head equivalent to 1 psi (6.895 kPa) of pressure is 144/62.408 or 2.31 ft (0.7 m).

VELOCITY HEAD

Flowing water possesses kinetic energy, that is, energy due to motion or velocity. The source of this energy in a pressurized water supply system is the potential energy available in the form of pressure or head on the system. To overcome the inertia of still water and produce velocity of flow, some of the potential energy or static head is converted to kinetic energy. The decrease in head corresponding to the velocity of flow is known as the *velocity head.*

In freely falling or flowing from a height, bodies are acted upon by the force of gravity and are accelerated at the rate of 32.2 ft/s^2. The height of fall and the velocity attained at any given moment are expressed in the following equations:

$$h = \tfrac{1}{2} g t^2 \qquad [5\text{-}9]$$

and

$$V = gt \qquad [5\text{-}10]$$

where h = height of fall, ft
g = gravitational acceleration, 32.2 ft/s^2
t = time, s
V = velocity, fps

The relationship between the height of fall and the velocity may be established by combining these two equations, substituting for the time factor t its equivalent V/g. This results in the following:

$$h = \frac{V^2}{2g} \qquad [5\text{-}11]$$

The term $V^2/2g$ is commonly called the *velocity head,* for it is the measure of the decrease in static head, in feet of water column, at any given point corresponding to the velocity of flow.

BERNOULLI'S THEOREM

In 1738, Daniel Bernoulli developed an equation to express the conservation of energy as applied to flow conditions in a frictionless, incompressible fluid. It is as follows:

$$Z+\frac{P}{w}+\frac{V^2}{2g}=\text{constant} \qquad [5\text{-}12]$$

where Z = vertical distance above the reference level, ft
P = pressure, lb/ft²
w = density of the fluid, lb/ft³
V = velocity of flow, fps
g = gravitational acceleration, 32.2 ft/s²

As the term P/w is equal to the head or height of liquid column, the term h may be substituted for it in the above equation. When applied to express the conditions of frictionless flow at different points in a piping system, the equation may be written as follows:

$$Z_1+h_1+\frac{V_1^2}{2g}=Z_2+h_2+\frac{V_2^2}{2g} \qquad [5\text{-}13]$$

where Z_1 and Z_2 = heights of points 1 and 2 above the reference level, ft
h_1 and h_2 = static heads at points 1 and 2, ft of fluid
V_1 and V_2 = velocities of flow at points 1 and 2, fps
g = gravitational acceleration, 32.2 ft/s²

This is illustrated in Fig. 5-1.

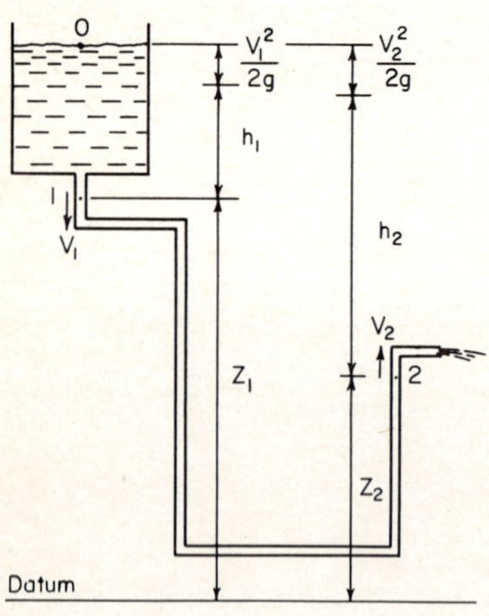

Fig. 5-1 Bernoulli's theorem—frictionless flow.

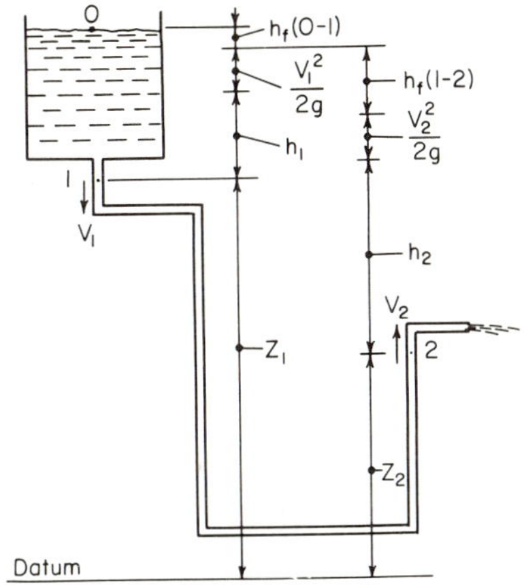

Fig. 5-2 Energy distribution with friction in flow.

FRICTION HEAD

Water flow in piping always produces friction due to the rubbing of water particles against each other and against the inner surfaces of pipe, valves, fittings, and equipment through which the water flows. Friction generates heat energy. In water piping, heat energy is dissipated in the form of temperature rise in both the water and piping. The temperature rise usually is inappreciable because just 1 Btu of heat energy is developed or interchanged from a potential energy loss of 778 ft·lb.

As friction occurs in piping whenever water flows, the pressure or head decreases along the line in the direction of flow. The decrease in head corresponding to the friction loss is called the *friction head.* This subject is more fully discussed in later parts of this chapter.

Figure 5-2 illustrates the energy conditions which prevail at two points in piping when friction occurs during flow. From this, it can be seen that the energy loss due to friction may be expressed as follows:

$$\left(Z_1 + h_1 + \frac{V_1^2}{2g}\right) - \left(Z_2 + h_2 + \frac{V_2^2}{2g}\right) = h_f = 778E_h \qquad [5\text{-}14]$$

where Z_1 and Z_2 = heights of points 1 and 2 above the reference level, ft

h_1 and h_2 = static heads at points 1 and 2, ft of fluid

V_1 and V_2 = velocities of flow at points 1 and 2, fps

g = gravitational acceleration, 32.2 ft/s²

h_f = head loss due to friction between points 1 and 2, ft of water

E_h = heat energy interchanged, Btu per lb of water flowing

TORRICELLI'S THEOREM

Torricelli, discoverer of the principle of the barometer and pupil of Galileo, engaged in experiments about 1640 to determine the velocity at which water escaped from an outlet in the side of a broad open tank filled with water to a considerable height above the level of the outlet.

The conclusion he drew from this experiment is known as the Torricelli theorem, which is as follows: "Except for minor frictional effects, the velocity of escape is the same as if the escaping liquid had fallen freely through the vertical distance measured from the water surface to the outlet." This is illustrated in Fig. 5-3.

The relationship between the height of free fall and the resulting velocity is given in the following equations:

$$h = \frac{V^2}{2g} \quad \text{and} \quad V = (2gh)^{1/2} \qquad [5\text{-}15]$$

where h = height of fall, ft
V = velocity attained, fps
g = gravitational acceleration, 32.2 ft/s^2

The maximum velocity that can be attained in discharge, owing to available pressure or head at the outlet during flow, is the same as that which may be attained as the result of free fall from a height equal to the head. Hence, this velocity may be termed the *ideal velocity*.

JET FLOW

The greatest velocity of flow from an outlet is found in the jet of water a short distance from the face of the outlet. At that point, the cross-sectional area of the jet contracts to a minimum, and the velocity becomes relatively uniform across the area of flow. This zone of the jet is known as the *vena contracta*, or contracted vein, as shown in Fig. 5-4. Beyond the contracted zone, the jet again expands in flow area.

The velocity attained by a body in frictionless free fall is the ideal velocity that may be achieved by a jet of water as the result of frictionless flow from an outlet due to the pressure or head available at the outlet during flow.

However, friction is always present with flow and prevents full development of ideal velocity. The amount of friction occurring during flow from an outlet varies with the available pressure or head, the size and shape of the outlet, and the smoothness of surfaces approaching and adjacent to the outlet. Hence, the actual velocity in practice is less than ideal, and the ratio of actual to ideal velocity is called the *coefficient of velocity*.

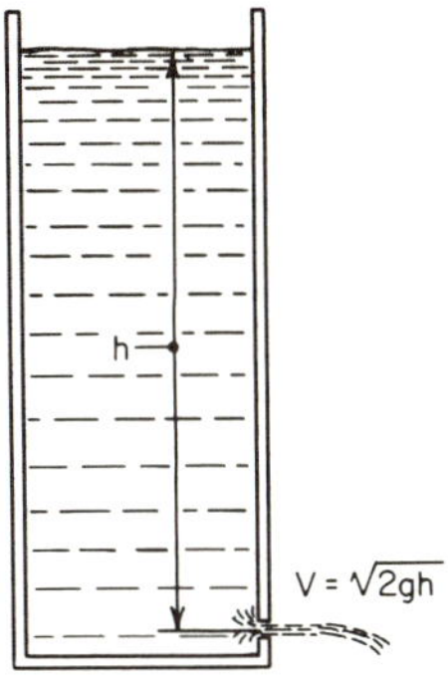

Fig. 5-3 Torricelli's theorem.

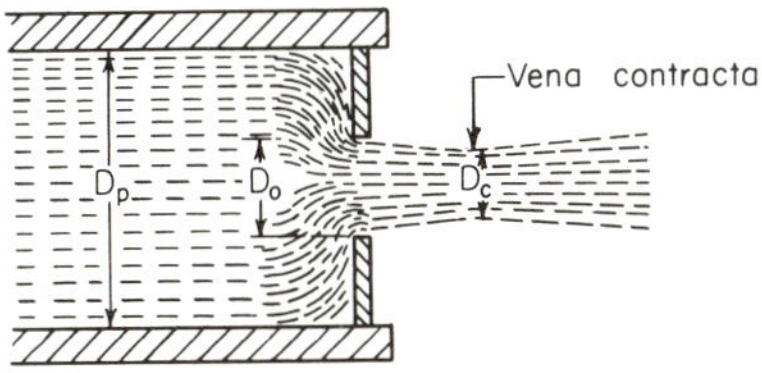

Fig. 5-4 Jet contraction.

For various kinds of outlets, coefficients of velocity have been determined experimentally. Standard square-edged circular orifices and well-designed, standard nozzles with smooth walls have coefficients ranging from about 0.95 to 0.99, the smaller values being applicable to small sizes and low pressures, while the larger values are applicable to larger sizes and higher pressures. In this regard, the coefficient increases as the result more of increase in size than of increase in pressure.

Short nipples connected to branch outlets of tees, in which the nipple enters about even with the inner face of the tee, have coefficients ranging from about 0.80 to 0.82 for low to high heads. Tank nipples, which project into tanks beyond their inner face, have lower coefficients ranging from about 0.72 to 0.78 for varying heads.

The ratio of the area of a jet at the vena contracta, the zone at which cross-sectional area is minimum, to the area of the outlet is called the *coefficient of contraction.* This may be determined by measuring the diameter of the jet at its narrowest point, calculating the cross-sectional area, and comparing it with that of the outlet.

It should be noted that for the many types of orifices, nozzles, and outlets, the coefficient of contraction varies from about 0.52 to 1.00 and has a wider range than is found for the coefficient of velocity. In addition, jet contraction is more sensitive to sharpness and slope of the edge of the outlet and the smoothness of the internal surfaces adjacent to the edge.

For standard square-edged circular orifices, the coefficient ranges from about 0.61 to 0.72, increasing with the decreasing size of the orifice and the decreasing head. A standard 2½-in orifice has a coefficient of about 0.61 at heads of 100 ft (30.48 m) or more, while 1- and ¾-in orifices have coefficients of 0.64 and 0.70, respectively, for similar heads. For nozzles, the coefficient varies with type, such as conical, straight tip, and convex, for which general values are, respectively, 0.80, 0.95, and 0.99.

Ordinary water supply system outlets vary so widely as to condition at the outlet edge that no definite coefficients may be given. For any particular outlet, its exact coefficient would have to be determined by test. However, a reasonable approximation may be assumed for a given circular outlet from values cited above for standard orifices and nozzles.

If a jet of water discharged from an outlet at ideal velocity and with no contraction, a condition which cannot occur in practice, the discharge would be at a rate termed the *ideal rate.* However, the actual rate of discharge is less than ideal owing to friction and jet contraction. The ratio of actual to ideal rate of discharge is called the *coefficient of discharge.*

In general, the discharge coefficient is simply the product of the velocity and contraction coefficients for a given orifice, nozzle, or outlet. This may be expressed as follows:

$$c_d = c_v c_c \qquad [5\text{-}16]$$

where c_d = coefficient of discharge
c_v = coefficient of velocity
c_c = coefficient of contraction

From the above, it may be recognized that the discharge coefficient for well-designed convex nozzles is relatively high, a good average value being about 0.98, while conical nozzles having a large vertex angle have a coefficient of about 0.76. Similarly, for standard orifices, the discharge coefficient may be found to range from about 0.59 to 0.70, depending upon the size and pressure at the orifice.

Ordinary pipe outlets, faucets, and corporation taps have a considerable range of discharge coefficients, depending upon various factors, including the shape and condition of the outlet edge, its size, and the available pressure during flow. But in view of the values given above for standard orifices, an average value of 0.67 may be assumed as a reasonable value for general application to such outlets.

QUANTITY RATE OF FLOW FROM OUTLETS

The velocity at which fluid flows from an outlet is due to the total energy available in the supply pipe at the outlet during flow. This total energy is the sum of the potential and kinetic energies of the moving fluid. Potential energy, in this instance, is the pressure or head exerted by the flowing fluid against the inside wall of the supply pipe and is termed *flow pressure.* It may be measured during flow by means of a pressure gage connected to the supply pipe near the outlet.

The amount of kinetic energy, or velocity head, in the supply pipe during flow usually is relatively small. Water piping ordinarily is designed so as to meet the demand at velocities limited to no more than 8 fps (2.4 m/s). As a result, the velocity head $V^2/2g$ equals 64/64.4 and amounts to only about 1 ft (0.3 m) of head or 0.433 psi (6.895 kPa) of pressure for maximum design velocity, and thus is evidently an insignificant factor by comparison with the flow pressure.

Consequently, the maximim rate of discharge from outlets of a water supply system may be determined with reasonable accuracy based just upon the flow pressure in the supply line and the diameter of the outlet. The discharge rate may be expressed as follows:

$$\begin{aligned} q_d &= c_d q_i = c_d(2.448 d_o^2 V_i) = c_d(2.448 d_o^2 \sqrt{2gh_m}) \\ &= c_d(19.65 d_o^2 \sqrt{h_m}) = c_d(29.87 d_o^2 \sqrt{p_m}) \end{aligned} \qquad [5\text{-}17]$$

where q_d = quantity rate of discharge from outlet in practice, gpm
q_i = ideal rate of discharge from outlet, gpm
c_d = coefficient of discharge for outlet
d_o = diameter of outlet, in
V_i = ideal velocity, fps
h_m = head measured in supply pipe during flow, ft of water
p_m = pressure measured in supply pipe during flow, psi

When the value of the coefficient of discharge c_d is assumed to be 0.67, a reasonable value for most small-diameter pipe outlets, the equation for the maximum rate of flow which may be obtained from an outlet in practice has the following simplified form:

$$\begin{aligned} q_d &= c_d(29.87 d_o^2 \sqrt{p_m}) = 0.67(29.87 d_o^2 \sqrt{p_m}) \\ &= 20 d_o^2 \sqrt{p_m} \end{aligned} \qquad [5\text{-}18]$$

where q_d = actual quantity rate of discharge from outlet, gpm
d_o = diameter of outlet, in
p_m = pressure measured in supply pipe during flow, psi

If the kinetic energy, or velocity head, in the supply pipe is considered in addition to the flow pressure, the only difference in the resulting equation for the quantity rate of discharge is the inclusion of an additional factor, known as the *correction factor,* for the velocity of approach to the outlet. When included, the equation appears as follows:

$$q_d = \frac{c_d(29.87 d_o^2 \sqrt{p_m})}{\sqrt{1 - (c_d)^2 (d_o/d_p)^4}} \qquad [5\text{-}19]$$

In this equation, the diameter of the orifice d_o and that of the pipe d_p are included as a ratio in the denominator. If this ratio were to be equal to one-half, then the correction factor would indicate a 1½ percent increase in the rate of discharge above that calculated solely on the basis of flow pressure. Hence, the correction factor by itself is not of appreciable significance, and the importance of assuming a proper value for the coefficient of discharge is evident.

QUANTITY RATE OF FLOW THROUGH SUBMERGED ORIFICES

From the equation for the maximum rate of flow which may be obtained in practice from an outlet discharging into the atmosphere, it is evident that the most critical factor affecting the rate of flow is the diameter of the outlet opening. As many plumbing fixtures are equipped with faucets and valve bodies supplied through tailpieces which are of much smaller internal diameter than the faucet spouts, it must be recognized that the amount of water dischargeable from a tailpiece into the atmosphere would not be increased by attaching the faucet thereto. Hence, the diameter of the faucet tailpiece, and similarly of any small-diameter orifice in the faucet or valve body or installed in the fixture supply pipe, is the most critical factor controlling the maximum rate of flow from a faucet or valve body.

Such tailpieces and orifices installed in supply piping are, in effect, submerged orifices through which jet flow occurs. The applicable coefficients of velocity, contraction, and discharge are practically the same as those for jets discharging into the atmosphere. However, in this case, the effective pressure under which flow proceeds through a submerged orifice is the difference between the measured pressure at its inlet and the back pressure at its outlet. This may be expressed as follows:

$$q_d = 20 d_o^2 \sqrt{(p_s - p_d)} \qquad [5\text{-}20]$$

where q_d = quantity rate of flow through submerged orifice, gpm
d_o = diameter of submerged orifice, in
p_s = pressure at supply side of submerged orifice during flow, psi
p_d = pressure at discharge side of submerged orifice during flow, psi

PRESSURE LOSS THROUGH METERS

The principal type of meter currently installed to measure the amount of water furnished to building water supply systems is the disk type of

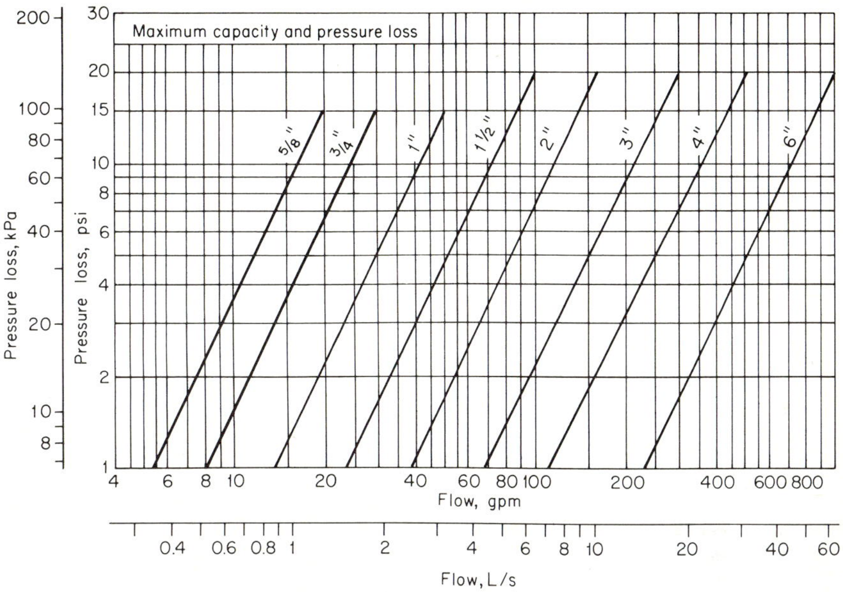

Fig. 5-5 Pressure loss in cold water meters, displacement type.

cold water meter. In this type, the flow of water displaces and imparts motion to a disk, which in turn operates a gear train so as to record on a dial a reading indicating the amount of water passed through the meter. The energy required to move the disk is converted from the potential or pressure energy of the water. Additional pressure energy is lost because of friction in flow through the meter.

Total pressure energy lost owing to flow through a meter is most conveniently expressed in terms of the pressure loss, in pounds per square inch, corresponding to the rate of flow. The maximum pressure drop permitted in standards established for new meters of the disk type at the capacity rates of flow designated for various sizes, and the flow capacities, in gallons per minute, for the different sizes are shown in Fig. 5-5.

Pressure loss through disk meters at flow rates less than capacity may be estimated from Fig. 5-5. For other types of meters, information regarding pressure losses pertaining thereto may be obtained directly from meter manufacturers.

PRESSURE LOSS DUE TO FRICTION IN PIPING

Whenever water flows in the supply system of a building, there is a continuous loss of pressure along the piping in the direction of flow.

As turbulent flow normally prevails in such a system, the pressure loss is due to friction generated between the moving water and the inner surface of the piping.

The amount of pressure or head lost because of friction is dependent upon a number of factors. Among these are physical properties of the fluid such as its viscosity, density, and temperature, roughness of interior pipe surface, length of the pipe, diameter of the pipe, and the velocity at which fluid flows through the pipe.

Experiments show that the head lost as a result of friction is directly proportional to the roughness of the interior surface of the pipe, to the length of the pipe, and to approximately the square of the velocity of flow, and also is inversely proportional to the diameter of the pipe. To express these relationships, numerous formulas have been devised by various experimenters. Some of these formulas are of the exponential type and refer to specific kinds of piping.

An adequate formula for general application in estimating head loss due to friction is that devised by Darcy many years ago. It is as follows:

$$h_f = \frac{f L V^2}{D 2g} \quad \text{or} \quad p_f = \frac{w f L V^2}{144 D 2g} \qquad [5\text{-}21]$$

where h_f = head loss due to friction, ft of fluid column
p_f = pressure loss due to friction, psi
w = density of fluid flowing, lb/ft^3
f = coefficient of friction corresponding to roughness of pipe surface and diameter of pipe
L = length of piping, ft
D = diameter of piping, ft
V = velocity of flow, fps
g = gravitational acceleration, 32.2 ft/s^2

In applying the above formula, proper selection should be made regarding the value of f, the coefficient of friction corresponding to the roughness of the pipe surface and the diameter of the pipe. Consideration should be given to changes in the condition of the pipe surface resulting from a reasonable period of service. As a guide for selecting appropriate values, one may use those given in Table 5-5, which have been determined from data, charts, and graphs on pressure loss contained in publications of authoritative agencies and numerous manufacturers of piping materials.

Evident from Table 5-5 is the fact that average values of f for fairly smooth pipes such as brass, copper, or plastic are about one-half the corresponding values for fairly rough pipe such as galvanized steel and wrought iron. In addition, the average value for all sizes listed for fairly

Table 5-5
AVERAGE VALUES FOR *f*, PIPE FRICTION COEFFICIENT
For Use in Darcy's Formula $h_f = f\frac{L}{D}\frac{V^2}{2g}$

Nominal size of pipe, in	Brass, copper, or plastic pipe, f	Galvanized-steel or wrought-iron pipe, f
⅜	0.023	0.058
½	0.022	0.044
¾	0.021	0.040
1	0.020	0.038
1¼	0.020	0.036
1½	0.019	0.035
2	0.018	0.033
2½	0.017	0.031
3	0.017	0.031
4	0.016	0.030

smooth pipes is about 0.02, while that for fairly rough pipes is about 0.04. Values of f shown in the table and the average values are applicable to turbulent flow conditions, such as exist in water supply systems during periods of peak demand.

However, as a matter of general interest, it may be noted that when streamline flow conditions prevail, such as during periods of very low demand when velocities of flow drop below the critical level, the value of f differs from, and is generally higher than, that for turbulent flow.

Osborne Reynolds established that the coefficient of friction during streamline flow was independent of the roughness of the pipe surface, and instead was dependent upon the viscosity of the fluid, velocity of flow, and diameter of the pipe. He developed a formula for determining the value of f applicable to streamline flow conditions. It is of significant use in calculations dealing with flow of very viscous fluids. The formula is as follows:

$$f = \frac{64\,\mu}{DVw} = \frac{64\,\nu}{DV} = \frac{64}{R_e} \qquad [5\text{-}22]$$

where f = coefficient of friction
D = diameter of pipe, ft
V = velocity of fluid flow, fps
μ = absolute viscosity of fluid, pdl • s/ft²
ν = kinematic viscosity of fluid, ft²/s
w = density of fluid, lb/ft³
R_e = Reynolds number

Calculations of pressure loss due to friction are performed generally in order to estimate pressures that may exist in piping systems based upon certain assumed conditions of flow. These assumptions should always provide a reasonable factor of safety. In view of this, it is permissible to apply as the coefficient of friction a value of 0.02 for brass, copper, and plastic piping, and a value of 0.04 for galvanized steel and wrought-iron piping, where water conditions are favorable.

To further simplify pressure loss calculations, it is generally advisable to combine the Darcy formula and the common flow formula and convert their terms to those more generally used in practice. This combination and conversion results in two different formulas, one for brass, copper, and plastic piping, and a second for galvanized-steel and wrought-iron piping. They may be used for direct calculations of estimated pressure loss due to friction in piping. They are as follows:

For brass, copper, or plastic piping:

$$h_f = 0.000623 q^2 \frac{L}{d^5} \qquad [5\text{-}23]$$

or

$$p_f = 0.00027 q^2 \frac{L}{d^5} \qquad [5\text{-}24]$$

For galvanized-steel and wrought-iron piping:

$$h_f = 0.001246 q^2 \frac{L}{d^5} \qquad [5\text{-}25]$$

or

$$p_f = 0.000539 q^2 \frac{L}{d^5} \qquad [5\text{-}26]$$

where h_f = head loss due to friction, ft of fluid column
p_f = pressure loss due to friction, psi
q = quantity rate of flow, gpm
L = length of piping, ft
d = diameter of piping, in

The preceding formulas may be rearranged so that they express the quantity rate of flow in terms of friction loss, pipe diameter, and pipe length, as in the following:

For brass, copper, or plastic piping:

$$q = 40.1\, d^{2\frac{1}{2}} \left(\frac{h}{L}\right)^{\frac{1}{2}} \qquad [5\text{-}27]$$

or

$$q = 60.8 d^{2\frac{1}{2}} \left(\frac{p}{L}\right)^{\frac{1}{2}} \qquad [5\text{-}28]$$

For galvanized-steel and wrought-iron piping:

$$q = 28.3d^{2\frac{1}{2}}\left(\frac{h}{L}\right)^{\frac{1}{2}} \qquad [5\text{-}29]$$

or

$$q = 43.0d^{2\frac{1}{2}}\left(\frac{p}{L}\right)^{\frac{1}{2}} \qquad [5\text{-}30]$$

In the above equations, the terms h/L and p/L represent, respectively, the loss of head, in feet of water column, and of pressure, in pounds per square inch, due to friction in just 1 ft of pipe length. Such terms may also be viewed as the uniform friction loss for piping in which the total friction loss is uniformly distributed throughout its length. The use of uniform friction loss as a basis for designing pipe sizes has definite advantages. For a given uniform friction loss, the quantity rates of flow may be calculated readily for any given pipe diameters by applying the above equations. The quantity rates of flow calculated in this way may be used as limits in sizing piping systems in accordance with the rates of flow to be conveyed while maintaining friction loss within the permissible limit established for any particular system.

PIPE FRICTION FORMULAS FOR VARIOUS SERVICE CONDITIONS

The most authoritative and generally accepted recommendations concerning appropriate pipe friction formulas for various conditions of service in water supply systems were presented by Roy B. Hunter in 1941 in the National Bureau of Standards Building Materials and Structures Report, BMS79, entitled, "Water-Distributing Systems for Buildings."

In designing water supply systems in buildings, it is not sufficient to provide simply for adequate supply when pipes are new. It is essential that suitable allowance be made for the decrease in capacity in service caused by deterioration and change of the interior surface of the pipes with age and service conditions. This involves the choice of piping materials, and the characteristics of the particular water to be conveyed. In addition, classifying the relative roughness of pipe surfaces simply as either smooth or rough may not be considered to achieve a sufficient degree of accuracy in design.

For greater accuracy, it was suggested that water supply pipe be considered in four classes as to hydraulic roughness:

1. Smooth pipe, in which class new copper or brass tubing with streamlined fittings and brass pipe may be placed

2. Fairly smooth pipe, in which class new butt-welded steel and wrought-iron pipe with threaded fittings may be placed

3. Fairly rough pipe, which represents a roughness intermediate between fairly smooth and rough pipe
4. Rough pipe, in which class any kind of badly corroded or badly caked pipe may be placed

The rational pipe friction formula used by Darcy is presented in Eq. (5-21). In it, the coefficient of friction f to be used must correspond to the roughness of the pipe surface and the diameter of the pipe. The value of f is a function of the Reynolds number and varies with the relative roughness of the internal pipe surface. Appropriate values may be found in hydraulics textbooks on graphs showing values of f versus values of Reynolds numbers and presenting special curves applicable to different degrees of relative roughness.

For different degrees of relative roughness, the relationship between f and the Reynolds number R_e may be represented by an equation in the mathematical model of

$$f = bR_e^{-a} \qquad [5\text{-}31]$$

Applying this model to special curves for the four classes of hydraulic roughness as suggested, Hunter developed the following equations:

For smooth pipe: $f = 0.32 R_e^{-0.25}$ [5-32]
For fairly smooth pipe: $f = 0.17 R_e^{-0.17}$ [5-33]
For fairly rough pipe: $f = 0.085 R_e^{-0.08}$ [5-34]
For rough pipe: $f = 0.054 R_e^{-0.0}$ [5-35]

A fair average temperature for cold water piping is 50°F (10°C). The kinematic viscosity of pure water at that temperature is 0.0000141 ft²/s as given in Table 5-2. For water at 50°F (10°C), the Reynolds number is as follows:

At 50°F: $$R_e = \frac{VD}{\nu} = \frac{VD}{0.0000141} \qquad [5\text{-}36]$$

where R_e = Reynolds number
ν = kinematic viscosity of fluid, ft²/s
V = velocity of fluid flow, fps
D = diameter of pipe, ft

The value of the Reynolds number as set forth in Eq. (5-36) may be substituted in Eqs. (5-32), (5-33), (5-34), and (5-35) to determine appropriate values for the coefficient of friction f to apply in the rational pipe friction formula for the four classes of hydraulic roughness considered for water supply piping.

To simplify pressure loss calculations further, the rational pipe friction formula and the common flow formula may be combined and their terms converted to those more generally used in practice. This combination and conversion results in a separate pipe friction formula for each of the four classifications of water supply pipe as to hydraulic roughness. The formulas are as follows:

For smooth pipe:	$q = 4.93\,p^{0.571} d^{2.714}$	[5-37]
For fairly smooth pipe:	$q = 4.57\,p^{0.546} d^{2.64}$	[5-38]
For fairly rough pipe:	$q = 4.29\,p^{0.521} d^{2.562}$	[5-39]
For rough pipe:	$q = 3.70\,p^{0.5} d^{2.5}$	[5-40]

where q = quantity rate of flow, gpm
p = pressure loss due to friction, psi/100 ft pipe length
d = diameter of pipe, in

The above formulas may be considered with regard to their application for various service conditions. For example, the smooth pipe classification includes new copper or brass tubing and brass pipe. The new condition should be considered as an "ideal" rather than a "service" condition. Similarly, the fairly smooth pipe classification includes new steel and wrought-iron pipe. Formulas applicable to "ideal" conditions should not be applied in design. Rather one should be guided by knowledge of actual conditions of service. Further recommendations on this subject may be found discussed in Chap. 6.

EQUIVALENT LENGTH OF FITTINGS AND VALVES

An important factor in pipe friction equations is the length of piping. For flow through a straight run of piping containing no fittings or valves, the length to be applied is simply the developed length measured along the centerline of the pipe.

However, where fittings or valves are included in the run, consideration must be given to the fact that they impose considerably more frictional resistance than pipe of the same developed length and size. Consequently, the proper length to apply in this case is the sum of the developed length of the piping plus an equivalent length which should be allowed as the additional resistance due to the fittings and valves in the piping.

The equivalent length of piping having the same frictional resistance as a given type of fitting or valve has been a subject pursued by many experimenters. Many data have been collected and published for general application. In some cases, the resistance is given in terms of additional lengths equal to multiples of the pipe diameter. In others, it is stated in the more convenient form of equivalent length of pipe of the same

Table 5-6*a*
EQUIVALENT LENGTH FOR FITTINGS AND VALVES—U.S. CUSTOMARY UNITS
Standard Pipe

Fitting or valve	Equivalent feet of pipe for various sizes							
	½ in	¾ in	1 in	1¼ in	1½ in	2 in	2½ in	3 in
45° elbow	1.2	1.5	1.8	2.4	3.0	4.0	5.0	6.0
90° elbow	2.0	2.5	3.0	4.0	5.0	7.0	8.0	10.0
T, run	0.6	0.8—	0.9	1.2	1.5	2.0	2.5	3.0
T, branch	3.0	4.0	5.0	6.0	7.0	10.0	12.0	15.0
Gate valve	0.4	0.5	0.6	0.8	1.0	1.3	1.6	2.0
Balancing valve	0.8	1.1	1.5	1.9	2.2	3.0	3.7	4.5
Plug-type cock	0.8	1.1	1.5	1.9	2.2	3.0	3.7	4.5
Check valve, swing	5.6	8.4	11.2	14.0	16.8	22.4	28.0	33.6
Globe valve	15.0	20.0	25.0	35.0	45.0	55.0	65.0	80.0
Angle valve	8.0	12.0	15.0	18.0	22.0	28.0	34.0	40.0

Table 5-6*b*
EQUIVALENT LENGTH FOR FITTINGS AND VALVES—SI UNITS
Standard Pipe

Fitting or valve	Equivalent meters of pipe for various sizes							
	12.7 mm	19.0 mm	25.4 mm	31.8 mm	38.1 mm	50.8 mm	63.5 mm	76.2 mm
45° elbow	0.36	0.45	0.54	0.73	0.91	1.21	1.52	1.82
90° elbow	0.60	0.76	0.91	1.21	1.52	2.13	2.43	3.04
T, run	0.18	0.24	0.27	0.36	0.45	0.60	0.76	0.91
T, branch	0.91	1.21	1.51	1.82	2.13	3.04	3.65	4.57
Gate valve	0.12	0.15	0.18	0.24	0.30	0.39	0.48	0.60
Balancing valve	0.24	0.33	0.45	0.57	0.67	0.91	1.12	1.37
Plug-type cock	0.24	0.33	0.45	0.57	0.67	0.91	1.12	1.37
Check valve, swing	1.70	2.56	3.41	4.26	5.12	6.82	8.53	10.24
Globe valve	4.57	6.09	7.62	10.66	13.71	16.76	19.81	24.38
Angle valve	2.43	3.65	4.57	5.48	6.70	8.53	10.36	12.19

Table 5-7*a*
EQUIVALENT LENGTH FOR FITTINGS AND VALVES—U.S. CUSTOMARY UNITS
Copper Water Tube

	Equivalent feet of tube for various sizes							
Fitting or valve	**½ in**	**¾ in**	**1 in**	**1¼ in**	**1½ in**	**2 in**	**2½ in**	**3 in**
45° elbow (wrought)	0.5	0.5	1.0	1.0	2.0	2.0	3.0	4.0
90° elbow (wrought)	0.5	1.0	1.0	2.0	2.0	2.0	2.0	3.0
T, run (wrought)	0.5	0.5	0.5	0.5	1.0	1.0	2.0	
T, branch (wrought)	1.0	2.0	3.0	4.0	5.0	7.0	9.0	
45° elbow (cast)	0.5	1.0	2.0	2.0	3.0	5.0	8.0	11.0
90° elbow (cast)	1.0	2.0	4.0	5.0	8.0	11.0	14.0	18.0
T, run (cast)	0.5	0.5	0.5	1.0	1.0	2.0	2.0	2.0
T, branch (cast)	2.0	3.0	5.0	7.0	9.0	12.0	16.0	20.0
Compression stop	13.0	21.0	30.0					
Globe valve				53.0	66.0	90.0		
Gate valve			1.0	1.0	2.0	2.0	2.0	2.0

Source: Excerpted from *Copper Tube Handbook,* 1965, by Copper Development Association.

Table 5-7*b*
EQUIVALENT LENGTH FOR FITTINGS AND VALVES—SI UNITS
Copper Water Tube

Fitting or valve	Equivalent meters of tube for various sizes							
	12.7 mm	19.0 mm	25.4 mm	31.8 mm	38.1 mm	50.8 mm	63.5 mm	76.2 mm
45° elbow (wrought)	0.15	0.15	0.30	0.30	0.60	0.60	0.91	1.21
90° elbow (wrought)	0.15	0.30	0.30	0.60	0.60	0.60	0.60	0.91
T, run (wrought)	0.15	0.15	0.15	0.15	0.30	0.30	0.60	
T, branch (wrought)	0.30	0.60	0.91	1.12	1.52	2.13	2.74	
45° elbow (cast)	0.15	0.30	0.60	0.91	1.12	1.52	2.43	3.35
90° elbow (cast)	0.30	0.60	1.21	1.52	2.43	3.35	4.26	5.48
T, run (cast)	0.15	0.15	0.15	0.30	0.30	0.60	0.60	0.60
T, branch (cast)	0.60	0.91	1.52	2.13	2.74	3.65	4.87	6.09
Compression stop	3.96	6.40	9.14					
Globe valve				16.15	20.11	27.43		
Gate valve			0.30	0.30	0.60	0.60	0.60	0.60

Source: Excerpted from *Copper Tube Handbook*, 1965, by Copper Development Association.

nominal diameter. Considerable variation in equivalent length for valves occurs depending upon the degree of valve closure, and consequently it is inappropriate to assign specific lengths to valves except for the fully open position.

Recommended equivalent lengths to be allowed for fittings and valves are presented in the customary English units in Tables 5-6*a* and 5-7*a*. To meet the current needs of many users and in the interest of standardization, the same two tables are presented also in metric units as Tables 5-6*b* and 5-7*b*.

PRESSURE LOSS THROUGH MISCELLANEOUS EQUIPMENT

Horizontal hot water storage tanks having bottom supply and top outlet connections may be reasonably considered to have frictional resistance equivalent to that of three 90° elbows of the same size as the supply and outlet piping. Vertical hot water storage tanks having a top outlet connection, and also a top supply connection from which water enters the lower portion of the tank via a dip tube of smaller diameter than the supply piping, produce much more frictional resistance because of the added resistance of the dip tube. The added resistance should be separately determined for the dip tube based upon its internal diameter, length, and the velocity of flow through it.

Submerged instantaneous domestic hot water coils vary considerably in frictional resistance because of the variations that exist in coil length and diameter and because of such design features as manifolding and coil finning or crimping. Information about pressure loss should be obtained directly from the manufacturer whose product is to be installed, especially in view of the fact that some coils are provided with flow control orifices to ensure that flow rates through the coils do not exceed performance ratings. These orifices may add a very great amount of resistance not otherwise included in the rated resistance of the coils.

Water softeners vary considerably in frictional resistance, depending upon design. For example, softeners having ¾-in piping connections produce resistance equivalent to from 50 to 200 ft of ¾-in piping length. More exact data on the equivalent length to be applied may be obtained from the manufacturer of any given type and model.

Water filters are the devices which generally provide the most frictional resistance of all used in water supply systems. The resistance varies considerably with the type of filter medium and the design of the apparatus. Data should be obtained directly from the manufacturer in each individual case.

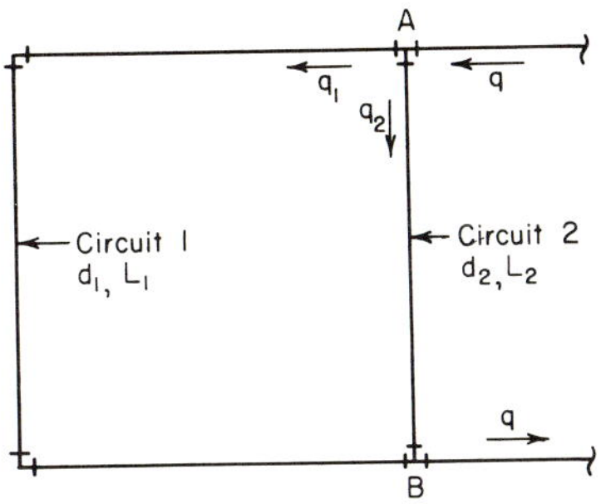

Fig. 5-6 Parallel pipe circuits.

DIVISION OF FLOW IN PARALLEL PIPE CIRCUITS

An arrangement of parallel pipe circuits is one in which flow from a single pipe divides at a branch fitting, and two or more approximately parallel pipe circuits convey separate portions of the flow to a second branch fitting where the total flow rejoins and proceeds again through a single pipe. This piping arrangement is illustrated in Fig. 5-6.

In the sketch, the total flow to fitting A and from fitting B is the same and is noted as q. The parallel pipe circuits between fittings A and B are noted as circuits 1 and 2, and their respective quantity rates of flow, diameters, and lengths are noted as q_1, q_2, d_1, d_2, L_1, and L_2.

It should be recognized that q is equal to the sum of q_1 and q_2 and that the pressure drop between the ends of circuits 1 and 2 is the same for each circuit and is equal to $h_A - h_B$. Hence, flow through each circuit may proceed only at a rate that will produce this pressure drop. This may be expressed as follows:

$$h_A - h_B = 0.000623(q_1)^2 \frac{L_1}{d_1^5} = 0.000623(q_2)^2 \frac{L_2}{d_2^5} \qquad [5\text{-}41]$$

The above expression applies specifically to brass, copper, or lead piping. For galvanized-steel or wrought-iron piping, a factor of 0.001246 should be applied instead of 0.000623.

To determine the amount of flow in circuit 1 as compared with that in circuit 2, the above equations may be rearranged to express this relationship as follows:

$$\frac{q_1}{q_2} = \sqrt{\frac{L_2}{L_1}\left(\frac{d_1}{d_2}\right)^5} \qquad [5\text{-}42]$$

WATER HAMMER

Hammering noises and severe shocks may occur in a pressurized water supply system when flow is halted abruptly by the rapid closure of a valve or faucet. This phenomenon is called *water hammer.* It occurs principally in long, straight runs of piping to which quick-closing valves or faucets are connected, and it is most prevalent in areas where water pressure is high.

The reason for this condition is that flowing water has kinetic energy due to its weight and velocity, and when flow is halted abruptly, this energy imparts to the water the effect of a battering ram inside the piping. If the pipe is relatively weak, it may burst or develop leaks. When the pipe is strong, the kinetic energy is converted to pressure energy which is developed in the process of compressing the water and distending the pipe. This reaction, compression, and distension sets up pressure waves which travel back and forth along the runs of piping with the speed of sound in water, about 4700 fps, and cause the pipe wall to expand and contract in rhythm with the rise and fall of pressure inside the piping.

The compressibility of water may be written as $1/K$, in which K is the coefficient of compressibility having a value of 43,200,000 lb/ft².

Brass, copper, iron, and steel are much less compressible than water, their coefficients ranging, respectively, from about 50 to 100 times as much as that for water. Consequently, most of the kinetic energy possessed by flowing water, when halted abruptly, is converted into pressure energy in compressing the water rather than in distending the pipe.

The kinetic energy of flowing water, per pound, is equal to its velocity head $V^2/2g$. In compressing the water, this energy is expended in generating pressure which rises to a maximum pounds per square foot P, the average pressure rise during the process being $P/2$ and the amount of compression produced being P/K. Work performed in compression is the product of the average pressure rise and the amount of compression, or $P^2/2K$ ft · lb. When this is divided by the density of water w the quotient is the amount of pressure energy developed per pound in compressing the water, the amount being equal to $P^2/2Kw$ ft of pressure head.

Maximum pressure developed in compressing the water may be found be equating the velocity head to the pressure head. This may be done as follows:

$$\frac{V^2}{2g} = \frac{P^2}{2Kw} \qquad [5\text{-}43]$$

$$P = V\sqrt{\frac{Kw}{g}} \quad \text{lb/ft}^2 \qquad [5\text{-}44]$$

$$p = \frac{V}{144}\sqrt{\frac{Kw}{g}} \quad \text{psi} \qquad [5\text{-}45]$$

where V = velocity of flow, fps
g = gravitational acceleration, 32.2 ft/s^2
P = maximum pressure in compression, lb/ft^2
p = maximum pressure in compression, psi
K = modulus of compressibility of water, 43,200,000 lb/ft^2
w = density of water, 62.4 lb/ft^3

By substituting the above values in the equation to determine the maximum pressure rise, in pounds per square inch, in terms of the velocity of flow, the result is as follows:

$$p = 63.6\,V \qquad \text{psi} \tag{5-46}$$

Under ideal conditions of instantaneous halting of flow and complete incompressibility of the piping, the extremely high pressure rise indicated by the above equation might occur. However, although such pressures may be approached in experiments, they rarely prevail in practice because of the time element required for valve closure, the small amount of elasticity of piping, and more important, the amount of dissolved gases contained in most potable water supplies. These factors tend to reduce the maximum pressure which may occur in practice.

Relatively high pressure rises are found to occur only under certain conditions related to the length of time required for valve closure, the length of the straight run of piping, and the velocity of sound in water. They may occur when the time required for valve closure, in seconds, is less than $2L/4700$, in which L is the length of the straight run of piping, in feet, and 4700 fps is taken as the velocity of sound in water.

Considerably lower but significant pressure rises may occur when the time required for valve closure, in seconds, is greater than $4L/4700$. Where the length L is great, the pressure rise can be significant and result in appreciable water hammer. In this circumstance, the time required for valve closure t in seconds may be applied to express the deceleration of flow as V/t, and the mass of water moving through the piping may be represented as wL/g. As force is the product of mass times acceleration, the average pressure applied in compressing the water is equal to wLV/gt. Consequently, maximum pressure developed in compression, being twice the average pressure, may be expressed as follows:

$$P = \frac{2wLV}{gt} \qquad \text{lb/ft}^2 \tag{5-47}$$

or

$$p = \frac{2wLV}{144gt} \qquad \text{psi} \tag{5-48}$$

By substituting the appropriate numerical values for the density of water and gravitational acceleration in the previous equation, it may be resolved into more convenient form, appearing as follows:

$$p = \frac{0.027LV}{t} \qquad [5\text{-}49]$$

where p = maximum pressure in compression, psi
L = length of the straight run of piping, ft
V = velocity of flow, fps
t = time required for valve closure, s

As an illustration of the pressure which may develop in a long run of piping because of the sudden shutoff of flow, take, for example, a straight run of piping which is 100 ft in length and through which water is flowing at 10 fps. Suddenly the flow is shut off by a valve that closes completely in 0.2 s. The maximum pressure developed may be determined as follows:

$$p = \frac{0.027LV}{t} = \frac{0.027 \times 100 \times 10}{0.2}$$
$$= 135 \text{ psi}$$

When this considerable pressure rise travels back through the piping as a pressure wave, it may produce severe shock or hammer noises and cause damage. In high-pressure areas, the total pressure in the piping may approach the limit of pipe or joint strength and cause leakage or bursting at some weak point.

Water hammer effects in piping systems may be almost completely eliminated or reduced to the degree that they are no longer audible. This may be accomplished in several ways. The most direct method is to remove the cause of the condition by providing slow-closing-type valves in place of quick-closing types. Any lengthening of the time required for valve closure causes a corresponding decrease in the pressure rise and shock effect accompanying deceleration of flow.

Other methods may be applied when quick-closing valves are installed. Air chambers or mechanical shock absorbers may be connected to the piping adjacent to quick-closing valves so as to absorb the kinetic energy of the flowing water at the point where it is halted abruptly. Such devices contain air, diaphragms, or springs which have much greater compressibility than the water and piping and, therefore, absorb most of the kinetic energy in compressing air, distending diaphragms, or compressing springs. Only a small fraction of the kinetic energy remains for dissipation by compression of the water and distension of the piping, and consequently water hammer effects may be reduced to the degree that they become unnoticeable.

VELOCITY EFFECTS IN PIPING

With turbulent flow, the energy lost in the form of pressure drop because of friction varies directly with approximately the square of the velocity. For example, when the velocity is increased from 4 to 10 fps, the pressure drop rises to 6.25 times as much as at the lower velocity. As this energy loss is directly related to roughness of the pipe surface, the abrasive or scouring effect upon the interior of the piping becomes more severe with increased velocity and can cause erosion of the pipe surface. The amount and intensity of erosion due to velocity depend upon the physical characteristics of the pipe material and the surface deposits or corrosion products lining the interior of the piping.

This energy loss may also result in the production of line noise, such as whistling, when the velocity of flow is high. As the water strikes against high spots in the pipe surface, energy is transferred into the pipe, causing it to move or vibrate. This effect may be dissipated by damping or absorbing the vibration. But when the piping arrangement permits resonance to develop, the vibration may gain sufficient amplitude so that excessive line noise results.

High velocities through fittings and valves, in which the direction of flow changes sharply, can produce the phenomenon known as *cavitation,* accompanied by severe noise effects. Where the direction of flow changes abruptly at a bend and the velocity of flow is high, the centrifugal force developed thereby may be so great that pressure at the inside of the bend drops below atmospheric to a pressure corresponding to the boiling point of the flowing water. Under this condition, a cavity or subatmospheric pressure zone forms at the inside of the bend, permitting some of the water there to flash into bubbles of vapor or steam. These bubbles collapse immediately upon flowing past the bend into a higher pressure zone. The rapid volumetric changes in the stream of flow, corresponding to the generation and collapse of flash steam bubbles, produce intense noise effects. Consequently, flow velocities should be limited so as to avoid severe noise effects coincident with the occurrence of cavitation.

SUCTION LIFT OF PUMPS

Pumps have only limited capacity to lift water by suction. Among the factors limiting suction lift, the two most important are (1) atmospheric pressure at the water surface and (2) vapor pressure of the water corresponding to its temperature.

Considering that a perfect vacuum, 0 psia pressure, were achieved in the pump suction pipe, the limit to which water could rise therein would be a height equivalent to atmospheric pressure. This varies with elevation, for atmospheric pressure is due to the weight of air bearing down from above at any given elevation. At sea level, this pressure is about 14.7 psia, and at 5200 ft above sea level it is only about 12.7 psia.

If a perfect vacuum were achieved in the pump suction pipe, upon entering the pipe some of the water would vaporize so as to establish a balance between the pressure in the suction pipe and the pressure at which the water boils. The vapor or boiling pressure of water varies with its temperature. Temperature-pressure conditions at which water boils are as follows: at 40°F slightly under 1 psia; at 150°F, about 4 psia; at 180°F, about 8 psia; and at 212°F (100°C), water boils at 14.7 psia, atmospheric pressure at sea level.

Under conditions of no flow in the suction pipe, the maximum height to which water can be lifted by pump suction is that equivalent to the difference between atmospheric pressure and the vapor pressure of the water corresponding to its temperature. In view of this, the maximum suction lift of a pump decreases sharply with increase in water temperature, becoming 0 at 212°F (100°C). Hence, where hot water is to be pumped, it is generally necessary to avoid any suction lift and to locate the pump at an elevation where positive head is available at the pump inlet.

When flow exists in the suction pipe, two other factors further tend to limit suction lift. They are (1) the amount of decrease in suction lift corresponding to the velocity head in the suction pipe and (2) the amount of decrease in suction lift due to friction in the suction pipe. Both these factors vary directly with the square of the velocity of flow in the suction pipe and thus are controllable in the selection of suction pipe diameter.

The factors noted above may be expressed as an equation for the theoretical suction lift a pump may develop. It is as follows:

$$h_s = h_a - h_b - \frac{V^2}{2g} - h_f \qquad [5\text{-}50]$$

where h_s = theoretical suction lift, ft of water head

h_a = atmospheric pressure, ft of water head

h_b = vapor or boiling pressure of the water corresponding to its temperature, ft of water head

V = velocity of flow in suction pipe, fps

h_f = loss due to friction in suction pipe, ft of water head

g = gravitational acceleration, 32.2 ft/s^2

Actual suction lift obtainable with any pump is less than the theoretical; for in addition to the abovementioned factors, there are others which are of influence in practice, such as characteristics of the water, total dynamic head, and pump speed, capacity, and impeller design. Abnormally high suction lifts can produce severe reduction in pump capacity and efficiency and may cause intense vibration and cavitation.

WORK DONE IN PUMPING

Of the total work done in pumping water, part is expended in raising water from one level to another and thereby increasing its potential energy, part is used to produce velocity of flow and is converted to kinetic energy, and part is dissipated in overcoming friction in the pumping system. Per pound of water raised, the amount of work done may be expressed as follows:

$$h_t = h_r + \frac{V^2}{2g} + h_f \qquad [5\text{-}51]$$

where h_t = total head delivered by the pump to the water, ft of water column
h_r = total height through which the water is raised, ft
V = velocity of flow in the pump discharge pipe, fps
g = gravitational acceleration, 32.2 ft/s^2
h_f = total head lost owing to friction, ft of water column

The power developed by a pump may be determined knowing the total head and the quantity rate at which water is delivered. This may be calculated in terms of horsepower by applying the following equation:

$$\text{hp output of pump} = \frac{q h_t (w/7.48)}{33{,}000} \qquad [5\text{-}52]$$

where q = quantity rate at which water is delivered, gpm
h_t = total head delivered by the pump to the water, ft of water column
w = density of water corresponding to water temperature, lb/ft^3

6

SIZING THE WATER SUPPLY SYSTEM

DESIGN OBJECTIVES

The most important design objective in sizing the water supply system is the satisfactory supply of potable water to all fixtures, at all times, and at proper pressure and flow rate for normal fixture operation. This may be achieved only if adequate sizes of pipes and appurtenances are provided. The sizes established must be large enough to prevent occurrence of negative pressures in any part of the system during periods of peak demand in order to avoid the hazard of water supply contamination due to backflow and back siphonage from potential sources of pollution. Hence, the sizing of building water supply systems is a matter of vital concern in protecting health and must be regulated by codes.

Other important objectives in the design of water supply systems are: (1) to achieve economical sizing of piping and eliminate overdesign; (2) to provide against potential supply failure due to gradual reduction of pipe bore with the passing of time, such as may result from deposits of corrosion or hard water scale in the piping; (3) to avoid erosion-corrosion effects and potential pipe failure or leakage conditions owing to corrosive characteristics of the water and/or to excessive design velocities of flow; and (4) to eliminate water hammer damage and objectionable whistling noise effects in the piping due to excessive design velocities of flow.

MINIMUM AVAILABLE PRESSURE

Design Pressures for System The water supply system should be designed in accordance with the minimum pressure available at the public water main, or other source of water supply pressure, and the minimum pressure required at all times at water outlets of the system.

Adequate Source of Water Supply Pressure Where the pressure available at the public main is insufficient to maintain the minimum required at the highest water outlet of the system, a pressure booster pump system, approved as to capacity and reliability, or an automatically controlled water supply tank of either the hydropneumatic pressure type or the elevated, gravity type should be provided.

Pressure Booster Pump System For a pressure booster pump system, the minimum pressure at the pump outlet designed to be maintained at all times should be considered the minimum pressure available at the source.

Automatically Controlled Water Supply Tanks Automatically controlled water supply tanks should be designed and installed so as to furnish the required minimum pressures. Where a pressure tank is provided, the minimum pressure at which it is designed to operate should be considered the minimum available at the source. Where a gravity tank is provided, the minimum pressure at the source should be considered as being the static pressure owing to the elevation of the low waterline at which the tank is designed to operate under peak demand conditions.

Water Outlets As a general rule, the minimum pressure required at ordinary faucets of plumbing fixtures should be deemed to be 8 psi (55 kPa). However, at direct supply-connected flush valves (Flushometers), the minimum should be 25 psi (172 kPa) for blow-out-type water closets and 15 psi (103 kPa) for other types of fixtures. At electrically operated supply valves of equipment, where higher than ordinary pressure frequently is recommended, the minimum should be that recommended by the manufacturer for satisfactory equipment performance.

DEMAND AT INDIVIDUAL WATER OUTLETS

Ordinary Pipe Outlets The demand at ordinary pipe outlets, flowing wide open at maximum rate of discharge, may be calculated approximately based on the flow pressure in the supply pipe and the diameter of the outlet. This may be done by applying the formula for the practical jet discharge from pipe outlets, which is as follows:

$$q_a = 20\,(d_o)^2\sqrt{p_m}$$

where q_d = actual quantity rate of discharge from pipe outlet, gpm
d_o = diameter of outlet, in
p_m = pressure measured in supply pipe during flow, psi

Electrically Operated Supply Valves Equipment having electrically operated supply valves varies considerably in demand requirements depending upon the type used and its performance characteristics. Information about demand may be obtained from data furnished by manufacturers of such equipment. In the absence of adequate information, the demand may be calculated based on the flow pressure required and the outlet diameter, just as for any ordinary pipe outlet. Usually, electrically operated supply valves are kept wide open for maximum flow rate.

Common Plumbing Fixtures Demand rates which have become generally accepted as normal, suitable rates of flow from individual water outlets at various plumbing fixtures and hose connections are given in Table 6-1. The demands shown for faucets which users may adjust manually according to their needs are not the maximum flow rates that may be obtained in each case. Rather, they are demands which are considered suitable for fixture usage without causing excessive splashing, a factor which is related to the shape and depth of the fixture.

Table 6-1
DEMAND AT INDIVIDUAL WATER OUTLETS

Type of outlet	Demand	
	gpm	L/s
Ordinary lavatory faucet	2.0	0.126
Self-closing lavatory faucet	2.5	0.158
Sink faucet, ⅜″ (9.52 mm) or ½″ (12.7 mm)	4.5	0.284
Sink faucet, ¾″ (19 mm)	6.0	0.378
Bath faucet, ½″ (12.7 mm)	5.0	0.315
Shower head, ½″ (12.7 mm)	5.0	0.315
Laundry faucet, ½″ (12.7 mm)	5.0	0.315
Ball cock in water closet flush tank	3.0	0.189
1″ (25.4 mm) flush valve [25 psi (172 kPa) flow pressure]	35.0	2.210
1″ (25.4 mm) flush valve [15 psi (103 kPa) flow pressure]	27.0	1.703
¾″ (19.0 mm) flush valve [15 psi (103 kPa) flow pressure]	15.0	0.946
Drinking fountain jet	0.75	0.047
Dishwashing machine (domestic)	4.0	0.252
Laundry machine [8 lb (3.6 kg) or 16 lb (7.3 kg)]	4.0	0.252
Aspirator (operating room or laboratory)	2.5	0.158
Hose bib or sill cock, ½″ (12.7 mm)	5.0	0.315

Control of Excessive Flow Rates at Water Outlets To avoid excessive flow rates and splashing at fixtures where the available pressure is considerably higher than the minimum required, it is generally necessary to provide some means of reducing the maximum flow rate to match normal demand. This is especially needed at fixtures located on the lower floors of high and tower-type buildings, for at such locations the pressure existing in water supply systems is many times the minimum required at water outlets.

Where the available pressure at water outlets is more than twice the minimum pressure required for satisfactory supply, it is recommended that means to control the rate of flow should be provided in the fixture supply pipe. For this purpose, individual regulating valves, variable orifice control devices, or fixed orifices may be provided.

Flush valves (Flushometers) are usually equipped with adjustment screws for regulating valve operation and are generally provided with a throttling valve at the valve inlet so as to reduce the available flow pressure to 25 or 15 psi (172 or 103 kPa), as may be required for satisfactory function of water closets, urinals, bedpan washers, flushing-rim slop sinks, or dishwashing machines.

For faucets and flush tank ball cocks, a throttling valve or flow control orifice may be installed in the fixture supply pipe to reduce the maximum flow rate to match normal demand. To control faucet flow, throttling valves should be adjusted so that flow matches demand when the faucet is wide open. Flow control orifices should be selected in accordance with desired demand and anticipated supply pressures.

Flow Control for Conservation of Water and Energy As a water conservation measure, flow rates at certain faucet outlets may be reduced slightly below those shown in Table 6-1, without causing any noticeably adverse effect in usage. Fixtures at which this may be applied and the minimum flow rates recommended are as follows: lavatory faucets in private bathrooms, 1.5 gpm (0.095 L/s); shower heads in private bathrooms, 2 gpm (0.126 L/s); and sink faucets at domestic kitchen sinks, 3 gpm (0.189 L/s).

As an energy conservation measure, the maximum flow rate for hot water lavatory faucets in restrooms to which the general public has access is recommended to be limited by design to ½ gpm (0.032 L/s), and the outlet temperature limited to 110°F (43.4°C). This requirement may be found in state energy conservation construction codes. In view of such requirement as to maximum flow rate for the hot water lavatory faucet in restrooms for the general public, it is recommended that the same limitation be observed for the cold water lavatory faucet.

ESTIMATING DEMAND

Need for a General Method for Estimating Demand The objectives in designing water supply systems for buildings are to ensure adequate water supply to all fixtures at all times and to achieve economical sizing of piping. To do this on a rational basis, it is necessary to estimate as accurately as possible the probable maximum rate of flow or the demand for which provision should be made in every portion of the system, including the water service, main supply lines, risers, and main branches.

Demand in building water supply systems cannot be determined exactly. Although some water outlets, such as those for watering gardens, washing sidewalks, irrigating lawns, and air-conditioning apparatus, impose continuous demand during periods of peak demand, the fact must be recognized that most plumbing fixtures in buildings are used intermittently, and the probability of simultaneous use of such fixtures cannot be definitely established. In addition, each type of plumbing fixture imposes its own singular loading effect on the system. This may be attributed to (1) average rate of supply required by a fixture for satisfactory service, (2) duration of fixture use, and (3) frequency of fixture use. Nevertheless, the demand imposed on the building water supply system by intermittently used fixtures is clearly related to the number, type, and probable simultaneous use of the fixtures to be supplied.

Many methods for estimating demand have been applied in the past. Most have been found to be unsatisfactory for one reason or another. Some resulted in considerable underestimates of demand; hence, undersizing of piping and consequent failure of supply or inadequate service to fixtures. By contrast, some produced extreme overestimates and uneconomical oversizing of systems. Still other methods were found unsuitable because they were adapted just to groups of fixtures of the same kind and were not applicable to large numbers of different types of fixtures. Some recent methods have been developed by plumbing engineering designers for specific building occupancy classifications, and they have been reportedly effective in achieving greater economy in such cases. However, such methods are appropriate only within their proved limits. They are not methods recommended for general application.

For a method to be generally acceptable for estimating the demand of a building water supply system, it must meet three basic requirements. First, it must produce estimates greater than the average demand for all fixtures on the system during periods of heaviest demand; otherwise failure will occur in the supply to some fixtures during maximum demand periods. Second, the method must yield reasonably accurate estimates of peak demand so as to avoid oversizing of piping and uneconomical waste. Third, the method must be adaptable for estimating the demand of groups of like fixtures as well as of different kinds of fixtures and building occupancy classifications.

Standard Method A generally accepted standard method for estimating the demand of building water supply systems has evolved and become recognized as meeting the above-mentioned three basic requirements. In 1923, a method of weighting plumbing fixtures in accordance with their water supply load producing effects was proposed by Roy B. Hunter of the National Bureau of Standards, U.S. Department of Commerce. This method was presented in the publication, "Recommended Minimum Requirements for Plumbing," Department of Commerce Building and Housing Series, BH13, 1924. After study of the application of the proposed method in the design of federal buildings over a period of years, the method was revised slightly and presented by the National Bureau of Standards in a 1940 publication, Building Materials and Structures Report, BMS65, "Methods of Estimating Loads in Plumbing Systems," by Roy B. Hunter.

In 1968, appropriate modifications were made in the method for modern plumbing fixtures as the result of research studies reported by Elmer Jones of the Agricultural Research Service, U.S. Department of Agriculture. These research studies were made on small housing developments and reflected the evident need posed by modern plumbing fixtures for changes in the demand values, in gallons per minute, corresponding to load values of from 1 to 60 water supply fixture units.

Up to the present, this method has filled the need for a reliable, rational way to estimate the demand in water supply systems for modern buildings of all types of occupancy and has become recognized as a generally accepted method.

Load Values (WSFUs) Assigned to Fixtures Demand in building water supply systems cannot be determined exactly. The demand imposed on a system by intermittently used fixtures is related to the number, type, and probable simultaneous use of the fixtures to be supplied.

In the standard method, fixtures using water intermittently under several conditions of service are assigned specific load values in terms of water supply fixture units. The water supply fixture unit (WSFU) is a factor so chosen that the load producing effects of different kinds of fixtures and their conditions of service can be expressed as multiples of that factor.

Values assigned to different kinds of fixtures are given in Table 6-14. For fixtures having both hot and cold water supplies, the values for separate hot and cold water demands should be taken as being three-fourths of the total value assigned to the fixture in each case. For example, the value assigned to a kitchen sink in a dwelling unit is 2 WSFU, while the separate demands on the hot and cold water piping thereto should be taken as being 1.5 WSFU.

Demand Corresponding to Fixture Load To determine the demand in gallons per minute or liters per second corresponding to any given load in water supply fixture units, reference may be made to two charts shown in Fig. 6-4 and also specifically mentioned later in this chapter under the subject, "Standard Code Regulations."

As an aid in this regard, tabulated values for the demand in gallons per minute or liters per second corresponding to given loads in water supply fixture units have been determined from the two curves and arranged conveniently in Table 6-2. In this way, easy conversion may be made from terms of water supply fixture units to gallons per minute or liters per second of flow, or vice versa. Intermediate values may be interpolated for loads between those shown in the table.

Table 6-2
TABLE FOR ESTIMATING DEMAND

Supply systems predominantly for flush tanks			Supply systems predominantly for Flushometer valves		
Load	Demand		Load	Demand	
Water supply fixture units (WSFU)	gpm	L/s	Water supply fixture units (WSFU)	gpm	L/s
1	3.0	0.19			
2	5.0	0.32			
3	6.5	0.41			
4	8.0	0.51			
5	9.4	0.59	5	15.0	0.95
6	10.7	0.68	6	17.4	1.10
7	11.8	0.74	7	19.8	1.25
8	12.8	0.81	8	22.2	1.40
9	13.7	0.86	9	24.6	1.55
10	14.6	0.92	10	27.0	1.70
12	16.0	1.01	12	28.6	1.80
14	17.0	1.07	14	30.2	1.91
16	18.0	1.14	16	31.8	2.01
18	18.8	1.19	18	33.4	2.11
20	19.6	1.24	20	35.0	2.21

Table 6-2 (*Continued*)

Supply systems predominantly for flush tanks			Supply systems predominantly for Flushometer valves		
Load	Demand		Load	Demand	
Water supply fixture units (WSFU)	gpm	L/s	Water supply fixture units (WSFU)	gpm	L/s
25	21.5	1.36	25	38.0	2.40
30	23.3	1.47	30	42.0	2.65
35	24.9	1.57	35	44.0	2.78
40	26.3	1.66	40	46.0	2.90
45	27.7	1.76	45	48.0	3.03
50	29.1	1.84	50	50.0	3.15
60	32.0	2.02	60	54.0	3.41
70	35.0	2.21	70	58.0	3.66
80	38.0	2.40	80	61.2	3.86
90	41.0	2.59	90	64.3	4.06
100	43.5	2.74	100	67.5	4.26
120	48.0	3.03	120	73.0	4.61
140	52.5	3.31	140	77.0	4.86
160	57.0	3.60	160	81.0	5.11
180	61.0	3.85	180	85.5	5.39
200	65.0	4.10	200	90.0	5.68
250	75.0	4.73	250	101.0	6.37
300	85.0	5.36	300	108.0	6.81
400	105.0	6.62	400	127.0	8.01
500	124.0	7.82	500	143.0	9.02
750	170.0	10.73	750	177.0	11.17
1000	208.0	13.12	1000	208.0	13.12
1250	239.0	15.08	1250	239.0	15.08
1500	269.0	16.97	1500	269.0	16.97
2000	325.0	20.50	2000	325.0	20.50
2500	380.0	23.97	2500	380.0	23.97
3000	433.0	27.32	3000	433.0	27.32
4000	525.0	33.12	4000	525.0	33.12
5000	593.0	37.41	5000	593.0	37.41

In viewing the charts and table, it may be noted that the demand corresponding to a given number of water supply fixture units of load is generally much higher for a system in which water closets are flushed predominantly by means of direct supply-connected flush valves (Flushometers) than for a system in which they are flushed predominantly by means of flush tanks. The difference in demand between the two systems gradually diminishes as the total number of fixture units of load rises, until at 1000 water supply fixture units the demand for both types is the same, 208 gpm (13.12 L/s).

Where a part of the system does not supply any water closets, such as is the case with hot water supply piping and some cold water supply branches, the demand corresponding to a given number of water supply fixture units may be determined from the values given for systems in which water closets are flushed predominantly by means of flush tanks. The demands determined in such cases undoubtedly are appropriate in view of the average of values assigned to fixtures other than water closets having direct supply-connected flush valves (Flushometers).

Total Demand Including Continuous Flow To estimate the total demand in gallons per minute or liters per second in any given water supply pipe which supplies outlets at which the demand is intermittent and also outlets at which demand is continuous, the demand for outlets which pose continuous demand during peak periods should be calculated separately and added to the demand for plumbing fixtures used intermittently. Examples of outlets which impose continuous demand are those for watering gardens, washing sidewalks, irrigating lawns, and for air-conditioning or refrigeration apparatus.

PRELIMINARY INFORMATION

Information for Establishing Design Basis for Sizing Obtain all information necessary for establishing a proper basis for sizing the building water supply system. Properness of the basis for sizing is contingent upon accuracy and reliability of the information applied. Thus, such information should be obtained from responsible parties and appropriate authorities recognized as sources of the necessary information.

Materials for System The kind of piping materials to be installed in the system should be determined specifically. This is a matter of selection by the owner of the building or an authorized representative, who may be the architect, engineer, or contractor.

Characteristics of the Water Supply The corrosivity of a given water supply with respect to various kinds of piping materials, and of its scale-forming tendency, is information which most officials, architects, engineers, and contractors in a water district normally have at their fingertips as a result of years of experience. For anyone without such experience and knowledge, the significant characteristics which may be applied to indicate the water supply's corrosivity and scale-forming tendency are its pH value, CO_2 content, dissolved air content, carbonate hardness, Langelier index, and Ryznar index. The most appropriate source of such information is the local water authority having jurisdiction over the system supplying the water, or over the wells from which water is pumped from the underground water table.

Location and Size of Water Supply Source The location and size of the public water main, where available, should be obtained from the local water authority. Where a private water supply source, such as a private well system, is to be used, the location and size as designed for the premises should be applied.

Developed Length of System Information should be obtained regarding the developed length of the piping run from the source of water supply to the service control valve of the building, i.e., the developed length of the water service pipe as shown on site plans. Also, determine the developed length of the piping run from the service control valve to the highest and most remote water outlet on the system. This may be established by measurement of the piping run on plans of the system.

Pressure Data Relative to Source of Supply Data as to pressure available in the public water main should be obtained from the local authority having jurisdiction over the public system. Maximum and minimum pressures available in the public main at all times should be obtained from that authority, as it is the only one recognized as the source of accurate and reliable information on this subject. Where a private well water supply system is to be used, the maximum and minimum pressures at which it will be adjusted to operate may be applied as appropriate in such cases.

Elevations The relative elevations of the source of water supply and the highest water supply outlets to be supplied in the building must be determined. In the case of a public water main, the elevation of the point where the water service connection is to be made to the public main should be obtained from the local water authority. It has the most authoritative record of elevations of the various parts of the public system, and such elevations are generally referred to datum as the reference level, and related to curb levels established for streets. Elevation of the curb level directly in front of the building should be obtained from the building plans, as such information is required to be shown on the building site plans. Elevations of each floor at which fixtures are to be supplied may be determined also from the building plans.

Minimum Pressure Required at Highest Water Outlets Information regarding the minimum pressure required at the highest water outlets for adequate, normal flow conditions, consistent with satisfactory fixture usage and equipment function, may be deemed to be as follows: 8 psi (55 kPa) for all water supply outlets at common plumbing fixtures, except for direct supply flush valves at which the minimum required flow pressure should be considered to be 15 psi (103 kPa) when supplying floor-outlet-type water closets and urinals, and as 25 psi (172 kPa) when supplying wall-hung siphon-jet or blowout-type water closets. For other types of water supplied equipment, the minimum flow pressure required should be obtained from the manufacturer.

Provision of Necessary Information on Plans Information necessary for establishing a proper basis for designing sizes of water supply piping should be provided on plans of the water supply system when submitted to plumbing plan examiners for proposed installations. Provision of such information permits the examiner to check easily and efficiently the adequacy of sizes proposed for the various parts of the building water supply system.

LIMITATION OF FRICTION

Consideration of Friction in Design Friction in the building water supply system must be limited by design to the degree necessary so that the highest water outlets on the system will have available during periods of peak demand at least the minimum pressure required at such outlets for satisfactory water supply conditions.

Maximum Permissible Friction Loss The limit to which pressure loss due to friction may be permitted to occur in the main water lines and risers supplying the highest water outlets during peak demand periods is the amount of static pressure available in excess of the minimum pressure required at such outlets when no-flow conditions exist. This may be calculated as the difference between the static pressure existing at the highest water outlets during no-flow conditions, and the minimum pressure required at such outlets for satisfactory supply conditions.

Where supplied by direct pressure from a public main, to calculate the static pressure at the highest outlet, deduct from the certified minimum pressure available in the public main the amount of static pressure loss corresponding to the height at which the outlet is located above the public main; i.e., deduct 0.433 psi pressure for each foot (or 9.79 kPa/m) of rise in elevation from the public main to the highest outlet.

Where supplied under pressure from a gravity water supply tank located at an elevation above the highest water outlet, the static pressure at that outlet is calculated as being equal to 0.433 psi for each foot (or 9.79 kPa/m) difference in elevation between the outlet and the water line in the tank. In this case, the minimum static pressure at the outlet should be determined as that corresponding to the level of the low-water line at which the tank is designed to operate.

Where supplied under pressure from a pressure booster pump system, to calculate the static pressure at the highest outlet, deduct from the minimum pressure designed to be maintained at all times at the pump outlet the amount of static pressure loss corresponding to the height at which the highest outlet is located above the level of the pump outlet, i.e., deduct 0.433 psi pressure for each foot (or 9.79 kPa/m) of rise in elevation from the pump outlet to the highest water outlet.

Basic Design Circuit Of the highest water outlets on a system, the one at which the least available pressure will prevail during periods of peak demand is that outlet which is supplied through the longest run of piping extending from the pressure source of water supply, as a general rule. As pipe friction loss is directly proportional to length of piping, the most extreme run of piping from the pressure source of water supply to the highest outlets on the system should be designated as the "basic design circuit" (BDC) for sizing the main water lines and risers in accordance with the limit to which pressure loss due to friction may be permitted to occur therein.

In most systems, the BDC may be found to consist of the run of cold water supply piping extending from the pressure source of water supply to the domestic hot water heating unit, and the run of hot water supply piping therefrom to the highest and most remote hot water outlet on the system. However, in systems having flush valve (Flushometer) supplied water closets at the topmost floor, the BDC may be found to be the run of cold water supply piping extending from the pressure source of water supply to the highest and most remote flush valve on the system.

Friction Loss in Equipment Where a water meter, water filter, water softener, fish trap or strainer, or instantaneous or tankless hot water heating coil is provided in the basic design circuit, the friction loss corresponding to the peak demand through such equipment must be determined and included in pressure loss calculations. Manufacturers' charts and data sheets on their products provide such information generally, and should be used as a guide in selecting the most appropriate type and size of equipment to use with consideration for the limit to which pressure loss due to friction may be permitted to occur in the basic design circuit. The rated pressure loss through such equipment should be deducted from the friction loss limit so as to establish the amount of pressure which may be permitted to be dissipated by friction in pipe, valves, and fittings of the basic design circuit.

Pressure Loss in Displacement-Type Cold Water Meters The American Water Works Association standard for cold water meters of the displacement type is designated AWWA C700–71. It covers displacement meters known as nutating- or oscillating-piston or disk meters, which are practically positive in action.

The standard establishes maximum capacity or delivery classifications for each meter size as follows:

⅝ in—20 gpm (15.9 mm—1.26 L/s)

¾ in—30 gpm (19 mm—1.89 L/s)

1 in—50 gpm (25.4 mm—3.1 L/s)

1½ in—100 gpm (38.1 mm—6.3 L/s)

2 in—160 gpm (50 mm—10.1 L/s)

3 in—300 gpm (75 mm—18.9 L/s)

4 in—500 gpm (100 mm—31.5 L/s)

6 in—1000 gpm (150 mm—63 L/s)

The standard also establishes the maximum pressure loss corresponding to the stated maximum capacities as follows:

15 psi (103 kPa) for the ⅝-in (15.9-mm), ¾-in (19.0-mm), and 1-in (25.5-mm) meter sizes

20 psi (138 kPa) for the 1½-in (38.1-mm), 2-in (50-mm), 3-in (75-mm), 4-in (100-mm), and 6-in (150-mm) sizes

For estimating pressure loss in displacement-type cold water meters, Fig. 5-5 is provided. Pressure loss in meters for flow at less than maximum rates for any given size of meter may be estimated from Fig. 5-5.

Uniform Pipe Friction Loss To facilitate calculation of appropriate pipe sizes corresponding to the permissible friction loss in pipe, valves, and fittings, it is recommended that the basic design circuit be designed in accordance with the principle of uniform pipe friction loss throughout its length. In this way the friction limit for the piping run may be established in terms of pounds per square inch per 100 ft (or Pa/m) of piping length. The permissible uniform pipe friction loss is calculated by dividing the permissible friction loss in pipe, valves, and fittings by the total equivalent length of the basic design circuit, and multiplying by 100.

Equivalent Length of Piping The total equivalent length of piping is its developed length plus the equivalent pipe length corresponding to the frictional resistance of all fittings and valves in the piping run. When sizes of fittings are known or have been established by sizing based upon appropriate limitation of velocity, corresponding equivalent lengths may be determined directly from available tables. Two such tables (Tables 5-6 and 5-7) are presented in Chap. 5.

As a general finding, experience indicates that the equivalent length to be allowed for fittings and valves as a result of such calculations is approximately 50 percent of the developed length of the basic design circuit in the case of copper water tube systems, and approximately 85 percent for standard threaded pipe systems.

ACCURATE PIPE FRICTION CHARTS FOR VARIOUS SERVICE CONDITIONS

Accurate pipe friction charts for various service conditions are presented herewith for each of the standard piping materials used for water supply systems in buildings. These charts were developed using the actual internal pipe diameters given in applicable standards for each of the different kinds of materials, rather than nominal diameters. All values calculated in developing the charts were made using the most authoritative pipe friction equations presented to date for application in accordance with the relative roughness of the internal pipe surface, i.e., roughness relative to pipe diameter. These equations were discussed in Chap. 5.

Flow rates corresponding to any given uniform pipe friction loss may be determined readily for each nominal size of the kind of pipe selected for the system. The appropriate chart to apply in any given case depends upon the kind of piping to be used and the effect that the water to be conveyed will produce within the piping after extended service. On each of the charts, the particular equation applied and the internal pipe surface condition to which it relates have been stated. Accordingly, appropriate selection of the proper chart to use in determining flow rates through various sizes of a given kind of piping may be made for a specific uniform pipe friction loss knowing the quality and characteristics of the water to be conveyed and the effects that the water will have upon the internal pipe surface in extended service.

Chart 6-1(a)
GALVANIZED IRON AND STEEL STANDARD WEIGHT PIPE (ASTM A72, A120)
Fairly Rough Surface Condition

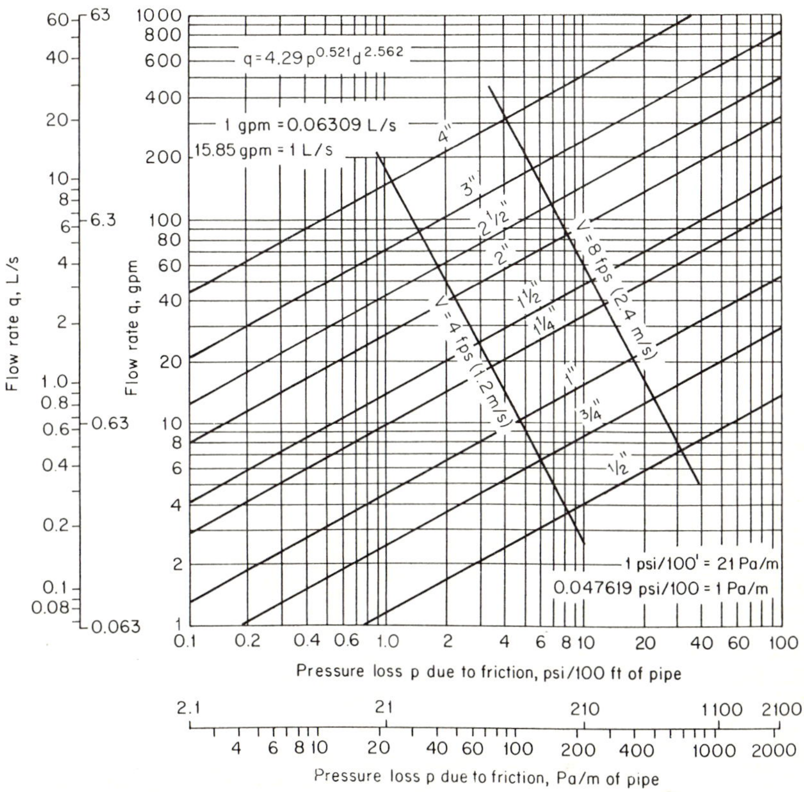

As a guide in selecting the appropriate chart to apply with regard to the quality and characteristics of the water to be conveyed, recommendations are as follows:

FOR RED BRASS AND COPPER PIPE, AND COPPER WATER TUBE

Where the water supply is relatively noncorrosive to the pipe material and is not scale-forming, apply the charts applicable to the "fairly smooth" surface condition.

Where the water supply is moderately scale-forming, apply the charts applicable to the "fairly rough" surface condition.

Where the water has an appreciable scale-forming tendency and would otherwise result in clogging of small-diameter pipes, and in reducing the effective diameter of other piping in the system,

Chart 6-1(b)
GALVANIZED IRON AND STEEL STANDARD WEIGHT PIPE (ASTM A72, A120)
Rough Surface Condition

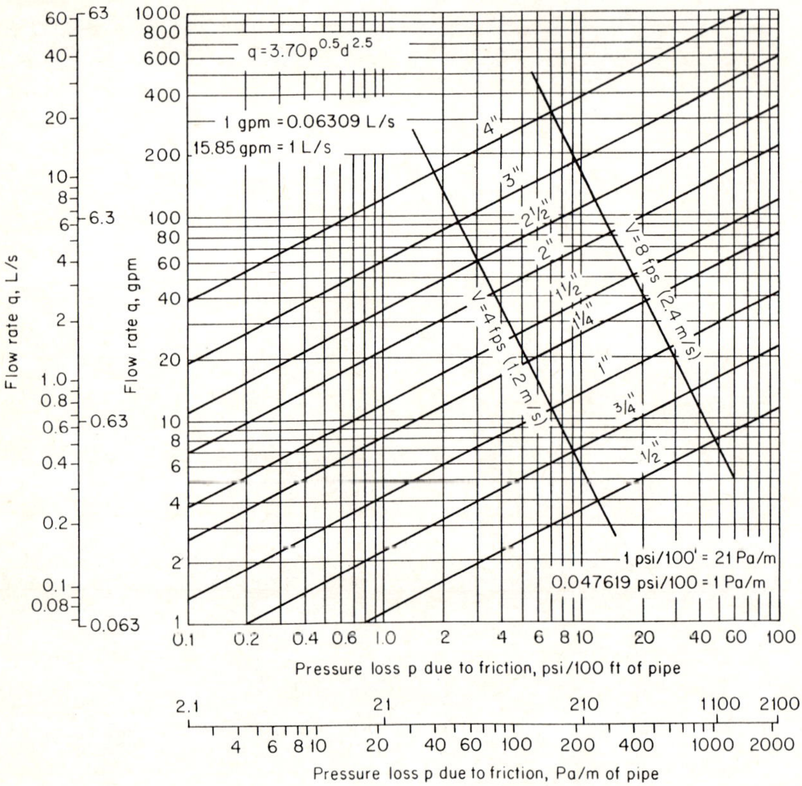

allow one standard pipe size larger than would be determined applying the "fairly smooth" chart.

FOR GALVANIZED IRON AND STEEL PIPE

Where the water supply is relatively noncorrosive to the pipe material and is not scale-forming, apply the chart applicable to the "fairly rough" surface condition.

Where the water supply is moderately corrosive or moderately scale-forming, apply the chart applicable to the "rough" surface condition.

Where the water supply is significantly corrosive to iron or steel or has an appreciable scale-forming tendency and would otherwise

Chart 6-2(a)
COPPER AND BRASS PIPE, S.P.S. (ASTM B42, B43)
Fairly Smooth Surface Condition

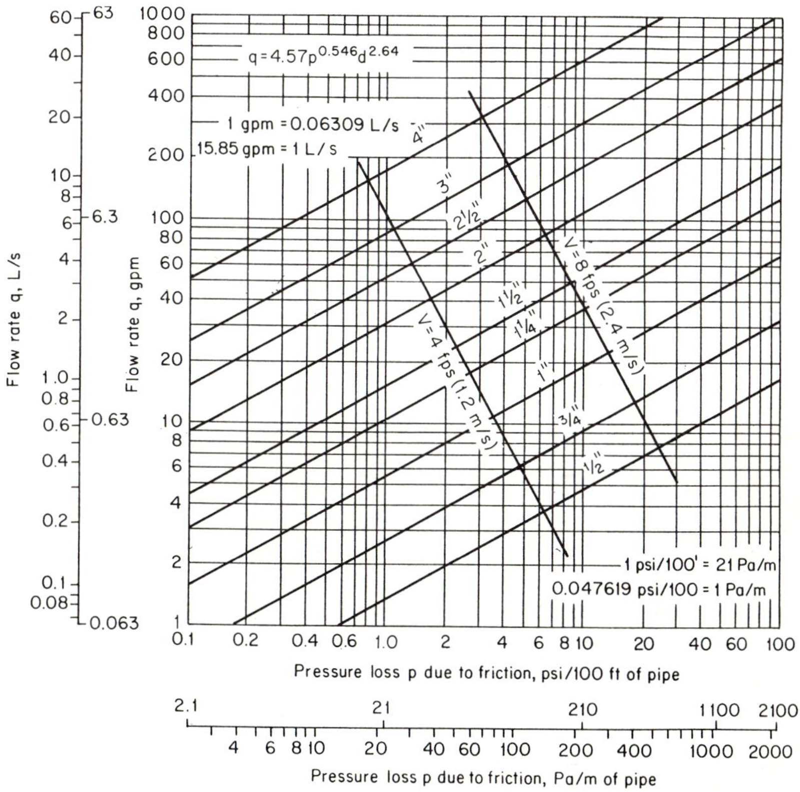

result in clogging of small-diameter pipes, and in reducing the effective diameter of other piping in the system, allow one standard pipe size larger than would be determined applying the "fairly rough" chart.

FOR PLASTIC PIPE (PE, ABS, AND PVC)

Where the water supply is relatively not scale-forming, apply the chart applicable to the "fairly smooth" surface condition.

Where the water supply is moderately scale-forming, apply the chart applicable to the "fairly rough" surface condition.

Where the water has an appreciable scale-forming tendency and would otherwise result in clogging of small-diameter pipes, and

Chart 6-2(b)
COPPER AND BRASS PIPE, S.P.S. (ASTM B42, B43)
Fairly Rough Surface Condition

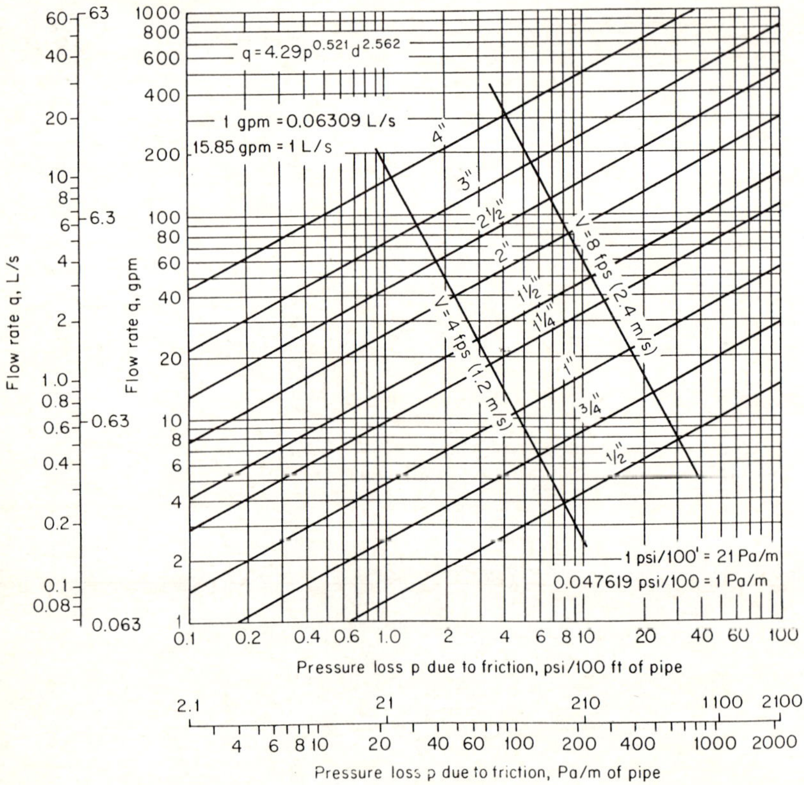

in reducing the effective diameter of other piping in the system, allow one standard pipe size larger than would be determined applying the "fairly smooth" chart.

LIMITATION OF VELOCITY

Consideration of Velocity in Design Velocity of flow through water supply piping during periods of peak demand is an important factor to consider in the design of building water supply systems. Limitation of water velocity should be observed in order to avoid objectionable noise effects in systems, shock damage to piping, equipment, tanks, coils and joints, and accelerated deterioration and eventual failure of piping from erosion-corrosion.

Chart 6-3(a)
THREADLESS RED BRASS AND COPPER PIPE (ASTM B302)
Fairly Smooth Surface Condition

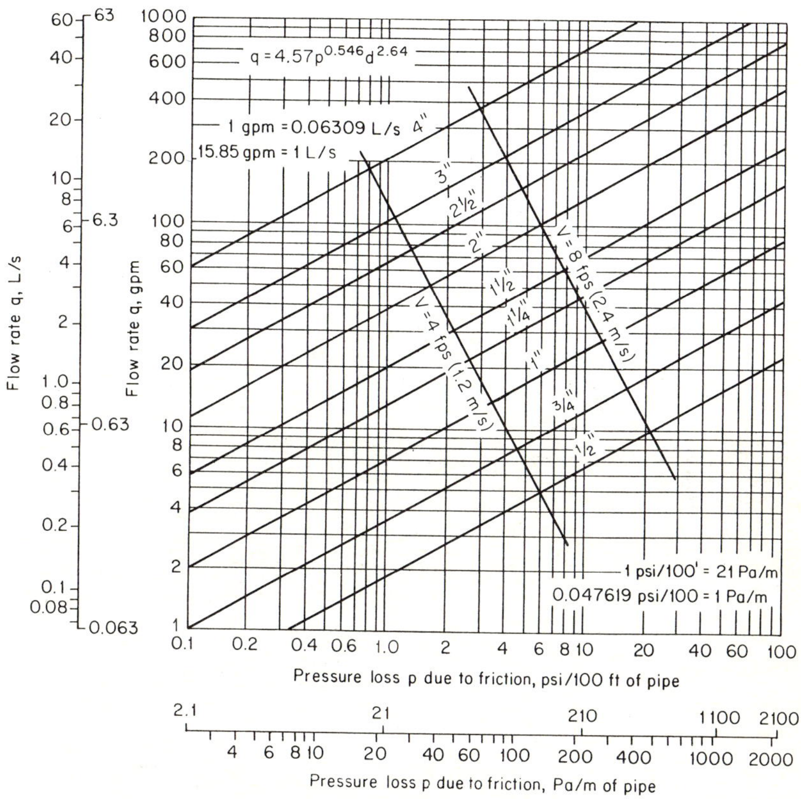

Good Engineering Practice In accordance with good engineering practice, it is recommended that maximum velocity in water supply piping be limited to no more than 8 fps (2.4 m/s). This is deemed essential in order to avoid such objectionable effects as the production of whistling line noise, the occurrence of cavitation, and associated excessive noise in fittings and valves.

It is also recommended that maximum velocity be limited to no more than 4 fps (1.2 m/s) in branch piping from mains, headers, and risers to water outlets at which the supply is controlled by means of quick-closing devices such as an automatic flush valve, solenoid valve, pneumatic valve, or a quick-closing valve or faucet of the self-closing, push-pull, push-button, or other similar type. This limitation is deemed necessary in order to avoid development of excessive and damaging

Chart 6-3(b)
THREADLESS RED BRASS AND COPPER PIPE (ASTM B302)
Fairly Rough Surface Condition

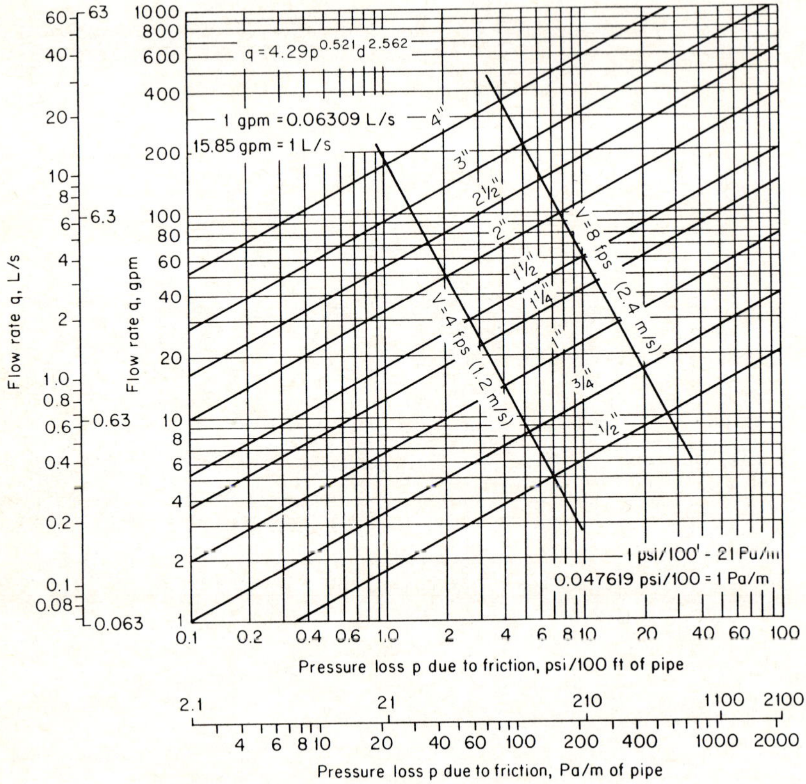

shock pressures in the piping and equipment when flow is suddenly shut off. Many items of plumbing equipment and systems are not designed to withstand the very high shock pressures which may occur as the result of sudden cessation of high-velocity flow in piping.

Manufacturers' Recommendations for Avoiding Erosion-Corrosion

Velocity limits recommended by pipe manufacturers to avoid accelerated deterioration of their piping materials due to erosion-corrosion should be observed. Recent research studies have shown that extreme turbulence accompanying high flow velocities is an important factor in causing erosion-corrosion, and that this is especially prone to occur where the water supply has a high carbon dioxide content (i.e., in excess of 10 ppm), and where it has been softened to zero hardness. Another important factor is very high water temperature (i.e., in excess of 150°F (66°C).

Chart 6-4(a)
COPPER WATER TUBE, TYPE K (ASTM B88)
Fairly Smooth Surface Condition

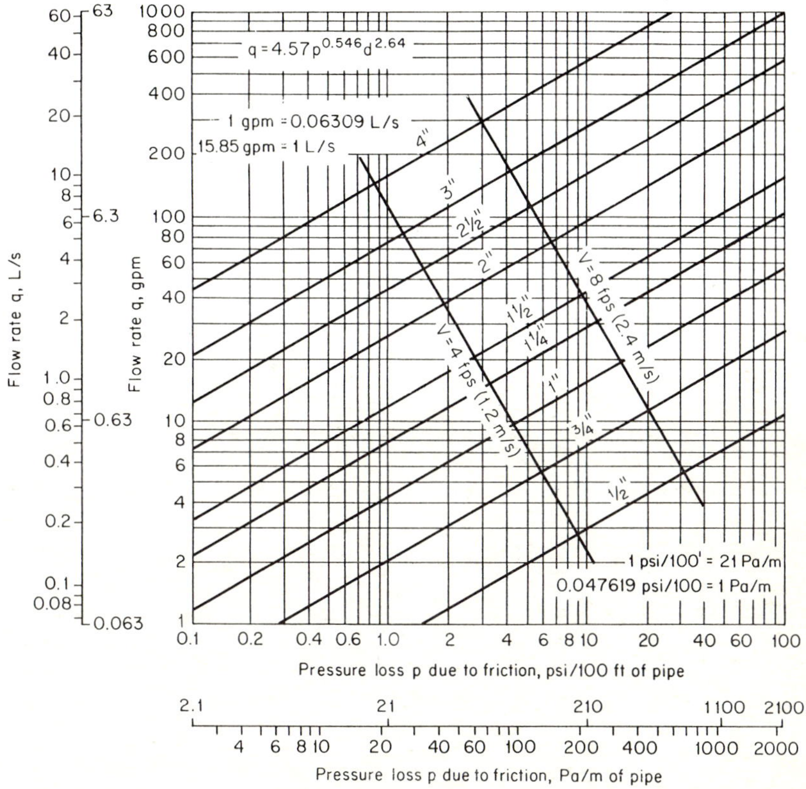

To control erosion-corrosion effects in copper water tube, and copper and brass pipe, pipe manufacturers' recommendations are as follows:

1. Where the water supply has a pH value higher than 6.9 and a positive scale-forming tendency, such as may be shown by a positive Langelier index, velocity should be limited to no more than 8 fps (2.4 m/s).
2. Where the water supply has a pH value lower than 6.9 and may be classified as aggressively corrosive or where the water supply has been softened to zero hardness by passage through a water softener, velocity should be limited to no more than 4 fps (1.2 m/s).
3. The 4 fps (1.2 m/s) velocity limit should be applied to all hot

Chart 6-4(b)
COPPER WATER TUBE, TYPE K (ASTM B88)
Fairly Rough Surface Condition

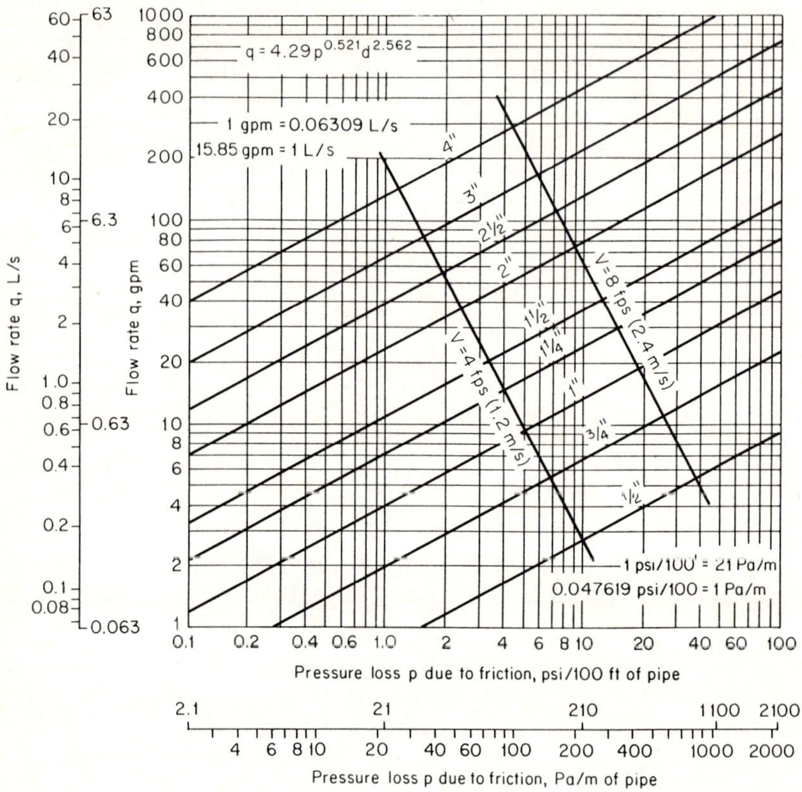

water supply piping conveying water at a temperature above 150°F (65.6°C) because of the accelerated rate of corrosion at such higher temperature.

Recommendations for Minimizing Cost of Pumping Velocity limitation is generally advisable and recommended in the sizing of inlet and outlet piping for water supply pumps. Frictional losses in such piping affect the cost of pumping, and consequently should be reduced to a reasonable minimum. The general recommendation in this instance is to limit velocity in both inlet and outlet piping for water supply pumps to no more than 4 fps (1.2 m/s). This may also be applied advantageously in the case of constant-pressure booster pump water supply systems.

Chart 6-5(a)
COPPER WATER TUBE, TYPE L (ASTM B88)
Fairly Smooth Surface Condition

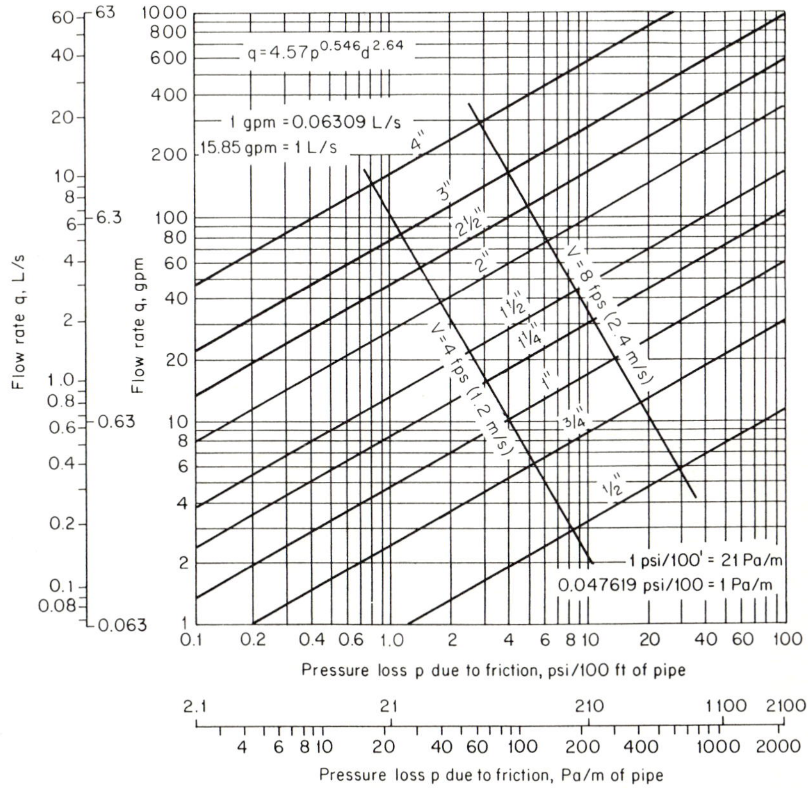

SIZING TABLES BASED ON VELOCITY LIMITATION

To provide a simplified guide for easy application by plumbing designers, and for plumbing plan examiners and inspectors to use in checking pipe sizes for building water supply systems, special sizing tables have been developed. They are presented here in Tables 6-3 through 6-8.

The tables show, for velocity limits of 4 fps (1.2 m/s) and 8 fps (2.4 m/s), the number of water supply fixture units of load permissible for each size and kind of pipe in common use. Hence, the tables may be used in much the same manner as tables currently applied for sizing drainage and vent systems.

An explanation of these tables and how they were developed may serve as an aid to users. There are two tables for each kind of pipe,

Chart 6-5(b)
COPPER WATER TUBE, TYPE L (ASTM B88)
Fairly Rough Surface Condition

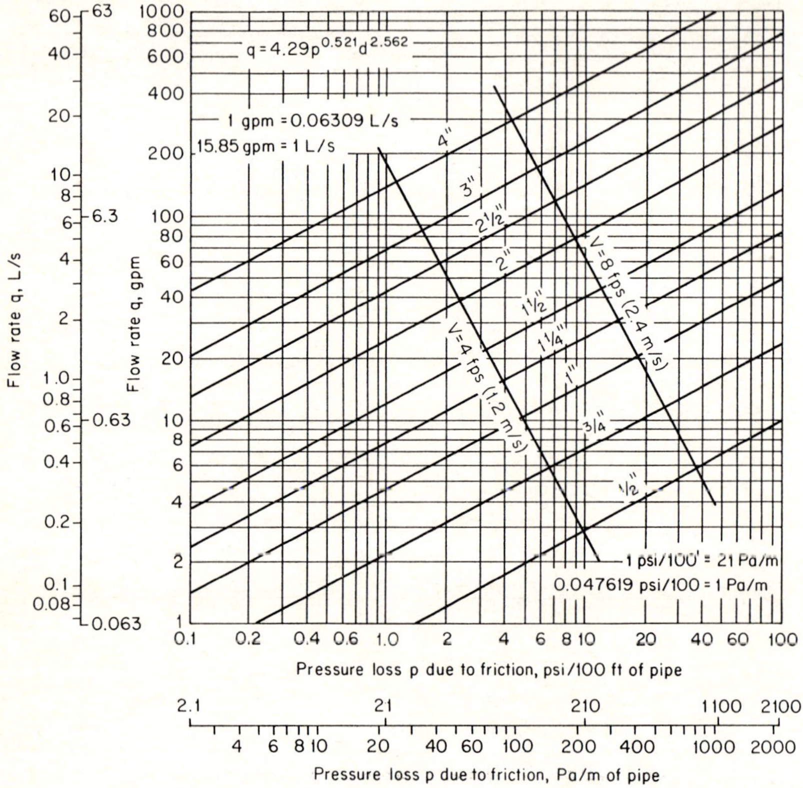

customary U.S. units and SI units. As an example, in Table 6-4*a* the first column lists the nominal pipe sizes, while the second column shows the actual internal pipe diameter as stated in the applicable ASTM standard for the pipe. Under the heading "Velocity = 4 fps" the column headed "Flow" shows the rate of flow, in gallons per minute, calculated for a velocity of 4 fps for each size of pipe based on the actual internal diameter. The column headed "Friction" shows the amount of pipe friction calculated for that rate of flow applying the formula noted beneath the table. Between these two columns are columns A and B, which show load values in terms of water supply fixture units. Column A applies to piping which does not supply flush valves (Flushometers). The load values shown in this column were determined as corresponding to the calculated flow rates using Table 6-2, selecting values as for supply

Chart 6-6(a)

SCHEDULE 40 PLASTIC PIPE: PE (ASTM D2104), ABS (ASTM D1527), PVC (ASTM D1785)

Fairly Smooth Surface Condition

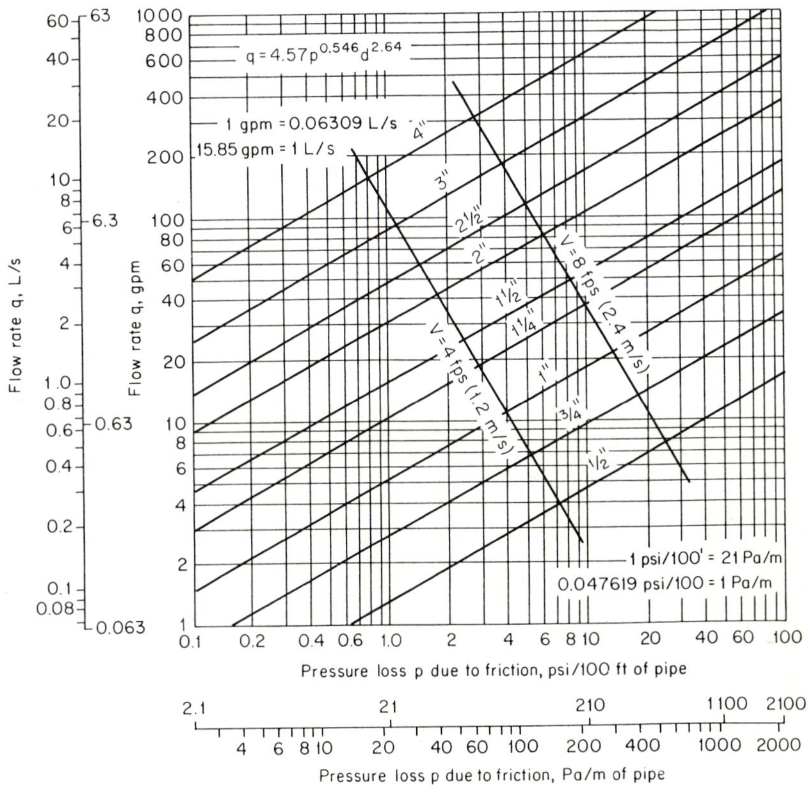

systems predominantly for flush tanks. Column B applies to piping which supplies flush valves. The load values shown therein were similarly determined as corresponding to the same calculated flow rates, using the same table, but selecting such values as for supply systems predominantly for flush valves.

Values shown in the tables for the quantity rate of flow were calculated using the adapted flow formula, which is as follows:

$$q = 2.448\, d^2 V$$

where q = quantity rate of flow, gpm
d = actual inside diameter of pipe, in
V = velocity of flow, fps

Chart 6-6(b)
SCHEDULE 40 PLASTIC PIPE: PE (ASTM D2104), ABS (ASTM D1527), PVC (ASTM D1785)
Fairly Rough Surface Condition

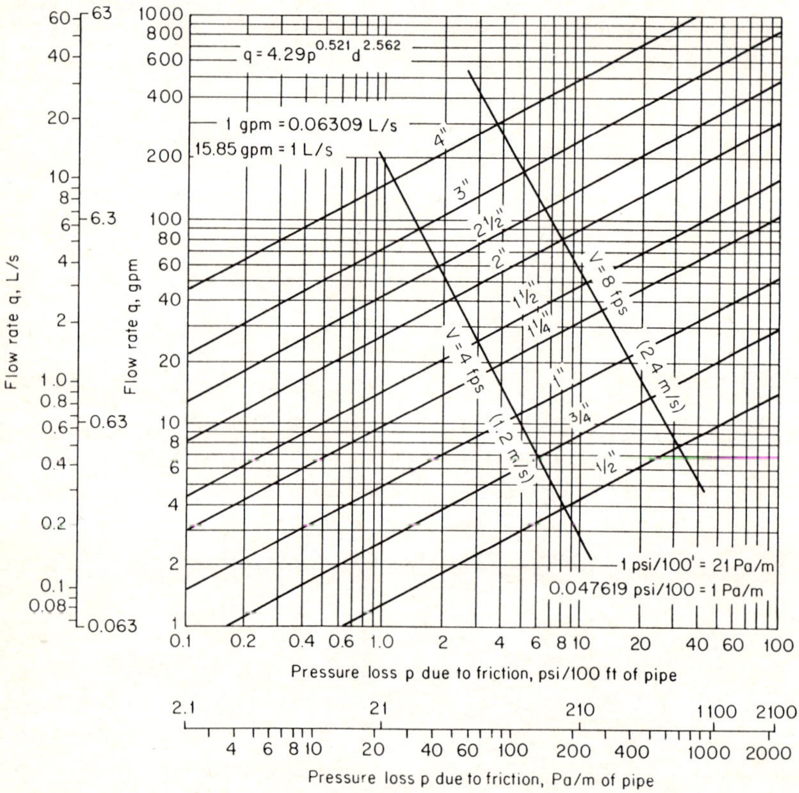

SIMPLIFIED METHOD FOR SIZING SYSTEMS

Application A simplified method for sizing building water supply systems in accordance with demand load, in terms of water supply fixture units, has been found in almost every case to constitute a complete and proper method for adequately sizing the water supply systems of a specific category of buildings. This category includes at least 95 percent of all residential buildings erected per year, and a significant percentage of buildings of other types of occupancy.

In this category are all buildings supplied from a source at which the minimum available water pressure is more than adequate for supplying the highest and most remote fixtures satisfactorily during peak demand, provided the water supply system is sized in accordance with

Table 6-3*a*

SIZING TABLE BASED ON VELOCITY LIMITATION—U.S. CUSTOMARY UNITS

Galvanized Iron and Steel Pipe, Standard Pipe Size

		Velocity = 4 fps				Velocity = 8 fps			
Nominal size, in	Actual ID, in	Flow q, gpm	Load WSFU (col. A)*	Load WSFU (col. B)†	Friction p, psi/100 ft‡	Flow q, gpm	Load WSFU (col. A)*	Load WSFU (col. B)†	Friction p, psi/100 ft‡
½	0.622	3.8	1.5		8.2	7.6	3.7		31.0
¾	0.824	6.7	3.0		6.0	13.4	8.4		22.5
1	1.049	10.8	6.1		4.6	21.6	25.3	7.7	17.2
1¼	1.380	18.6	17.5	6.0	3.4	37.2	77.3	23.7	12.8
1½	1.610	25.4	37.0	9.3	2.9	50.8	132.3	52.0	10.8
2	2.067	41.8	93.0	29.8	2.2	83.6	293.0	171.6	8.4
2½	2.469	59.8	174.0	75.6	1.8	119.6	477.0	361.0	6.8
3	3.068	92.0	335.0	209.0	1.4	184.0	842.0	806.0	5.4
4	4.026	158.6	688.0	615.0	1.1	317.0	1930.0	1930.0	4.1

* Col. A applies to piping which does not supply flush valves.

† Col. B applies to piping which supplies flush valves.

‡ Friction loss p, corresponding to flow rate q, for piping having fairly smooth surface condition after extended service applying the formula

$$q = 4.57\, p^{0.546}\, d^{2.64}$$

Table 6-3*b*
SIZING TABLE BASED ON VELOCITY LIMITATION—SI UNITS
Galvanized Iron and Steel Pipe, Standard Pipe Size

Nominal size, mm	Actual ID, mm	Velocity = 1.2 m/s				Velocity = 2.4 m/s			
		Flow q, L/s	Load WSFU (col. A)*	Load WSFU (col. B)†	Friction p, Pa/m‡	Flow q, L/s	Load WSFU (col. A)*	Load WSFU (col. B)†	Friction p, Pa/m‡
12.7	15.8	0.23	1.5		172.3	0.47	3.7		651.5
19.0	20.9	0.42	3.0		126.1	0.84	8.4		472.8
25.4	26.6	0.68	6.1		96.7	1.36	25.3	7.7	361.5
31.8	35.1	1.17	17.5	6.0	71.5	2.34	77.3	23.7	269.0
38.1	40.9	1.60	37.0	9.3	60.9	3.20	132.3	52.0	227.0
50.8	52.5	2.63	93.0	29.8	46.2	5.27	293.0	171.6	176.5
63.5	62.7	3.77	174.0	75.6	37.8	7.54	477.0	361.0	142.9
76.2	77.7	5.80	335.0	209.0	29.4	11.60	842.0	806.0	113.5
102.0	102.3	10.00	688.0	615.0	23.1	20.01	1930.0	1930.0	86.2

* Col. A applies to piping which does not supply flush valves.

† Col. B applies to piping which supplies flush valves.

‡ Friction loss p, corresponding to flow rate q, for piping having fairly smooth surface condition after extended service applying the formula

$$q = 4.57\, p^{0.546}\, d^{2.64}$$

Table 6-4*a*

SIZING TABLE BASED ON VELOCITY LIMITATION—U.S. CUSTOMARY UNITS

Copper and Brass Pipe, Standard Pipe Size

		Velocity = 4 fps				Velocity = 8 fps			
Nominal size, in	Actual ID, in	Flow q, gpm	Load WSFU (col. A)*	Load WSFU (col. B)†	Friction p, psi/100 ft‡	Flow q, gpm	Load WSFU (col. A)*	Load WSFU (col. B)†	Friction p, psi/100 ft‡
½	0.625	3.8	1.5		6.8	7.0	3.7		24.2
¾	0.822	6.6	3.0		5.1	3.2	8.4		18.0
1	1.062	11.0	6.3		3.7	22.0	26.4	8.0	13.3
1¼	1.368	18.3	16.8	6.4	2.8	36.6	75.0	22.7	10.0
1½	1.600	25.2	36.3	9.3	2.3	50.4	130.0	51.0	8.4
2	2.062	41.6	92.0	29.5	1.7	83.2	291.0	170.0	6.2
2½	2.500	61.2	181.0	80.0	1.4	122.4	492.0	376.0	4.9
3	3.062	92.0	335.0	209.0	1.1	184.0	842.0	807.0	3.9
4	4.000	158.0	685.0	611.0	0.8	316.0	1920.0	1920.0	2.9

* Col. A applies to piping which does not supply flush valves.

† Col. B applies to piping which supplies flush valves.

‡ Friction loss p corresponding to flow rate q for piping having fairly smooth surface condition after extended service applying the formula

$$q = 4.57\, p^{0.546}\, d^{2.64}$$

Table 6-4*b*
SIZING TABLE BASED ON VELOCITY LIMITATION—SI UNITS
Copper and Brass Pipe, Standard Pipe Size

		Velocity = 1.2 m/s				Velocity = 2.4 m/s			
Nominal size, mm	Actual ID, mm	Flow q, L/s	Load WSFU (col. A)*	Load WSFU (col. B)†	Friction p, Pa/m‡	Flow q, L/s	Load WSFU (col. A)*	Load WSFU (col. B)†	Friction p, Pa/m‡
12.7	15.9	0.24	1.5		143.0	0.48	3.7		508.6
19.0	20.9	0.42	3.0		107.2	0.83	8.4		378.3
25.4	27.0	0.69	6.3		71.8	1.39	26.4	8.0	279.5
31.8	34.7	1.15	16.8	6.4	58.8	2.31	75.0	22.7	210.2
38.1	40.6	1.59	36.3	9.3	48.3	3.18	130.0	51.0	176.5
50.8	52.4	2.62	92.0	29.0	35.7	5.25	291.0	170.0	130.3
63.5	63.5	3.86	181.0	80.0	29.4	7.72	492.0	376.0	103.0
76.2	77.8	5.80	335.0	209.0	23.1	11.61	842.0	807.0	82.0
102.0	101.6	9.97	685.0	611.0	16.8	19.94	1920.0	1920.0	61.0

* Col. A applies to piping which does not supply flush valves.

† Col. B applies to piping which supplies flush valves.

‡ Friction loss p corresponding to flow rate q for piping having fairly smooth surface condition after extended service applying the formula

$$q = 4.57\, p^{0.546}\, d^{2.64}$$

Table 6-5*a*

SIZING TABLE BASED ON VELOCITY LIMITATION—U.S. CUSTOMARY UNITS

Threadless Copper and Red Brass Pipe (TP)

		Velocity = 4 fps				Velocity = 8 fps			
Nominal size, in	Actual ID, in	Flow q, gpm	Load WSFU (col. A)*	Load WSFU (col. B)†	Friction p, psi/100 ft‡	Flow q, gpm	Load WSFU (col. A)*	Load WSFU (col. B)†	Friction p, psi/100 ft‡
½	0.710	4.9	2.0		5.9	9.8	5.3		20.8
¾	0.920	8.3	4.2		4.4	16.6	13.2	5.7	15.5
1	1.185	13.7	9.0		3.3	27.4	44.0	10.5	11.7
1¼	1.530	22.9	28.9	8.3	2.4	45.8	110.0	40.0	8.5
1½	1.770	30.6	55.0	14.5	2.1	61.2	181.0	80.0	7.2
2	2.245	49.4	126.0	48.5	1.6	98.8	369.0	240.0	5.6
2½	2.745	74.0	245.0	125.0	1.3	148.0	631.0	537.0	4.4
3	3.334	109.0	421.0	305.0	1.0	218.0	1081.0	1081.0	3.5
4	4.286	180.0	816.0	774.0	0.8	360.0	2318.0	2318.0	2.6

* Col. A applies to piping which does not supply flush valves.

† Col. B applies to piping which supplies flush valves.

‡ Friction loss p corresponding to flow rate q for piping having fairly smooth surface condition after extended service applying the formula

$$q = 4.57\, p^{0.546}\, d^{2.64}$$

Table 6-5*b*
SIZING TABLE BASED ON VELOCITY LIMITATION—SI UNITS
Threadless Copper and Red Brass Pipe (TP)

		Velocity = 1.2 m/s				Velocity = 2.4 m/s			
Nominal size, mm	Actual ID, mm	Flow q, L/s	Load WSFU (col. A)*	Load WSFU (col. B)†	Friction p, Pa/m‡	Flow q, L/s	Load WSFU (col. A)*	Load WSFU (col. B)†	Friction p, Pa/m‡
12.7	18.0	0.31	2.0		124.0	0.62	5.3		437.1
19.0	23.4	0.52	4.2		92.5	1.05	13.2	5.7	325.7
25.4	30.1	0.86	9.0		69.4	1.73	44.0	10.5	245.9
31.8	38.9	1.44	28.9	8.3	50.4	2.89	110.0	40.0	178.6
38.1	47.0	1.91	55.0	14.5	44.1	3.86	181.0	80.0	151.3
50.8	57.0	3.11	126.0	48.5	33.6	6.23	369.0	240.0	117.7
63.5	69.7	4.6	245.0	125.0	27.3	9.34	631.0	537.0	92.5
76.2	84.7	6.8	421.0	305.0	21.0	13.75	1081.0	1081.0	73.6
102.0	108.9	11.4	816.0	774.0	16.8	22.71	2318.0	2318.0	54.6

* Col. A applies to piping which does not supply flush valves.

† Col. B applies to piping which supplies flush valves.

‡ Friction loss p corresponding to flow rate q for piping having fairly smooth surface condition after extended service applying the formula

$$q = 4.57\, p^{0.546}\, d^{2.64}$$

Table 6-6*a*
SIZING TABLE BASED ON VELOCITY LIMITATION—U.S. CUSTOMARY UNITS
Copper Water Tube, Type K

		Velocity = 4 fps				Velocity = 8 fps			
Nominal size, in	Actual ID, in	Flow q, gpm	Load WSFU (col. A)*	Load WSFU (col. B)†	Friction p, psi/100 ft‡	Flow q, gpm	Load WSFU (col. A)*	Load WSFU (col. b)†	Friction p, psi/100 ft‡
½	0.527	2.7	0.75		8.5	5.4	2.3		31.0
¾	0.745	5.5	2.3		5.6	11.0	6.3		20.2
1	0.995	9.7	5.3		4.1	19.4	19.5	5.8	14.4
1¼	1.245	15.2	10.8	5.0	3.1	30.4	54.0	14.0	11.1
1½	1.481	21.5	25.0	7.8	2.6	43.0	98.0	34.0	9.2
2	1.959	37.6	78.0	24.0	1.8	75.2	251.0	130.0	6.5
2½	2.435	58.2	166.0	69.0	1.4	116.4	460.0	340.0	5.2
3	2.907	82.8	289.0	161.0	1.2	165.6	725.0	663.0	4.2
4	3.857	146.0	609.0	528.0	0.8	292.0	1705.0	1705.0	3.0

* Col. A applies to piping which does not supply flush valves.

† Col. B applies to piping which supplies flush valves.

‡ Friction loss p corresponding to flow rate q for piping having fairly smooth surface condition after extended service applying the formula

$$q = 4.57\, p^{0.546}\, d^{2.64}$$

Table 6-6*b*
SIZING TABLE BASED ON VELOCITY LIMITATION—SI UNITS
Copper Water Tube, Type K

		Velocity = 1.2 m/s				Velocity = 2.4 m/s			
Nominal size, mm	Actual ID, mm	Flow q, L/s	Load WSFU (col. A)*	Load WSFU (col. B)†	Friction p, Pa/m‡	Flow q, L/s	Load WSFU (col. A)*	Load WSFU (col. B)†	Friction p, Pa/m‡
12.7	13.4	0.17	0.75		178.6	0.34	2.3		651.5
19.0	18.9	0.34	2.3		117.7	0.69	6.3		424.5
25.4	25.3	0.61	5.3		86.2	1.22	19.5	5.8	302.6
31.8	31.6	0.95	10.8	5.0	65.1	1.91	54.0	14.0	233.3
38.1	37.6	1.35	25.0	7.8	54.6	2.71	98.0	34.0	193.3
50.8	49.8	2.37	78.0	24.0	37.8	4.74	251.0	130.0	136.6
63.3	61.8	3.67	166.0	69.0	29.4	7.34	460.0	340.0	109.3
76.2	73.8	5.22	289.0	161.0	25.2	10.44	725.0	663.0	88.3
102.0	98.0	9.21	609.0	528.0	16.8	18.42	1705.0	1705.0	63.0

* Col. A applies to piping which does not supply flush valves.

† Col. B applies to piping which supplies flush valves.

‡ Friction loss p corresponding to flow rate q for piping having fairly smooth surface condition after extended service applying the formula

$$q = 4.57\, p^{0.546}\, d^{2.64}$$

Table 6-7*a*
SIZING TABLE BASED ON VELOCITY LIMITATION—U.S. CUSTOMARY UNITS
Copper Water Tube, Type L

		Velocity = 4 fps				Velocity = 8 fps			
Nominal size, in	Actual ID, in	Flow q, gpm	Load WSFU (col. A)*	Load WSFU (col. B)†	Friction p, psi/100 ft‡	Flow q, gpm	Load WSFU (col. A)*	Load WSFU (col. B)†	Friction p, psi/100 ft‡
½	0.545	2.9	1.0		8.2	5.8	2.5		29.0
¾	0.785	6.0	2.5		5.2	12.0	7.3		18.7
1	1.025	10.3	5.5		3.9	20.6	22.5	7.0	13.7
1¼	1.265	15.7	11.5	5.0	3.0	31.4	58.0	15.5	10.7
1½	1.505	22.8	28.5	8.0	2.5	45.6	109.0	38.0	8.7
2	1.985	38.6	82.0	26.0	1.8	77.2	261.0	138.0	6.3
2½	2.465	59.5	172.0	75.0	1.4	119.0	474.0	356.0	4.9
3	2.945	85.0	300.0	178.0	1.1	170.0	750.0	692.0	4.0
4	3.905	149.0	636.0	544.0	0.8	298.0	1759.0	1759.0	2.8

* Col. A applies to piping which does not supply flush valves.

† Col. B applies to piping which supplies flush valves.

‡ Friction loss p corresponding to flow rate q for piping having fairly smooth surface condition after extended service applying the formula

$$q = 4.57\, p^{0.546}\, d^{2.64}$$

Table 6-7*b*
SIZING TABLE BASED ON VELOCITY LIMITATION—SI UNITS
Copper Water Tube, Type L

		Velocity = 1.2 m/s				Velocity = 2.4 m/s			
Nominal size, mm	Actual ID, mm	Flow q, L/s	Load WSFU (col. A)*	Load WSFU (col. B)†	Friction p, Pa/m‡	Flow q, L/s	Load WSFU (col. A)*	Load WSFU (col. B)†	Friction p, Pa/m‡
12.7	13.8	0.18	1.0		172.3	0.36	2.5		609.4
19.0	19.9	0.37	2.5		109.3	0.75	7.3		392.9
25.4	26.0	0.64	5.5		80.7	1.29	22.5	7.0	287.9
31.8	31.1	0.99	11.5	5.0	63.0	1.98	58.0	15.5	224.9
38.1	38.2	1.43	28.5	8.0	52.5	2.87	109.0	38.0	182.8
50.8	50.4	2.43	82.0	26.0	37.8	4.87	261.0	138.0	132.4
63.5	62.6	3.75	172.0	75.0	29.4	7.50	474.0	356.0	102.9
76.2	74.8	5.36	300.0	178.0	23.1	10.72	750.0	692.0	84.0
102.0	99.2	9.40	636.0	544.0	16.8	18.80	1759.0	1759.0	58.8

* Col. A applies to piping which does not supply flush valves.

† Col. B applies to piping which supplies flush valves.

‡ Friction loss p corresponding to flow rate q for piping having fairly smooth surface condition after extended service applying the formula

$$q = 4.57\, p^{0.546}\, d^{2.64}$$

Table 6-8*a*
SIZING TABLE BASED ON VELOCITY LIMITATION—U.S. CUSTOMARY UNITS
Schedule 40 Plastic Pipe (PE, PVC, and ABS)

		Velocity = 4 fps				Velocity = 8 fps			
Nominal size, in	Actual ID, in	Flow q, gpm	Load WSFU (col. A)*	Load WSFU (col. B)†	Friction p, psi/100 ft‡	Flow q, gpm	Load WSFU (col. A)*	Load WSFU (col. B)†	Friction p, psi/100 ft‡
½	0.622	3.8	1.5		6.8	7.6	3.7		24.2
¾	0.824	6.7	3.0		5.1	13.4	8.4		18.0
1	1.049	10.8	6.1		3.7	21.6	25.3	7.7	13.2
1¼	1.380	18.6	17.5	6.0	2.8	37.2	77.3	23.7	9.6
1½	1.610	25.4	37.0	9.3	2.3	50.8	132.3	52.0	8.2
2	2.067	41.8	93.0	29.8	1.7	83.6	293.0	171.6	6.1
2½	2.469	59.8	174.0	75.6	1.4	119.6	477.0	361.0	4.8
3	3.068	92.0	335.0	209.0	1.1	184.0	842.0	806.0	3.8
4	4.026	158.6	688.0	615.0	0.8	317.2	1930.0	1930.0	2.8

* Col. A applies to piping which does not supply flush valves.

† Col. B applies to piping which supplies flush valves.

‡ Friction loss p corresponding to flow rate q for piping having fairly smooth surface condition after extended service applying the formula

$$q = 4.57\, p^{0.546}\, d^{2.64}$$

Table 6-8*b*
SIZING TABLE BASED ON VELOCITY LIMITATION—SI UNITS
Schedule 40 Plastic Pipe (PE, PVC, and ABS)

		Velocity = 1.2 m/s				Velocity = 2.4 m/s			
Nominal size, mm	Actual ID, mm	Flow q, L/s	Load WSFU (col. A)*	Load WSFU (col. B)†	Friction p, Pa/m‡	Flow q, L/s	Load WSFU (col. A)*	Load WSFU (col. B)†	Friction p, Pa/m‡
12.7	15.8	0.23	1.5		142.9	0.47	3.7		508.6
19.0	20.9	0.42	3.0		107.2	0.84	8.4		378.3
25.4	26.6	0.68	6.1		77.8	1.36	25.3	7.7	277.4
31.8	35.1	1.17	17.5	6.0	58.8	2.34	77.3	23.7	201.7
38.1	40.9	1.60	37.0	9.3	48.3	3.20	132.3	52.0	172.3
50.8	52.5	2.63	93.0	29.8	35.7	5.27	293.0	171.6	128.2
63.5	62.7	3.77	174.0	75.6	29.4	7.54	477.0	361.0	100.9
76.2	77.7	5.80	335.0	209.0	23.1	11.60	842.0	806.0	79.9
102.0	102.3	10.00	688.0	615.0	16.8	20.01	1930.0	1930.0	58.8

* Col. A applies to piping which does not supply flush valves.

† Col. B applies to piping which supplies flush valves.

‡ Friction loss p corresponding to flow rate q for piping having fairly smooth surface condition after extended service applying the formula

$$q = 4.57\, p^{0.546}\, d^{2.64}$$

the velocity limitations which should be observed. Included are almost all one- and two-family dwellings, most multiple dwellings up to at least three stories in height, and a considerable proportion of commercial and industrial buildings of limited height and area, where supplied from a source at which the minimum available pressure is no less than 40 psi (276 kPa). Under such conditions, the available pressure generally is more than enough for overcoming static head and ordinary pipe friction losses, and pipe friction is not an additional factor to consider in sizing.

Simplified Method Based on Velocity Limitation This method is based solely on the application of velocity limitations that are (1) recognized as good engineering practice; (2) authoritative recommendations issued by manufacturers of piping materials regarding proper use of their products in order to achieve durable performance and avoid failure in service, especially in water areas where the supply is aggressively corrosive; and (3) general recommendations for minimizing the cost of pumping where water supply pumps are provided. These limitations have been detailed in the section "Limitation of Velocity."

STEP-BY-STEP PROCEDURE FOR THE SIMPLIFIED SIZING METHOD

For sizing water supply systems in relatively low buildings, a simplified method has been developed in a step-by-step procedure which may be applied in the design of modern buildings. The procedure consists of seven steps. They are as follows:

1. Obtain all information necessary for establishing a proper basis for sizing the system. Properness of the basis for sizing is contingent upon accuracy and reliability of the information applied. Such information should be obtained from responsible parties and appropriate local authorities recognized as sources of the necessary information. See the preceding discussion under the subject "Preliminary Information."

2. Provide a schematic elevation of the complete water supply system. Show all piping connections in proper sequence and all fixture supplies. Identify all fixtures and risers by means of appropriate letters, numbers, or combinations thereof. Specially identify all piping conveying water at a temperature above 150°F (66°C), and all branch piping to such water outlets as automatic flush valves, solenoid valves, pneumatic valves, or quick-closing valves or faucets. Provide on the schematic elevation all the

necessary information obtained as per step 1, and as discussed under the subject "Preliminary Information."

3. Mark on the schematic elevation, for each section of the complete system, the hot and cold water loads conveyed thereby in terms of water supply fixture units in accordance with Table 6-14.
4. Mark on the schematic elevation, adjacent to all fixture units notations, the demand in gallons per minute or liters per second corresponding to the various fixture unit loads in accordance with Table 6-2.
5. Mark on the schematic elevation, for appropriate sections of the system, the demand in gallons per minute or liters per second for outlets at which demand is deemed continuous, such as outlets for watering gardens, irrigating lawns, air-conditioning apparatus, refrigeration machines, and other similar equipment using water at a relatively continuous rate during peak demand periods. Add the continuous demand to the demand for intermittently used fixtures, and show the total demand at those sections where both types of demand occur.
6. Size all individual fixture supply pipes to water outlets in accordance with the minimum sizes permitted by regulations. Minimum fixture supply pipe sizes for common plumbing fixtures are given in Table 6-15.
7. Size all other parts of the water supply system in accordance with velocity limitations recognized as good engineering practice, with velocity limitations recommended by pipe manufacturers for avoiding accelerated deterioration and failure of their products under various conditions of service, and with velocity limitations generally recommended for minimizing the cost of pumping where water supply pumps are provided. [Sizing tables (Tables 6-3 to 6-8) based on such velocity limitations and showing permissible loads in terms of water supply fixture units for each size and kind of piping material have been provided and may be applied advantageously in this step.]

APPLICATION OF SIMPLIFIED METHOD TO ILLUSTRATIVE PROBLEMS

Problem 1

Draw a schematic elevation, and size the piping of the following water distributing system using the simplified sizing method:

A one-family dwelling, two stories and cellar in height, is to be supplied by direct pressure from an 8-in public water main in which the certified minimum pressure available is 50 psi. Top floor fixture outlets are 20 ft above the public main and require 8 psi flow pressure for satisfactory operation.

Authoritative reports indicate that the public water supply has a pH of 7.2, carbon dioxide content of 7 ppm, and a solids content of 90 ppm. Records show no significant corrosion of copper by the water up to 150°F.

Copper water tube with wrought copper fittings has been selected, and is to be of type K for the water service and of type L for inside the building.

Water supply for the premises is to be metered at the point of entry by a disk-type meter. The system is to be of the upfeed riser type. A 52-gal automatic hot water storage heater, having a rated pressure loss of 1.5 psi at peak demand, is to provide 140°F tank controlled hot water supply.

The most extreme run of piping from the public main to the highest and most remote outlet is 180 ft in developed length, consisting of the following: 100 ft of water service, 45 ft of cold water piping from the water service valve to the hot water storage heater, and 35 ft therefrom to the top floor hot water outlet at the shower. Plans of the entire water supply system are available.

Fixtures provided on the system are as follows:

> Cellar—an automatic laundry washing machine, two hose bib outlets (only one to be used at any one time) at the outside of the building for lawn watering, and one valved outlet for supplying the house heating boiler
>
> First floor—a kitchen equipped with a sink and a domestic dishwasher, and a powder room containing one lavatory and one water closet with flush tank
>
> Second floor—two bathroom groups, one containing a lavatory, bathtub with shower above, and a water closet with flush tank, and a second containing a lavatory, shower stall, and a water closet with flush tank

Solution to Problem 1

Step 1 All information required for establishing a proper design basis has been obtained from appropriate sources.

Step 2 A schematic elevation of the building water supply system is shown in Fig. 6-1. This drawing was developed using the plans of the

DESIGN BASIS

PIPING:
Copper water tube with wrought copper fittings, type K for water service, type L for inside building.

PUBLIC SUPPLY SYSTEM:
8" public main located in street in in front of building. Certified minimum available pressure is: 50 psi.

WATER CHARACTERISTICS:
Records show no significant corrosion of copper by the water up to 150 °F.

WATER ANALYSIS:

pH	7.2
CO_2	7 ppm
Dissolved solids	90 ppm

ELEVATIONS:

Curb (as datum)	0.00'
Public main	-4.00'
Cellar floor	-7.00'
First floor	+2.00'
Second floor	+12.00'
Highest outlet	+15.00'

DEVELOPED LENGTHS OF PIPING FROM PUBLIC MAIN TO HIGHEST AND FARTHEST OUTLET:
Sections: PM-A=100'; A-B=40'; B-C=5'; C-D=5'; D-E=10'; E-F=10'; F-G=10'; TOTAL DL=180'.

HOT WATER TEMPERATURE:
140°F system, tank control.

DESIGN VELOCITY LIMITS:
8 fps for all piping, except 4 fps for branches to quick-closing valves (noted by *).

KEY FOR NOTATIONS MADE IN SIZING PROCEDURE
Load in water supply fixture units is shown unenclosed.
Load in gpm of flow is shown in ( ).
Continuous load in gpm of flow is shown in (_).
Sizes selected are shown in ["].

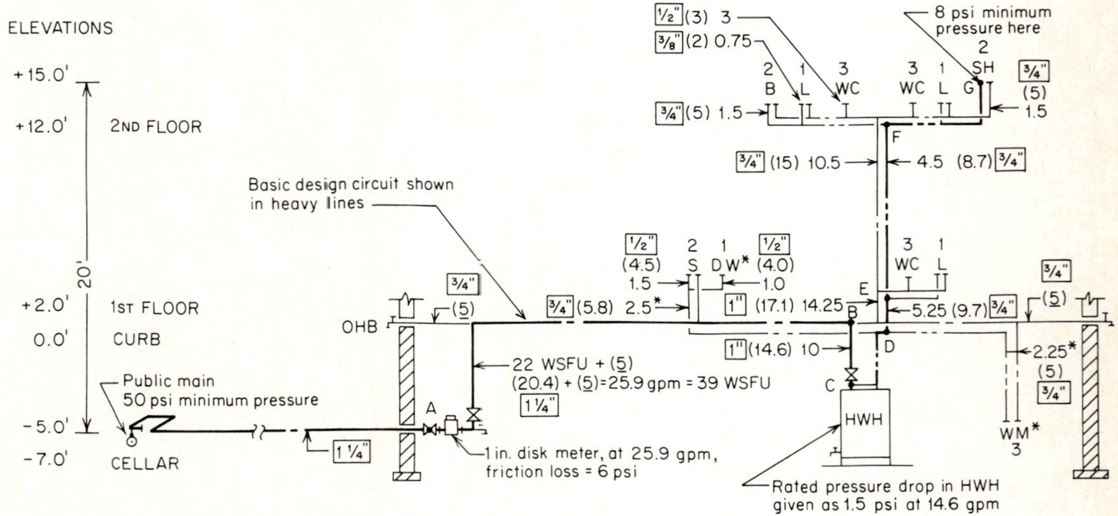

Fig. 6-1 Water supply system design sheet for Prob. 1.

system. All piping connections have been shown in proper sequence as per the plans, and the developed lengths of each section of the basic design circuit determined therefrom. All information required for establishing a proper design basis has been shown on the left side of the design sheet. Fixtures have been identified by letters, and those fixtures and branches having quick-closing valve outlets have been specially identified by means of an asterisk.

Step 3 For each section of the system, notations have been made showing the hot and cold water loads conveyed thereby in terms of water supply fixture units. Fixture unit values have been shown unenclosed by parentheses.

Step 4 Adjacent to fixture unit load notations, the demand in gallons per minute corresponding to such loads has been shown in parentheses. The demand in gpm was determined from Table 6-2, applying the values shown therein under the heading, "Supply Systems Predominantly for Flush Tanks," in view of the fact that no flush valves (Flushometers) are involved in this water supply system.

Step 5 The continuous demand posed by the two outside hose bibs, only one of which is to be used at a time, has been shown on the design sheet. This demand has been specially underlined and included in parentheses as gallons per minute where added to fixture units of load. The normal demand posed by a hose bib was obtained from Table 6-1.

Step 6 All individual fixture supply pipes to water outlets have been sized on the design sheet in accordance with the minimum sizes shown under the subject, Standard Code Regulations, in Table 6-15.

Step 7 All other parts of the system have been sized in accordance with the velocity limitations established for this system as the proper basis for design, i.e., 8 fps for all piping, except 4 fps for branches to quick-closing valves as noted by asterisks on the design sheet. Sizing was done in accordance with total fixture units of load corresponding to total demand in each section. For those sections of the cold water header in the cellar which convey both demand of intermittently used fixtures and continuous demand of a hose bib, the total demand in gallons per minute was converted to equivalent water supply fixture units of load and proper sizes were determined therefor, although proper sizing could also have been done simply on the basis of demand rates in gallons per minute. Sizing was done using Tables 6-3 through 6-8, and specifically those tables dealing with copper water tube, type K, for sizing the water service pipe, and with copper water tube, type L, for sizing piping inside the building.

SUPPLEMENTARY CHECK RE PIPE FRICTION LOSSES IN BDC

To prove the adequacy of the simplified sizing method as applied in this problem, a supplementary check has been made regarding the total pipe friction losses in the main lines and risers for the longest run of piping from the public water main to the highest and most remote water outlet. This run of piping was shown in heavy lines on the design sheet and letters marked thereon to denote the various sections of the run, the BDC.

The certified minimum pressure available at the public main was given as 50 psi. After appropriate deductions therefrom, the maximum pressure available for pipe friction losses during peak demand in the BDC was determined to be 25.8 psi. The appropriate deductions were as follows: 8.7 psi for static head loss due to rise from the public main to the highest outlet, 8.0 psi minimum pressure required at the top outlet, 6.0 psi friction loss through the 1-in disk meter at 25.9 gpm flow rate, and 1.5 psi friction loss rated for 14.6 gpm flow rate through the 52-gal hot water storage heater.

Table 6-9 shows the losses determined for each section of the circuit. Total pipe friction loss for the BDC was determined to be 18.6 psi, a total of 7.2 psi less than the 25.8 psi previously determined as the maximum pressure available for pipe friction losses during peak demand in the BDC. This proves that the sizes determined in this problem on the basis of velocity limitations exclusively are adequate.

Problem 2

Draw a schematic elevation, and size the piping of the following water distributing system using the simplified sizing method:

A nine-family multiple dwelling, three stories and cellar in height, fronts on a public street and is to be supplied by direct street pressure from a 10-in public water main in which the maximum pressure is 55 psi (379.5 kPa) and the minimum pressure available is 40 psi (276 kPa) as certified by a hydrant flow test. Top floor fixture outlets are 29 ft (87 m) above the public main and require 8 psi (55.2 kPa) flow pressure for satisfactory operation.

Water analysis of the public supply is as follows:

pH 7.1	Hardness 23 ppm
Iron 0.07 ppm	Alkalinity 13 ppm

Table 6-9
PIPE FRICTION LOSSES IN BASIC DESIGN CIRCUIT FOR PROB. 1
Copper Water Tube

Piping section	Tube type	Tube size, in	Developed length, ft	Equivalent length of fittings and valves, ft	Total equivalent length of section, ft	Flow rate q gpm	Friction (Charts 6-4*a*, 6-5*a*), psi/100′	Friction loss, psi
PM–A	K	1¼	100	1, 1¼″ tap at 1.9′ 1, 1¼″ gate valve at 1.0′ 4, 1¼″ elbows at 2.0′	110.9	25.9	8.4	9.3
A–B	L	1¼	40	1, 1¼″ gate valve at 1.0′ 4, 1¼″ tees, run at 0.5′ 4, 1¼″ elbows at 2.0′	51.0	25.9	7.4	3.8
B–C	L	1	5	1, 1″ gate valve at 1.0′ 4, 1″ elbows at 1.0′	10.0	14.6	7.2	0.7
C–D	L	1	5	2, 1″ tees, run at 0.5′ 4, 1″ elbows at 1.0′	10.0	14.6	7.2	0.7
D–E	L	¾	10	1, ¾″ tee, run at 0.5′ 3, ¾″ elbows at 1.0′	13.5	9.7	12.0	1.6
E–F	L	¾	10	1, ¾″ tee, branch at 2′	12.0	8.7	10.0	1.2
F–G	L	¾	10	1, ¾″ globe valve at 20′ 1, ¾″ tee, run at 0.5′ 4, ¾″ elbows at 1.0′	34.5	5.0	3.7	1.3
Totals			180 (100%)		241.9 (134%)			18.6

Chlorides 3 ppm
Dissolved salts 40 ppm
Dissolved air Supersaturated
Ryznar index +12.1 at 60°F and +10.5 at 140°F
Sulfates 9 ppm
Free CO_2 2 ppm
Langelier index −2.5 at 60°F and −1.7 at 140°F

Records show no significant corrosion of copper by the water up to 150°F.

Copper water tube with wrought copper fittings has been selected, and is to be of type K for the water service and of type L for inside the building.

Water supply for the premises is to be metered at the point of entry by a disk-type meter. The system is to be of the upfeed riser type. A horizontal hot water storage tank is to provide hot water to the entire building, and is to be equipped with automatic tank control of water temperature set for 140°F.

The most extreme run of piping from the public main to the highest and most remote outlet is 137 ft in developed length, consisting of the following: 50 ft of water service, 12 ft of cold water piping from the water service valve to the individual branch to the horizontal hot water storage tank, and 75 ft therefrom to the top floor hot water outlet at the kitchen sink. Plans of the entire water supply system are available.

The building has a full cellar and three above-grade stories, each 10 ft (3 m) in height from floor to floor. The first floor is 2 ft (0.6 m) above curb level in front of the building. The public water main is located beneath the street, 5 ft (1.5 m) from the curb and 4 ft (1.2 m) below curb level.

On each of the above-grade stories, there are three dwelling units. Each dwelling unit has a kitchen containing a countertop sink and an automatic dishwasher, and a bathroom equipped with a close-coupled water closet and flush tank combination, a lavatory, and a bathtub with shower head above.

In the cellar are the following: a laundry room containing two automatic laundry washing machines and a slop sink, an employees' toilet room equipped with a lavatory and a close-coupled water closet and flush tank combination, and a porter's room in which a slop sink is provided.

On the building exterior are two hose bibs, one located at the front wall and another at the rear wall. They are remote from each other and are not to be used simultaneously for lawn watering.

Solution to Problem 2

Step 1 All information required for establishing a proper design basis has been obtained from appropriate sources.

Step 2 A schematic elevation of the building water supply system is provided in Fig. 6-2. This drawing was developed using the plans of the system. All piping connections have been shown in proper sequence as per the plans, and the developed lengths of each section of the basic design circuit have been determined therefrom. Fixtures and risers have been identified by combinations of letters and numbers, and those fixtures and branches having quick-closing valve outlets have been specially identified by means of an asterisk. All information required for establishing a proper design basis has been shown on the left side of the design sheet.

Step 3 For each section of the system, notations have been made showing the hot and cold water loads conveyed thereby in terms of water supply fixture units. Fixture unit values have been shown unenclosed by parentheses.

Step 4 Adjacent to fixture unit load notations, the demand in gallons per minute corresponding to such loads has been shown in parentheses. The demand in gallons per minute was determined from Table 6-2, applying the values shown therein under the heading, "Supply Systems Predominantly for Flush Tanks," in view of the fact that no flush valves are involved in this water supply system.

Step 5 The continuous demand posed by the two outside hose bibs, only one of which is to be used at a time, has been shown on the design sheet. This demand has been specially underlined and included in parentheses as gallons per minute, or otherwise designated CL, where added to fixture units of load. The normal demand posed by a hose bib was obtained from Table 6-1.

Step 6 All individual fixture supply pipes to water outlets have been sized on the design sheet in accordance with the minimum sizes shown under Standard Code Regulations in Table 6-15.

Step 7 All other parts of the system have been sized in accordance with the velocity limitations established for this system as the proper basis for design, i.e., 8 fps for all piping, except 4 fps for branches to quick-closing valves as noted by asterisks on the design sheet. Sizing was done in accordance with total water supply fixture units of load corresponding to total demand in each section. For those sections of the cold water header in the cellar which convey both demand of intermittently used fixtures and continuous demand of a hose bib, the total demand in gallons per minute was converted to equivalent water supply fixture units of load and proper sizes were determined therefor, although

DESIGN BASIS

PIPING:
Copper water tube with wrought copper fittings, type K for water service, type L for inside building.

PUBLIC SUPPLY SYSTEM:
10" public main located in street in front of building. Certified available pressure is: 55 psi maximum, and 40 psi minimum.

WATER CHARACTERISTICS:
Records show no significant corrosion of copper by the water up to 150°F.

WATER ANALYSIS:

pH	7.1
Hardness	23.00 ppm
Iron	0.07 ppm
Alkalinity	13.00 ppm
Chlorides	3.00 ppm
Sulphates	9.00 ppm
Dissolved salts	40.00 ppm
Free CO_2	2.00 ppm

Dissolved air–Supersaturated.

Langelier index:
−2.5 at 60°F; −1.7 at 140°F.

Ryznar index:
+12.1 at 60°F; +10.5 at 140°F.

ELEVATIONS:

Curb (as datum)	0.00'
Public main	−4.00
Cellar floor	−8.00
First floor	+2.00
Second floor	+12.00
Third floor	+22.00
Highest outlet	+25.00

HIGHEST OUTLET PRESSURE:
8 psi required.

DEVELOPED LENGTHS OF PIPING FROM PUBLIC MAIN TO HIGHEST AND FARTHEST OUTLET:
Sections: PM-A = 50'; A-B = 12'; B-C = 8'; C-D = 4'; D-E = 7'; E-F = 6.5'; F-G = 12.5'; G-H = 12'; H-I = 3'; I-J = 10'; J-K = 10'; K-L = 2'.
TOTAL = 137'.

HOT WATER TEMPERATURE:
140°F System, tank control.

DESIGN VELOCITY LIMITS:
8 fps for all piping, except 4 fps for branches to quick-closing valves (noted by*).

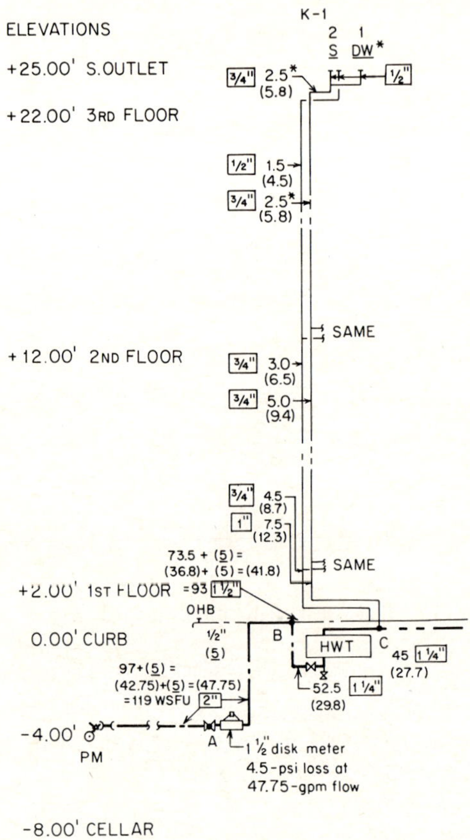

Fig. 6-2 Water supply system design sheet for Prob. 2.

KEY FOR NOTATIONS MADE IN SIZING PROCEDURE

Load in water supply fixture units is shown unenclosed.
Load in gpm of flow is shown in ( ).
Continuous load in gpm of flow is shown in (_).
Sizes selected are shown in ["].

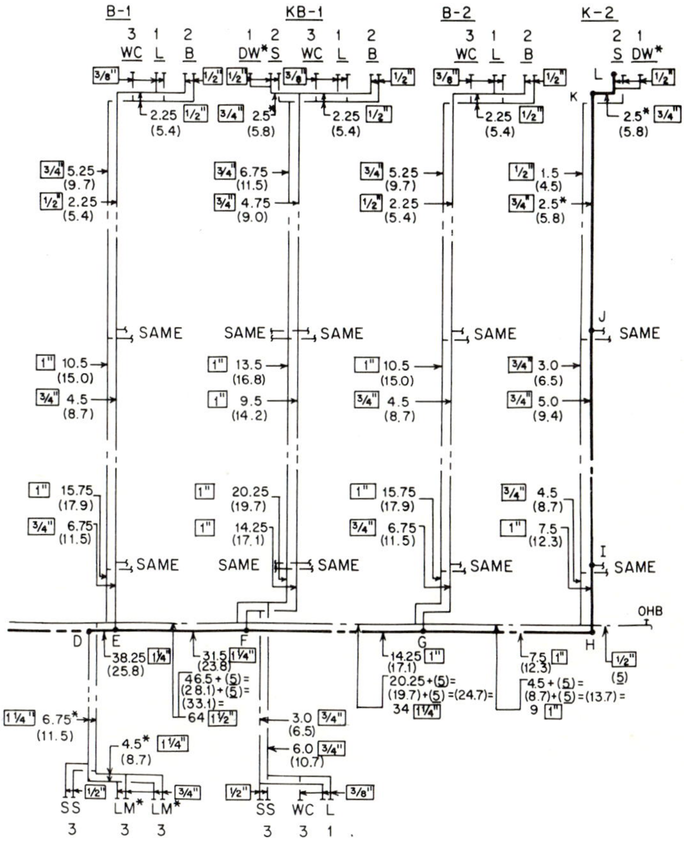

proper sizing could also have been done simply on the basis of demand rates in gallons per minute. Sizing was done using Tables 6-6*a* and *b* and 6-7*a* and *b*.

SUPPLEMENTARY CHECK RE PIPE FRICTION LOSSES IN BDC

To prove the adequacy of the simplified sizing method as applied in this second problem, a supplementary check has been made regarding the total pipe friction losses in the main lines and risers for the longest run of piping from the public water main to the highest and most remote water outlet. This run of piping was shown in heavy lines on the design sheet and letters marked thereon to denote the various sections of the run, the BDC.

The certified minimum pressure available at the public main was given as 40 psi. After appropriate deductions therefrom, the maximum pressure available for pipe friction losses during peak demand in the BDC was determined to be 14.2 psi. The appropriate deductions were as follows: 12.6 psi for static head loss due to rise from the public main to the highest outlet, 8.0 psi minimum pressure required at the top outlet, 4.5 psi friction loss through the 1½-in disk meter at 47.75 gpm flow rate, and 0.7 psi friction loss rated at 29.8 gpm flow rate through the horizontal hot water storage tank.

Table 6-10 shows the losses determined for each section of the circuit. Total pipe friction loss for the BDC was determined to be 11.3 psi, a total of 2.9 psi less than the 14.2 psi previously determined as the maximum pressure available for pipe friction losses during peak demand in the BDC. This proves that the sizes determined in this second problem on the basis of velocity limitation exclusively are adequate.

APPLICATION TO SYSTEMS IN HIGH BUILDINGS

This simplified method of sizing, based upon the velocity limitations which should be observed in design of building water supply systems, has much broader application than just to systems in one-, two-, and three-story buildings where ample excess pressure is available at the source of supply. These velocity limitations should be observed in all building water supply systems. Thus, the sizes determined by this method are the minimum sizes recommended for use in any case. Where pipe friction is an additional limiting factor to be considered in design, larger sizes may be required with respect to the sizing of main lines and risers.

DETAILED METHOD FOR SIZING SYSTEMS IN BUILDINGS OF ANY HEIGHT

For sizing water supply systems in buildings of any height, a detailed method has been developed in a step-by-step procedure which may be

PIPE FRICTION LOSSES IN BASIC DESIGN CIRCUIT FOR PROB. 2
Copper Water Tube

Piping section	Tube type	Tube size, in	Developed length, ft	Equivalent length of fittings and valves, ft	Total equivalent length of section, ft	Flow rate q gpm	Friction (Charts 6-4*a*, 6-5*a*), psi/100′	Friction loss, psi
PM–A	K	2	50	1, 2″ tap at 2.0′ 1, 2″ gate valve at 2.0′ 4, 2″ elbows at 2.0′	62.0	47.75	2.9	1.8
A–B	L	2	12	3, 2″ elbows at 2.0′ 2, 2″ Ts, branch at 7′	32.0	47.75	2.6	0.83
B–C	L	1¼	8	1, 1¼″ gate valve at 1.0′ 5, 1¼″ elbows at 2.0′ 2, 1¼″ Ts, branch at 4′	27.0	29.8	9.5	2.57
C–D	L	1¼	4	1, 1¼″ T, run at 0.5′	4.5	27.7	8.4	0.38
D–E	L	1¼	7	1, 1¼″ T, run at 0.5′	7.5	25.8	7.2	0.54
E–F	L	1¼	6.5	1, 1¼″ T, run at 0.5′	7.0	23.8	6.4	0.45
F–G	L	1	12.5	1, 1″ T, run at 0.5′	13.0	17.1	9.6	1.25
G–H	L	1	12	1, 1″ elbow at 1.0′	13.0	12.3	5.4	0.70
H–I	L	1	3	1, 1″ T, run at 0.5′	3.5	12.3	5.4	0.19
I–J	L	¾	10	1, ¾″ T, run at 0.5′	10.5	9.4	11.5	1.21
J–K	L	¾	10	1, ¾″ elbow at 1.0′	11.0	5.8	4.8	0.53
K–L	L	¾	2	1, ¾″ T, run at 0.5′ 1, ¾″ globe valve at 15′ 2, ¾″ elbows at 0.5′	18.5	5.8	4.8	0.89
Totals			137 (100%)		209.5 (152.9%)			11.34

applied in the design of modern buildings. The procedure consists of 16 steps. They are as follows:

1. Obtain all information necessary for establishing a proper basis for sizing the system. Properness of the basis for sizing is contingent upon accuracy and reliability of the information applied. Such information should be obtained from responsible parties and appropriate local authorities recognized as sources of the necessary information. See the preceding discussion under "Preliminary Information."

2. Provide a schematic elevation of the complete water supply system. Show all piping connections in proper sequence and all fixture supplies. Identify all fixtures and risers by means of appropriate letters, numbers, or combinations thereof. Specially identify all piping conveying water at a temperature above 150°F (66°C), and all branch piping to such water outlets as automatic flush valves, solenoid valves, pneumatic valves, or quick-closing valves or faucets. Provide on the schematic elevation all the necessary information obtained as per step 1, and as discussed under "Preliminary Information."

3. Mark on the schematic elevation, for each section of the complete system, the hot and cold water loads conveyed thereby in terms of water supply fixture units in accordance with Table 6-14.

4. Mark on the schematic elevation, adjacent to all fixture unit notations, the demand in gallons per minute or liters per second corresponding to the various fixture unit loads in accordance with Table 6-2.

5. Mark on the schematic elevation, for appropriate sections of the system, the demand in gallons per minute or liters per second for outlets at which demand is deemed continuous, such as outlets for watering gardens, irrigating lawns, air-conditioning apparatus, refrigeration machines, and other similar equipment using water at a relatively continuous rate during peak demand periods. Add the continuous demand to the demand for intermittently used fixtures, and show the total demand at those sections where both types of demand occur.

6. Size all individual fixture supply pipes to water outlets in accordance with the minimum sizes permitted by regulations. Minimum fixture supply pipe sizes for common plumbing fixtures are given in Table 6-15.

7. Size all other parts of the water supply system in accordance with velocity limitations recognized as good engineering practice, with velocity limitations recommended by pipe manufacturers for avoiding accelerated deterioration and failure of their products under various conditions of service, and with velocity limitations generally recommended for minimizing the cost of pumping where water supply pumps are provided. [Sizing tables based on such velocity limitations and showing permissible loads in terms of water supply fixture units for each size and kind of piping material have been provided (Tables 6-3 through 6-8) and may be applied advantageously in this step.]

8. Assuming conditions of no flow in the system, calculate the amount of pressure available at the topmost fixture in excess of the minimum pressure required at such fixture for satisfactory supply conditions. The calculated excess pressure is the limit to which friction losses may be permitted for flow during peak demand in the system. (1 ft of water column = 0.433 psi pressure, and 1 m of water column = 9.795 kPa pressure.)

9. Determine which piping circuit of the system is the basic design circuit (BDC) for which pipe sizes in main lines and risers should be designed in accordance with friction loss limits. This circuit is the most extreme run of piping through which water flows from the public main, or other pressure source of supply, to the highest and most distant water outlet. The basic design circuit (BDC) should be specially identified on the schematic elevation of the system.

10. Mark on the schematic elevation the rated pressure loss due to friction corresponding to the demand through any water meter, water softener, or instantaneous or tankless hot water heating coil that may be provided in the basic design circuit.

11. Calculate the amount of pressure remaining and available for dissipation as friction loss during peak demand through the pipe, valves, and fittings in the basic design circuit. Deduct from the excess static pressure available at the topmost fixture (determined in step 8) the rated friction losses for any water meters, water softeners, or instantaneous or tankless hot water heating coils provided in the basic design circuit (determined in step 10).

12. Calculate the total equivalent length of the basic design circuit. Pipe sizes established on the basis of velocity limitations in step 7 for main lines and risers must be considered just tenta-

tive at this stage but may be deemed appropriate for determining corresponding equivalent lengths of fittings and valves in this step.

13. Calculate the permissible uniform pressure loss for friction in piping of the basic design circuit. The amount of pressure available for dissipation as friction loss due to pipe, fittings, and valves in the circuit (determined in step 11) should be divided by the total equivalent length of the circuit (determined in step 12). This establishes the pipe friction limit for the circuit in terms of pressure loss, in psi/ft (Pa/m) for the total equivalent pipe length. Multiply this value by 100 in order to express the pipe friction limit in terms of psi per 100 ft of length.

14. Set up a sizing table showing the rates of flow, for various sizes of the kind of piping to be used, corresponding to the permissible uniform pressure loss for pipe friction calculated for the basic design circuit (determined in step 13). Such rates may be determined from an accurate pipe friction chart appropriate for the kind of piping to be used and for the effects upon the piping of the quality of water to be conveyed thereby for extended service.

15. Size all parts of the basic design circuit, and all other main lines and risers which supply water upward to the highest water outlets on the system, in accordance with the sizing table set up in step 14. Where sizes determined in this step are larger than those previously established in step 7 (based just on velocity limitations), the increased size is applicable for limitation of friction.

16. Due consideration must be given to the action of the water on the interior of the piping, and proper allowance must be made where necessary as a design consideration, such as where the kind of piping selected and the characteristics of the water conveyed are such that an appreciable buildup of corrosion products or hard-water scale may be anticipated to cause a significant reduction in bore of the piping system and inadequate capacity for satisfactory supply conditions during the normal service life of the system. A reasonable allowance in such cases may be considered to be provision of at least one standard pipe size larger than the sizes determined in the preceding steps. Where the water supply is treated in such manner as to avoid buildup of corrosion products or hard-water scale, no allowance need be made in sizing piping conveying such treated water.

APPLICATION OF DETAILED METHOD TO ILLUSTRATIVE PROBLEMS

Problem 3

Draw a schematic elevation, and size the piping of the following water distributing system using the detailed sizing method:

A 102-family multiple dwelling, seven stories and basement in height, fronts on a public street and is to be supplied by direct street pressure from an 8-in public water main located beneath the street in front of the building. The public system is of cast iron and a hydrant flow test indicates a certified minimum available pressure of 75 psi. Top floor fixture outlets are 65 ft 8 in above the public main and require 8 psi flow pressure for satisfactory operation.

Authoritative water analysis reports show that the public water supply has a pH of 6.9, carbon dioxide content of 3 ppm, dissolved solids content of 40 ppm, and is supersaturated with air. Reports show that the public water supply has no significant corrosion effect on red brass for temperatures up to 150°F.

Cement-lined cast iron, class B, corporation water pipe, valves, and fittings have been selected for the water service pipe. Red brass pipe, standard pipe size, has been selected for the water distributing system inside the building.

Water supply for the building is to be metered at the point of entry by a compound meter installed in the basement. The system is to be of the upfeed riser type. A horizontal hot water storage tank is to provide hot water to the entire building, and is to be equipped with automatic tank control of water temperature set for 140°F. The tank is to have a submerged heat exchanger.

The most extreme run of piping from the public main to the highest and most remote outlet is 420 ft in developed length, consisting of the following: 83 ft of water service, 110 ft of cold water piping from the water service valve to the hot water storage tank, and 227 ft of hot water piping from the tank to the top floor hot water outlet at the kitchen sink. Plans of the entire water supply system are available.

The building has a basement and seven above-grade stories. The basement floor is 3 ft 8 in below curb level, the first floor is 5.0 ft above curb level, and the public water main is 5.0 ft below curb level. Each of the above-grade stories is 9 ft 4 in in height from floor to floor. The highest fixture outlet is 3 ft above floor level.

Fixtures provided on the system for the occupancies are as follows:

1. There are 17 dwelling units on each of the second, third, fourth, fifth, sixth, and seventh floors; and each dwelling unit is provided with a sink and domestic dishwashing machine in the kitchen,

and a close-coupled water closet and flush tank combination, a lavatory, and a bathtub with shower head above in a private bathroom.

2. The first floor is occupied for administrative and general purposes, and has the following provisions for such occupancy: one flush-valve supplied water closet and one lavatory in an office toilet room; one flush-valve supplied water closet, one flush-valve supplied urinal and one lavatory in a men's toilet room; two flush-valve supplied water closets and one lavatory in each of two women's toilet rooms; a sink and domestic dishwashing machine in a demonstration kitchen; one sink in an office kitchen; one sink in a craft room; and two drinking fountains in the public hall.
3. The basement is occupied for building equipment rooms, storage, utility, laundry, and general purposes and has the following provisions for such occupancy: one flush-valve supplied water closet and one lavatory in a women's toilet room; one flush-valve supplied water closet, one lavatory, and one shower stall in a men's toilet room; one service sink and six automatic laundry washing machines in a general laundry room; one faucet above a floor drain in the boiler room; and one valve-controlled primary water supply connection to the building heating system.
4. At each story and in the basement, a service sink is provided in a janitor's closet in the public hall.
5. Four outside hose bibs (only two to be used at any time) are provided for lawn watering at appropriate locations on the exterior of the building.

Fixture arrangements are typical on the six upper floors of the building, and 24 sets of risers are provided. Of these, 5 sets are for back-to-back bathrooms, 2 sets are for back-to-back kitchens, 4 sets are for back-to-back kitchen and bathroom groups, 9 sets are for separate kitchens, 3 sets are for separate bathrooms, and one set is for a service sink on each floor above the basement. Fixtures on the first floor are connected to adjacent risers. Basement fixtures are connected to overhead mains, which also supply directly the four outside hose bibs.

Solution to Problem 3

Step 1 All information required for establishing a proper design basis has been obtained from appropriate sources.

Step 2 A schematic elevation of the building water supply system is provided in Fig. 6-3. This drawing was developed using the plans of the system. All piping connections have been shown in proper sequence as per the plans, and the developed lengths of each section of the basic design circuit have been determined therefrom. Fixtures and risers have been identified by combinations of letters and numbers, and those fixtures and branches having quick-closing valve outlets have been specially identified by means of an asterisk. All information required for establishing a proper design basis has been shown on the left side of the design sheet.

Step 3 For each section of the system, notations have been made showing the hot and cold water loads conveyed thereby in terms of water supply fixture units. Fixture unit values have been shown unenclosed by parentheses.

Step 4 Adjacent to fixture unit load notations, the demand in gallons per minute corresponding to such loads has been shown in parentheses. The demand in gallons per minute was determined from Table 6-2, applying the values shown therein under the heading, "Supply Systems Predominantly for Flush Tanks," for all piping except for short branch piping which supplies water to water closets and urinals equipped with flush valves on the first floor and in the basement.

Step 5 The continuous demand posed by the four outside hose bibs, only two of which are to be used at any time, has been shown on the design sheet. This demand has been specially underlined and included in parentheses as gallons per minute, or otherwise designated CL, where added to fixture units of load. The normal demand posed by a hose bib was obtained from Table 6-1.

Step 6 All individual fixture supply pipes to water outlets have been sized on the design sheet in accordance with the minimum sizes shown under Standard Code Regulations in Table 6-15.

Step 7 All other parts of the system have been sized in this step in accordance with the velocity limitations established for this system as the proper basis for design, i.e., 8 fps for all piping, except 4 fps for branches to quick-closing valves as noted by asterisks on the design sheet. Sizing was done in accordance with total fixture units of load corresponding to total demand in each section. For those sections of the cold water header in the basement which convey both demand of intermittently used fixtures and continuous demand of hose bibs, the

DESIGN BASIS

PIPING:
Cement-lined cast iron, class B, corporation water pipe, valves and fittings for water service pipe. Red brass pipe, standard pipe size, for water distributing system within the building.

PUBLIC SUPPLY SYSTEM:
8" cast iron public water main located in street in front of building. Hydrant flow test indicates a certified minimum available pressure of 75 psi.

WATER CHARACTERISTICS:
Authoritative reports show that the public water supply has no significant corrosion effect on red brass for temperatures up 150°F.

WATER ANALYSIS:

pH	6.9
CO_2	3.0 ppm
Dissolved solids	40.0 ppm
Dissolved air – saturated.	

ELEVATIONS:

Curb (as datum)	0.00'
Public main	-5.00'
Basement floor	-3.67'
First floor	+5.00'
Second floor	+14.33'
Third floor	+23.00'
Fourth floor	+31.67'
Fifth floor	+40.33'
Sixth floor	+49.00'
Seventh floor	+57.67'
Highest outlet	+60.67'

HIGHEST OUTLET PRESSURE:
8 psi required.

DEVELOPED LENGTHS OF PIPING FROM PUBLIC MAIN TO HIGHEST AND FARTHEST OUTLET:
Sections: PM-A = 83'; A-B = 30'; B-C = 10'; C-D = 10'; D-E = 10'; E-F = 10'; F-G = 20'; G-H = 30'; H-I = 15'; I-J = 15'; J-K = 15'; K-L = 12'; L-M = 12'; M-N = 12'; N-O = 12'; O-P = 12'; P-Q = 12'; Q-R = 12'; R-S = 12'; S-T = 12'; T-U = 16.65'; U-V = 8.67' V-W = 8.67'; W-X = 8.67'; X-Y = 8.67'; Y-Z = 12.67'
Total developed length of basic design circuit = 420'.

HOT WATER TEMPERATURE:
140°F system, tank control.

DESIGN VELOCITY LIMITS:
8 fps for all piping, except 4 fps for branches to quick-closing valves (noted by*).

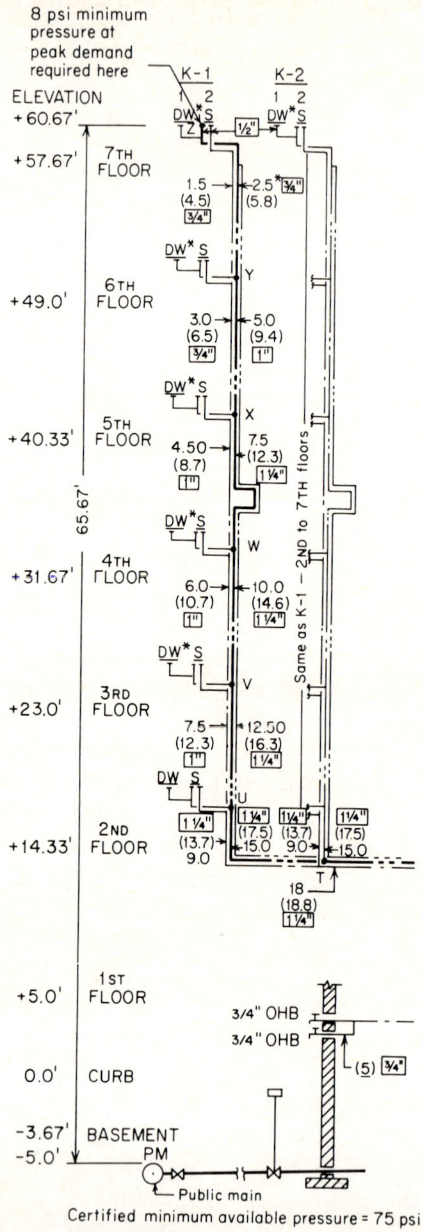

Fig. 6-3 Water supply system design sheet for Prob. 3.

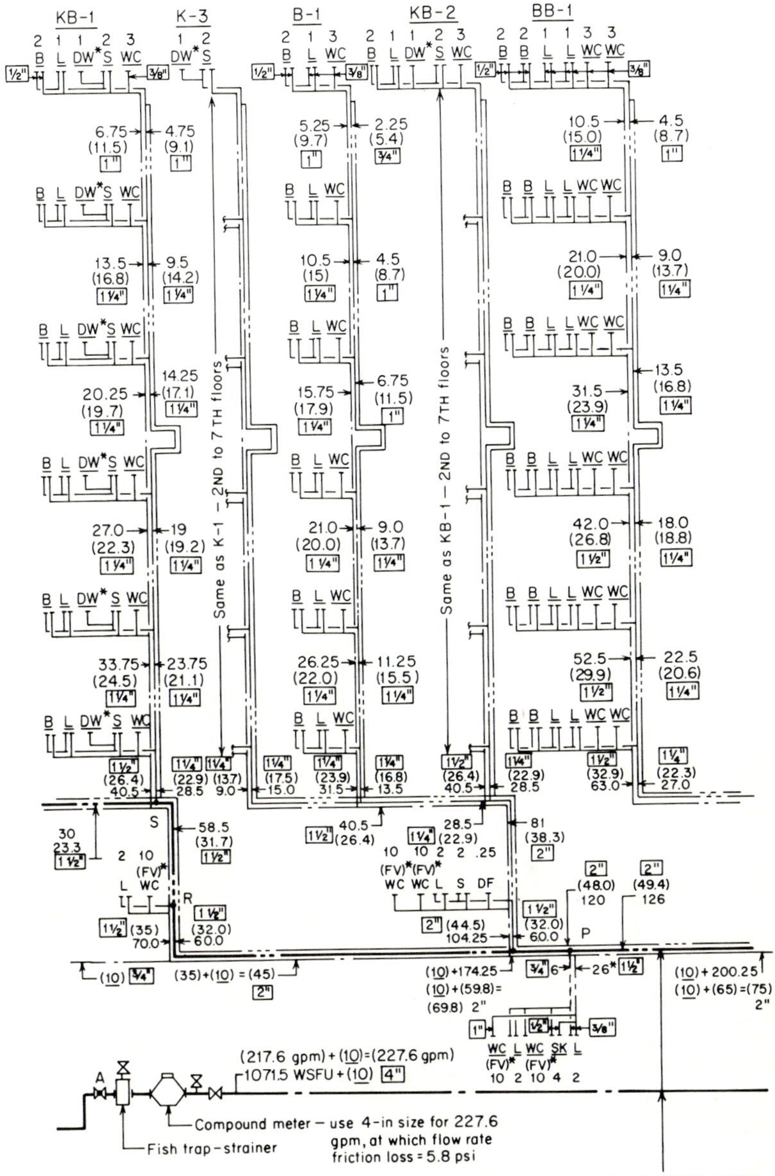

BASIC DESIGN CIRCUIT SHOWN IN HEAVY LINES

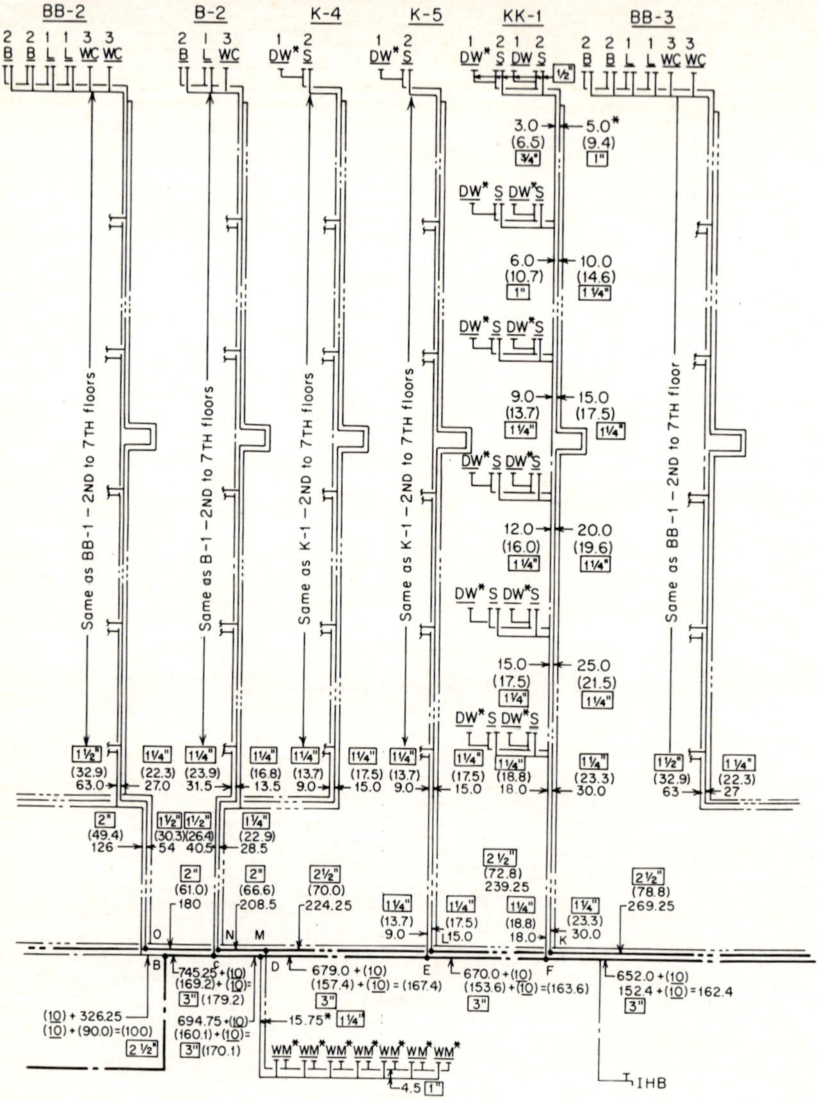

Fig. 6-3 (*Continued*)

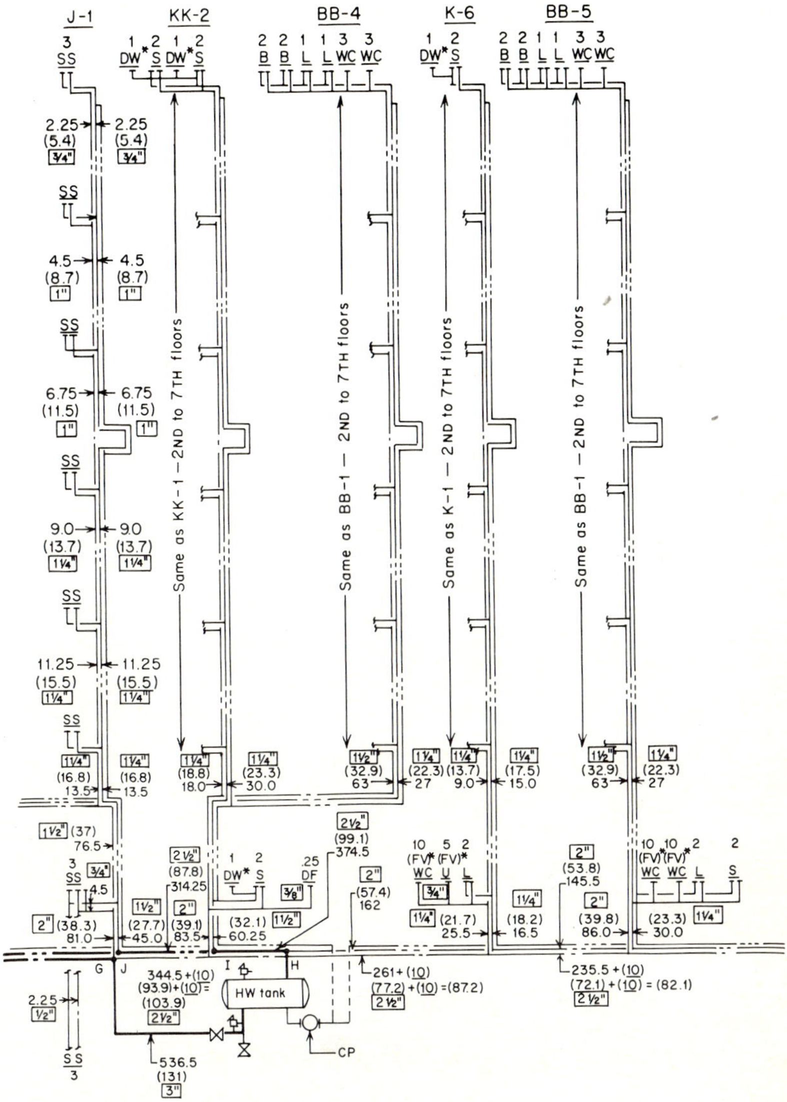
J-1
KK-2
BB-4
K-6
BB-5
3
SS
1 2 1 2
DW* S DW* S
2 2 1 1 3 3
B B L L WC WC
1 2
DW* S
2 2 1 1 3 3
B B L L WC WC
2.25
(5.4)
3/4"
SS
4.5
(8.7)
1"
SS
6.75
(11.5)
1"
SS
9.0
(13.7)
1¼"
SS
11.25
(15.5)
1¼"
SS
1¼"
(16.8)
13.5
Same as KK-1 —2ND to 7TH floors
Same as BB-1 — 2ND to 7TH floors
Same as K-1 — 2ND to 7TH floors
Same as BB-1 —2ND to 7TH floors
1¼"
(18.8)
18.0
1¼"
(23.3)
30.0
1½"
(32.9)
63
1¼"
(22.3)
27
1¼"
(13.7)
9.0
1¼"
(17.5)
15.0
1½"
(32.9)
63
1¼"
(22.3)
27
1½" (37)
76.5
3
SS
3/4"
4.5
2" (38.3)
81.0
1½"
(27.7)
45.0
2½"
(87.8)
314.25
2"
(39.1)
83.5
1 2
DW* S
.25
DF
3/8"
(32.1) 1½"
60.25
2½"
(99.1)
374.5
2"
(57.4)
162
10 5 2
(FV)* (FV)*
WC U L
3/4"
1¼" (21.7)
25.5
1¼"
(18.2)
16.5
2"
(53.8)
145.5
2"
(39.8)
86.0
10 10 2 2
(FV)* (FV)*
WC WC L S
(23.3) 1¼"
30.0
G
J
I
H
344.5 + (10)
(93.9) + (10) =
(103.9)
2½"
HW tank
261 + (10)
(77.2) + (10) = (87.2)
2½"
235.5 + (10)
(72.1) + (10) = (82.1)
2½"
2.25
½"
SS
3
536.5
(131)
3"
CP

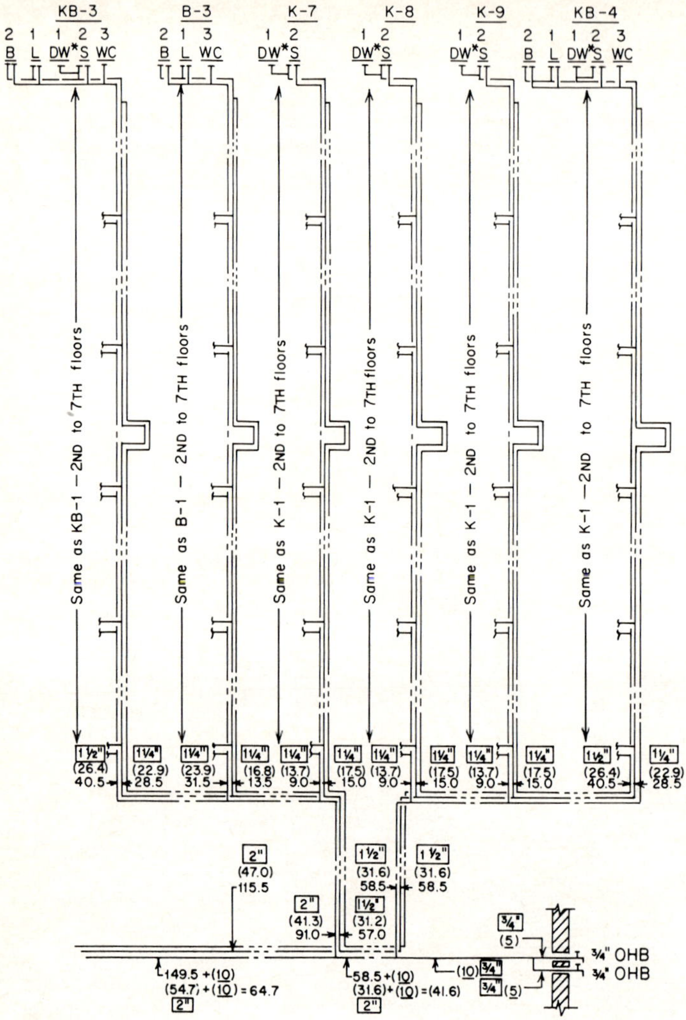

Fig. 6-3 (*Concluded*)

total demand in gallons per minute was converted to equivalent water supply fixture units of load and proper sizes were determined therefor, although proper sizing could also have been done simply on the basis of demand rates in gallons per minutes. Sizing was done using Tables 6-3 through 6-8 specifically Tables 6-4*a* and 6-4*b*, for sizing piping inside the building.

Step 8 Assuming conditions of no flow in the system, the amount of excess pressure available at the topmost fixture in excess of the minimum pressure required at the fixture for satisfactory supply conditions was determined as follows:

$$\text{Excess pressure available} = 75 \text{ psi} - 8 \text{ psi} - (65.67 \times 0.433) = 38.6 \text{ psi}$$

Step 9 The basic design circuit of the water supply system for the building was specially identified and shown in heavy lines on the schematic elevation provided in step 2. For each of the 26 sections of the circuit,

Table 6-11
PRESSURE CALCULATIONS FOR BASIC DESIGN CIRCUIT

Minimum at public main	75.0 psi
Loss in rise to top outlet (65.67 ft × 0.433)	−28.4 psi
Static pressure at top outlet	46.6 psi
Minimum pressure at top outlet	− 8.0 psi
Excess static pressure at top outlet available for friction loss	38.6 psi
Friction loss through 4-in compound meter at 227 gpm flow rate (manufacturer's charts)	− 5.8 psi
	32.8 psi
Friction loss through horizontal hot water storage tank assumed for rated flow at 8 fps	− 0.7 psi
Maximum pressure remaining for friction in pipe, valves, and fittings	32.1 psi
Developed length of circuit from public main to top outlet	420 ft
Equivalent length for valves and fittings in circuit (based on sizes established on velocity limitation basis)	363 ft
Total equivalent length of circuit	783 ft
Maximum uniform pressure loss for friction in basic design circuit (32.1 psi/783 ft)	0.04 psi/ft or 4.0 psi/100 ft

the developed length was given on the design sheet as determined from the plans of the system.

Step 10 The rated pressure loss through the compound water meter selected for this system was determined from appropriate meter data to be 5.8 psi for peak demand flow rate of 227.6 gpm. This has been noted on the design sheet. The rated pressure loss for flow through the horizontal hot water storage tank, i.e., entrance and exit losses, may be assumed to be approximately 1.6 ft head or 0.7 psi.

Step 11 The amount of pressure available for dissipation as friction loss during peak demand through pipe, valves, and fittings in the basic design circuit is

$$38.6 - 5.8 - 0.7 = 32.1 \text{ psi}$$

Step 12 In step 7, tentative pipe sizes for the main lines and risers were established based on velocity limitations. Using such tentative sizes for the basic design circuit, corresponding equivalent lengths for valves and fittings were determined and added to the developed length to calculate the total equivalent length of the circuit. The equivalent length for valves and fittings was found to be 363.2 ft, which when added to the 420 ft developed length resulted in a total equivalent length for the basic design circuit of 783.2 ft.

Step 13 The maximum uniform pressure loss for friction in the basic design circuit is

$$32.1 \text{ psi}/783.2 \text{ ft} = 0.04 \text{ psi/ft or } 4.0 \text{ psi}/100 \text{ ft}$$

This is the pipe friction limit for the basic design circuit. It is to be applied for sizing all the main lines and risers supplying water to fixtures on upper floors of the building.

Step 14 In Table 6-12 flow rates have been tabulated, through various sizes of red brass pipe of standard pipe size, that correspond to the velocity limits of 4 and 8 fps, and to the friction limit of 4.0 psi/100 ft of total equivalent piping length. The values shown therein for velocity limitations were taken from the tables cited in step 7. The values shown therein for friction limitation were taken directly from Chart 6-2*a*, one of the accurate pipe friction charts presented earlier in this chapter. The chart applied to red brass pipe of standard pipe size and was appropriate in view of the water supply conditions and surface condition, "fairly smooth."

Table 6-12
SIZING TABLE FOR SYSTEM IN PROB. 3
Red Brass Pipe, Standard Pipe Size

Nominal pipe size, in	Velocity limit flow rate at $V=4$ fps: WSFU (col. A)	$V=4$ fps: gpm	$V=8$ fps: WSFU (col. A)	$V=8$ fps: gpm	Friction limit flow rate at 4.0 psi/100 ft, gpm
½	1.5	3.8	3.7	7.6	2.8
¾	3.0	6.6	8.4	13.2	5.8
1	6.3	11.1	26.4	22.0	11.7
1¼	16.8	18.3	75.0	36.6	22.5
1½	36.3	25.2	130.0	50.4	33.0
2	92.0	41.6	291.0	83.2	66.0
2½	181.0	61.2	492.0	122.4	112.0
3	335.0	92.0	842.0	184.0	288.0
4	685.0	158.0	1920.0	316.0	380.0

Note: Apply the column headed "Velocity limit, $V=4$ fps," to size branches to quick-closing valves. Apply the column headed "Velocity limit, $V=8$ fps," to all piping other than individual fixture supplies. Apply the column headed "Friction limit," just for sizing piping that conveys water to top floor outlets. Where two columns apply and two different sizes are indicated, select the larger size.

Step 15 All the main lines and risers on the design sheet have been subjected to sizing in accordance with the friction limitation for the basic design circuit. Where sizes determined in this step were larger than those previously determined in step 7 (based on velocity limitation), the increased size was noted directly on the design sheet. Increased sizes were made in all risers and in some parts of the main lines in this system. As an example, in the basic design circuit the sizes of many sections were increased and may be specifically cited as follows: sections J–K, K–L, and L–M were increased from 2-in to 2½-in; sections O–P and P–Q were increased from 1½-in to 2-in; sections Q–R, R–S, and S–T were increased from 1¼-in to 1½-in; sections T–U, U–V, and V–W were increased from 1-in to 1¼-in; section W–X was increased from ¾-in to 1¼-in; and section X–Y was increased from ¾-in to 1-in.

Step 16 From the characteristics of the water supply stated in the problem, it is recognized that the water is relatively noncorrosive and nonscaling. Consequently, there is no need for additional allowance in sizing in this case.

Problem 4

Provide a sizing table showing flow rates corresponding to the appropriate velocity and friction limits for the following water distributing system:

An office building, 20 stories in height, is designed to be supplied with water by means of two gravity water supply tanks elevated at the same height above the building roof. The tanks are to supply a standpipe system through connections at the tank bottoms and the building's potable water distributing system through connections 2 ft 3 in above the tank bottoms. Three pumps, one of which is reserved for standby service, draw water under direct pressure from two public water mains and, at the additional head required, supply the tanks at a rate double the peak demand on the system. The tanks and pumping equipment are designed to maintain as a normal operating condition a minimum tank water line 41 ft above the highest fixture outlets at the top story. The minimum pressure required at these outlets is 8 psi.

Authoritative water analysis reports show that the public water supply has a pH of 7.0, carbon dioxide content of 4 ppm, dissolved solids content of 55 ppm, and is supersaturated with air. Records show no significant corrosion of copper by the water up to 150°F.

Copper water tube, type L, with wrought copper fittings has been selected for this system. Joints are to be made with solder consisting of 95 percent tin and 5 percent antimony.

A horizontal hot water storage tank with submerged heater is provided in the cellar and is to be equipped with an automatic tank control set for 140°F hot water temperature.

The most extreme run of piping measured from the water supply tank connection to the highest and most remote outlet is 600 ft in developed length, consisting of the following: 210 ft of downfeed cold water riser from the supply tank connections down to the cellar, 75 ft of cold water piping to the horizontal hot water storage tank in the cellar, 140 ft of hot water piping from the tank to the base of the most extreme riser in the cellar, and 175 ft of hot water riser piping to the highest hot water outlet.

Fixtures are provided on each floor for the occupancies. Toilet rooms for men and for women are provided on each floor. Water closets, urinals, and lavatories are provided in men's toilet rooms, and water closets and lavatories are provided in women's toilet rooms. Drinking fountains are provided on each floor at appropriate locations. Janitor's closets on each floor contain mop sinks. All water closets, urinals, lavatories, and drinking fountains are supplied by means of quick-closing devices, including automatic flush valves, self-closing faucets, and push-button quick-closing valves.

Solution to Problem 4 (Steps 8 through 16)

Step 8 Assuming conditions of no flow in the system, the amount of excess pressure available at the topmost fixture in excess of the minimum pressure required at the fixture for satisfactory supply conditions is as follows:

$$\begin{aligned}\text{Excess pressure available} &= (41\ \text{ft} \times 0.433\ \text{psi/ft}) - 8\ \text{psi}\\ &= 17.75 - 8 = 9.75\ \text{psi}\end{aligned}$$

Step 9 The problem states that the most extreme run of piping measured from the water supply tank connection to the highest and most remote outlet is 600 ft in developed length. This is the "basic design circuit' of the system.

Step 10 The rated pressure loss for flow through the horizontal hot water storage tank, i.e., entrance and exit losses, may be assumed for flow at 8 fps to be approximately 1.6 ft head or 0.7 psi.

Step 11 The amount of pressure available for dissipation as friction loss during peak demand through pipe, valves, and fittings in the basic design circuit is

$$9.75\ \text{psi} - 0.7\ \text{psi} = 9.05\ \text{psi}$$

Step 12 Total equivalent length of the basic design circuit, established tentatively based upon its developed length plus 50 percent as an appropriate assumed allowance for equivalent length of copper water tube fittings and valves, is as follows:

$$\begin{aligned}600 + (0.50 \times 600) &= 600 + 300\\ &= 900\ \text{ft}\end{aligned}$$

Step 13 The maximum permissible uniform pressure loss for friction in the basic design circuit is

$$9.05\ \text{psi}/900\ \text{ft} = 0.01006\ \text{psi/ft or } 1.0\ \text{psi}/100\ \text{ft}$$

Step 14 In Table 6-13 flow rates have been tabulated, through various sizes of copper water tube, type L, that correspond to the velocity limit of 4 fps, and to the friction limit of 1.0 psi/100 ft of total equivalent piping length. The values shown therein for the velocity limit of 4 fps were taken from the tables cited in step 7 of the detailed method. The values shown therein for the friction limit of 1.0 psi/100 ft were taken directly from Chart 5-5*a*, one of the accurate pipe friction charts pre-

Table 6-13
SIZING TABLE FOR SYSTEM IN PROB. 4
Copper Water Tube, Type L

Nominal pipe size, in	Velocity limit flow rate at $V=4$ fps		Friction limit flow rate at 1.0 psi/100 ft, gpm
	WSFU (col. B)	gpm	
¾		6.0	2.4
1		10.3	4.8
1¼	5.0	15.7	8.6
1½	8.0	22.8	13.5
2	26.0	38.6	28.0
2½	75.0	59.5	49.0
3	178.0	85.0	80.0
4	544.0	149.0	170.0

Note: Apply the column headed "Velocity limit" to size all branches from main lines and risers. Apply the column headed "Friction limit" to size main lines and risers supplying upper floor outlets.

sented earlier in this chapter. This chart was appropriate in view of the water supply conditions and the surface condition "fairly smooth." All branches from main lines and risers should be sized in accordance with the velocity limit. All main lines and risers should be sized in accordance with the friction limit.

STANDARD CODE REGULATIONS

Water Service Pipe The water service pipe shall be of sufficient size to furnish an adequate supply of water to meet the requirements of the building at peak demand. No water service pipe shall be smaller than ¾ in nominal size, except that where direct supply flush valves are supplied at pressure from the water service pipe, such pipe shall be at least 1¼ in nominal size.

Minimum Available Pressure The distributing system shall be designed on the basis of the minimum available pressure at the water main or other source of water supply.

Demand Load The water supply demand load in the distributing system shall be based upon the number, type, and probable simultaneous use of the fixtures to be supplied.

Table 6-14
DEMAND LOAD OF FIXTURES

Fixture	Occupancy	Type of supply control	Load values assigned, water supply fixture units		
			Cold	Hot	Total
Water closet	Public	Flush valve	10		10
Water closet	Public	Flush tank	5		5
Urinal	Public	1″ (25.4 mm) flush valve	10		10
Urinal	Public	¾″ (19 mm) flush valve	5		5
Urinal	Public	Flush tank	3		3
Lavatory	Public	Faucet	1.5	1.5	2
Bathtub	Public	Faucet	3	3	4
Showerhead	Public	Mixing valve	3	3	4
Service sink	Offices, etc.	Faucet	2.25	2.25	3
Kitchen sink	Hotel, restaurant	Faucet	3	3	4
Drinking fountain	Offices, etc.	⅜″ (9.52 mm) valve	0.25		0.25
Water closet	Private	Flush valve	6		6
Water closet	Private	Flush tank	3		3
Lavatory	Private	Faucet	0.75	0.75	1
Bathtub	Private	Faucet	1.5	1.5	2
Shower stall	Private	Mixing valve	1.5	1.5	2
Kitchen sink	Private	Faucet	1.5	1.5	2
Laundry trays (1 to 3)	Private	Faucet	2.25	2.25	3
Combination fixture	Private	Faucet	2.25	2.25	3
Dishwashing machine	Private	Automatic		1	1
Laundry machine [8 lb (3.6 kg)]	Private	Automatic	1.5	1.5	2
Laundry machine [8 lb (3.6 kg)]	Public or general	Automatic	2.25	2.25	3
Laundry machine [16 lb (7.3 kg)]	Public or general	Automatic	3	3	4

Note: For fixtures not listed, loads should be assumed by comparing the fixture with one listed using water in similar quantities and at similar rates. The assigned loads for fixtures with both hot and cold water supplies are given for separate hot and cold water loads and for total load, the separate hot and cold water loads being three-fourths of the total load for the fixture in each case.

Sizing Procedure The procedure for sizing the piping of the water supply distributing system shall conform to good engineering practice. Adequate and approved design factors shall be used to determine pipe sizes.

The size of piping, other than individual fixture supply pipes, should be such that the velocity of water flow during maximum demand will not exceed 8 fps (2.4 m/s) for piping aboveground, within dwelling portions of buildings.

In determining the available pressure at outlets, proper allowance should be made for the pressure loss owing to friction in the piping.

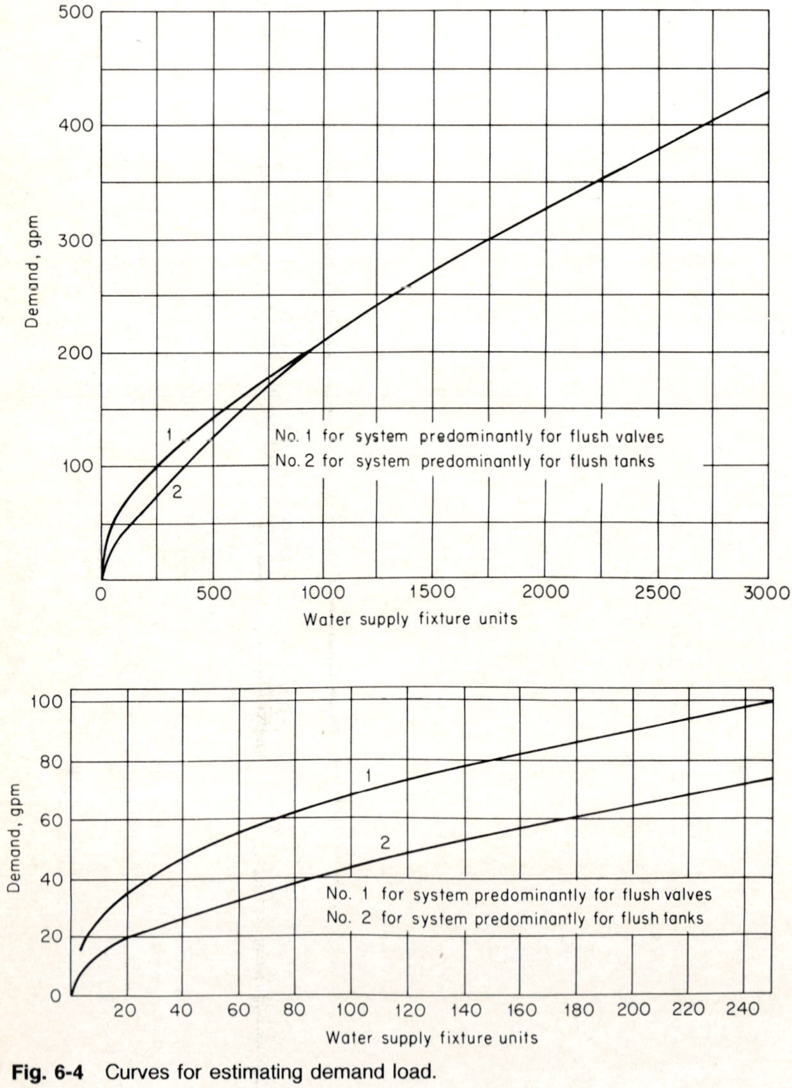

Fig. 6-4 Curves for estimating demand load.

Estimated Water Supply Demand Load For purposes of estimating the water supply load, the demand load values, in terms of water supply fixture units, for different plumbing fixtures under several conditions of service, are given in Table 6-14.

The estimated demand load for fixtures used intermittently on any supply pipe, in gallons per minute corresponding to the total number of water supply fixture units, is given in two charts in Fig. 6-4.

To estimate the total demand in gallons per minute, the demands for outlets such as hose connections and air-conditioning apparatus, which impose continuous demand during periods of heavy use, should be calculated separately and added to the demand for fixtures used intermittently.

Fixture Supply Pipes The minimum size of fixture supply pipes shall be as shown in Table 6-15. The fixture supply pipe shall terminate not more than 30 in (762 mm) from the point of connection to the fixture and in every instance shall extend into the room through the floor or wall adjacent to the fixture.

Risers The minimum size of water supply risers shall be not less than ½ in (12.7 mm) when no flush valves are supplied directly therefrom, nor less than

Table 6-15
MINIMUM SIZE OF FIXTURE SUPPLY PIPES

Fixture or device	Size in	Size mm
Bathtub	½	12.7
Combination sink and laundry tray	½	12.7
Drinking fountain	⅜	9.5
Dishwashing machine (domestic)	½	12.7
Kitchen sink (domestic)	½	12.7
Kitchen sink (commercial)	¾	19.0
Lavatory	⅜	9.5
Laundry tray (1, 2, or 3 compartments)	½	12.7
Shower (single head)	½	12.7
Sink (service, slop)	½	12.7
Sink (flushing rim)	¾	19.0
Urinal [1″ (25.4 mm) flush valve]	1	25.4
Urinal [¾″ (19.0 mm) flush valve]	¾	19.0
Urinal (flush tank)	½	12.7
Water closet (flush tank)	⅜	9.5
Water closet (flush valve)	1	25.4
Hose bib	½	12.7
Wall hydrant or sill cock	½	12.7

Note: For fixtures not listed in the above table, the minimum size of fixture supply pipes shall be the same as given in the table for comparable fixtures.

1¼ in (31.7 mm) when one or two flush valves are supplied directly therefrom, nor less than 1½ in (38.1 mm) when three or more flush valves are supplied directly therefrom.

Allowance for Water Characteristics Where the piping is subject to excessive corrosion or deposits because of characteristics of the water, pipe sizes shall be at least one standard pipe size larger than the minimums specified in this section.

7

HOT WATER SUPPLY SYSTEMS

OBJECTIVES OF STANDARD DESIGN

Standard design of hot water supply systems has the following objectives: (1) compliance with applicable regulations; (2) safe, satisfactory quality of service performance; (3) efficient utilization of available, economical sources of heat; (4) economy and durability of installation; and (5) economy and convenience of operation and maintenance of the completed installation.

In the latter portion of this chapter are presented provisions dealing with hot water supply systems which have become relatively common in standard code regulations throughout the United States. Very often these regulations are not found in just a single code, but rather in several different codes or sets of ordinances. They may be found scattered frequently amongst the provisions of building codes, plumbing codes, and sanitary or health codes. However, all applicable regulations should be observed in designing hot water supply systems.

The service performance of systems should be both safe and of satisfactory quality. Safety must be built into the systems by provisions to protect against the hazards of excessive pressure and excessive temperature so as to prevent injury to persons and damage to property which might otherwise occur. Satisfactory quality of service performance may be attained by designing the system so as to satisfy the demand for hot water at properly controlled volume, pressure, and temperature conditions.

Hot water temperature should be suitable for the service conditions required for safe use and proper function at fixtures. For most hot water supplied fixtures, at which persons come in contact with the water, the hot water temperature should be limited to a range of 130 to 140°F

(54 to 60°C) in order to avoid scalding the skin of users. But, for commercial dishwashing machines and commercial laundry washing machines at which persons do not come in contact with the water, the temperature should be limited to a range of 180 to 190°F (82 to 88°C), for sanitizing. In another type of machine use, domestic automatic laundry machines, manufacturers frequently recommend that hot water be supplied at 165°F (74°C).

Energy conservation in supplying hot water is a significant design consideration in large buildings. Satisfactory performance generally may be achieved with hot water supply systems set at slightly lower than maximum operating temperature range, such as 120°F (49°C). For example, lavatories in restrooms of public facilities should be equipped with outlet devices which limit the flow of hot water to a maximum of 0.5 gpm (0.032 L/s), and the outlet temperature should be limited to 110°F (43°C).

The most positive energy conservation measure is to insulate effectively all the hot water supply piping, recirculation piping, hot water tanks, and heaters so as to retain heat therein and minimize heat waste and loss to the building environment. The most advantageous time to do this is in original construction of the building.

Automatically operated and controlled systems are a practical necessity in order to maintain temperature control within the range established as standard for hot water supply systems. Nonautomatic systems generally are considered substandard as they are unreliable for maintaining proper control of hot water supply temperature. The use of nonautomatic systems is not recommended except in certain exceptional circumstances.

Efficient utilization of available, economical sources of heat is a most important consideration in deciding upon the type of equipment to install in any given building. In areas where one particular kind of fuel or energy is much cheaper than all others, the matter of savings in fuel or energy cost may be the decisive factor in the selection of water heating equipment. Where an abundance of waste steam or high-temperature steam condensate is available, suitable heat reclaiming equipment may be utilized. Similarly, innumerable other sets of conditions may govern the selection of both the method and equipment to be used for water heating.

The economy and durability of hot water supply systems are features which may be designed into such installations. Original cost of the various component parts of the system is but one of the factors to be considered. Durability of each component part should be consistent with that of the entire system, or special arrangements for ease of replacement of the less durable parts should be provided. The standard recommendation is that systems should be designed and installed so as to be durable, without need for frequent repairs or major replacements, and so as to minimize service interruptions.

Economy and convenience very frequently go hand in hand in matters of operation and maintenance of hot water supply systems. These are essentially features which are provided economically only in the design of original installations. To make such provisions afterward is almost always very costly and inconvenient for one reason or another. By following the recommendations of standard design as contained herein, reasonable economy of operation may be provided and convenient operation and maintenance can be achieved.

SAFETY DEVICES

Standard plumbing regulations require that hot water supply systems be provided with safety devices to relieve hazardous pressures and excessive temperatures. These regulations were enacted so as to prevent scalding, explosion or bursting of tanks, and injury to persons and damage to property, many instances of which have occurred in cases where systems were unprotected.

Pressures are considered hazardous when they exceed the water working pressures which the equipment and piping are designed to withstand. Although many individual items of equipment are designed for high pressures, the fact remains that most available standard items of equipment, including hot water storage tanks, are designed ordinarily for no more than 125 psi. Greater pressures may burst weak parts of systems and cause personal injury and property damage. In addition, deterioration of parts of systems as a result of corrosion and long periods of high-temperature operation increases the incidence of eventual failure and resultant effects.

Water expands as it is heated, the total amount of expansion depending upon the volume of water heated and the degree of temperature rise imparted. The density of water at 40°F (4.4°C) is 62.422 lb/ft^3, and at 140°F (60°C) it is 61.387 lb/ft^3. If a cubic foot of water is heated from 40 to 140°F (4.4 to 60°C), its volume under constant pressure conditions increases to 62.422/61.387 ft^3, or 1.0168 ft^3. Therefore, for this 100°F (55.6°C) temperature rise, it can be seen that the water expands 1.68 percent, or 1/60th of its original volume.

As an illustration, this may be applied to the case of a 60-gal hot water storage tank. The total expansion which results from heating it from a cold start, 40°F (4.4°C) to the normal hot water supply temperature, 140°F (60°C), is 1 gal (3.75 L). This is calculated by multiplying 60 gal by 1/60th.

Water is relatively incompressible. One pound per square inch of pressure is required to be applied to water to compress it just 1/300,-000th in volume. If all the water in a hot water supply system were heated through a 100°F (55.6°C) temperature rise under constant volume, confined conditions, the theoretical pressure rise would be 1/60th divided by 1/300,000th, or 5000 psi (34,500 kPa).

Although this maximum pressure rise is purely theoretical and would not be attained in practice because of distension of pipe walls and equipment, potentially hazardous pressure conditions can occur in any closed domestic hot water supply system, including a tankless heater system in which the hot water storage capacity of the heater coil may be just a fraction of the total storage capacity of the hot water supply system.

Where there is no check valve in the cold water supply line to water heating equipment, the pressure rise may automatically relieve itself as the hot water expands back into the cold water supply piping. This often is objectionable and may damage nonmetallic parts of cold water meters. In view of the fact that check valves may be installed on any system, and that pressure-reducing valves generally are provided on systems in high-water-pressure areas, positive means are necessary to relieve pressure before it rises to an excessive and hazardous amount.

A pressure relief valve should be installed in an effective location on every domestic hot water supply system so as to prevent development of hazardous pressures. The relief valve should be set to relieve at a pressure 25 psi (172 kPa) higher than the maximum service pressure under which the system may operate at any time, but in no case should the working pressure exceed 125 psi (862 kPa), the maximum working pressure at which many items of standard equipment are designed to serve, unless every part of the system is specifically designed for higher-pressure service.

Pressure relief valves should be of approved type and conform to recognized standards. They should be designed to shut off automatically after relieving excessive pressure and should be equipped with test levers so that they may be inspected and tested periodically. In selecting pressure relief valves, careful consideration should be given to ascertaining that their relief ratings are adequate in every case to match the equipment which they are intended to protect.

For constant effective service, pressure relief valves should be installed reasonably close to the heater or tank. It is mandatory that no intervening valve, check or other type, be installed between the pressure relief valve and the heater or tank. Preferably, a pressure relief valve should be installed in the cold water supply line to the heater or tank. This location tends to minimize the incidence of building up corrosion products and scale deposits on the relief valve seat and also permits cold rather than hot water to be discharged when the valve relieves excessive pressure.

However, in hard water areas, it frequently is found more advisable to locate the valve in the hot water supply piping about 3 to 4 ft from the heater or tank outlet. In this location, it has been reported that there is less incidence of hard water scale deposits building up on the valve seat and impairing valve operation.

Water temperatures in the hot water supply system are considered excessive and hazardous when they exceed 210°F (98.9°C). Under atmospheric pressure, water boils at 212°F (100°C). Under service pressure, water remains in liquid state until its temperature reaches the particular boiling point corresponding to the pressure at which it is confined in the system. Boiling points corresponding to 30, 50, and 70 psig pressure are, respectively, 274°F (134°C), 297°F (147°C), and 316°F (157°C).

When water is heated in a closed pressurized system to temperatures above 212°F (100°C), it becomes superheated. This can be extremely hazardous in view of the fact that when superheated water is discharged into the atmosphere, all its superheat above 212°F (100°C) is suddenly released and a portion of the water vaporizes or flashes into steam at atmospheric pressure.

This may result in the severe scalding of persons using faucets, as scalding may be produced by the application of moist heat at temperatures as low as 120°F (48.9°C), if continued for a sufficient period of time.

Sudden, severe pressure reductions in the system may cause rumbling noises because of the formation of small bubbles of flash steam in tanks, heaters, and piping when water temperatures are excessive. If a small leak develops in a hot water storage tank which contains water at a temperature above 212°F (100°C), the leakage will flash into a jet of steam. If the tank wall is weakened sufficiently, it may tear open or rupture with a violent explosive reaction.

The discharge of superheated water into the atmosphere may occur under uncontrolled conditions, as may exist when a hot water storage tank ruptures instantaneously. The steam released thereby may jet or drive the tank off its setting and, in some cases, may even cause it to fly through floors, walls, and roofs of the buildings. The number of instances where severe personal injury and property damage have occurred in the past are too numerous to mention.

The amount of energy released in this manner may be extremely great, varying directly with the degree of superheat above 212°F (100°C) in the water, and with the total weight of superheated water. The rate at which the energy release occurs varies directly with the rate at which superheated water is discharged to atmospheric conditions.

A description can be given of the amount of energy involved when water is superheated to 225°F (107°C). Each pound of water heated to 225°F (107°C) possesses 13 Btu of superheat (225 − 212°F). In a 60-gal hot water storage tank filled with water at 225°F (107°C), the total amount of superheat energy would be equal to 13 Btu × 60 gal × 8.3 lb/gal, or 6474 Btu. If this heat energy were converted to mechanical energy (1 Btu equals 778 ft/lb), it would be equal to 6474 Btu × 778 ft · lb/Btu, or 5,036,772 ft · lb.

This tremendous amount of energy that could be released at atmospheric pressure from water heated to 225°F (107°C) in a 60-gal hot water storage tank illustrates theoretically the great hazard inherent wherever superheated water can develop and be stored in significant volume.

Consequently, on every domestic hot water supply system equipped with a hot water storage tank, it is necessary to provide in an effective location an approved safety device to prevent water temperature from rising to a hazardous degree [i.e., no more than 210°F (98.9°C)].

A temperature relief valve or an energy shutoff device should be installed in an effective location in every hot water supply system so as to prevent development of hazardous water temperatures. Temperature relief valves should be designed to relieve water at a temperature of not more than 210°F (98.9°C), and energy shutoff devices should be designed to shut off the flow of energy to the heater at a water temperature of not more than 210°F (98.9°C).

Temperature relief valves should be of approved type and conform to recognized standards. They should be designed to shut off automatically after relieving excessive temperature and should be equipped with test levers so that they may be inspected and tested periodically. In selecting temperature relief valves, careful consideration should be given to ascertaining whether their relief ratings are adequate in every case to match the equipment which they are intended to protect.

Energy shutoff devices should be of approved type and conform to recognized standards. They should be designed to shut off the flow of energy to the heater when the water temperature becomes excessive. In selecting energy shutoff devices, careful attention should be given to ascertaining whether their performance ratings are adequate in every case to match the equipment which they are intended to protect.

For constant effective service of temperature relief valves and energy shutoff devices, their temperature-sensitive elements must be in contact with the hottest water in the system. It is mandatory that no intervening valve, check or other type, be installed between a temperature relief valve or energy shutoff device and the tank or heater. For temperature relief valves, the temperature-sensitive element should be located within the upper 6 in (150 mm) of the tank of an underfired hot water storage heater, above the hot water inlet of a tank equipped with a sidearm heater, and above the topmost heating element of an electric water heater. For energy shutoff devices, the temperature-sensitive element of an immersion type should be immersed in the hottest water in the system; in the case of a strap-on type, the temperature-sensitive element should be mounted on the tank wall sufficiently high so as to be responsive to the highest water temperature within the tank.

In small water heating equipment installations, combination pressure-temperature relief valves of adequate rated relief capacity are available and may be used in lieu of separate pressure and temperature relief valves. Both pressure and temperature elements of the valve should have adequate relief ratings to match the equipment served. The location of combination pressure-temperature relief valves should conform to the recommendations for temperature relief valves, that is, so that the temperature-sensitive element is in contact with the hottest water in the system.

In the case of a tankless heater submerged in a low-pressure steam or hot water heating system boiler—which is oil- or gas-fired and is equipped with the typical complement of automatic high limit controls—the possibility of heating the domestic hot water supply to 210°F (98.9°C) is relatively remote, if the boiler water temperature is 212°F (100°C) or less. This is true especially in view of the heat transfer rates and efficiencies involved.

The high limit controls on the boiler serve as safety devices to limit energy input of the boiler which may be transferred into water in the tankless heater by indirect heating. In addition, the volume of water heated and stored in the tankless heater is relatively small.

Although such protection may prevent the development of superheated water in tankless heaters, there is still the danger of persons being scalded by excessively hot water at outlets. To prevent this, it is a practical necessity to provide a tempering valve on hot water supply piping at the tankless heater outlet.

No piping from relief valve outlets should be directly connected with drainage or vent piping. Such connections should not be permitted for they constitute potential sources of pollution to the potable water supply system, and they conceal instances of continuous discharge from relief valves which do not shut off tightly. Where pipes from such relief outlets discharge into plumbing fixtures, an air gap should be provided in conformity with the minimum required air gaps for potable water outlets at plumbing fixtures. The relief outlets of relief valves should be connected with proper piping so as to discharge into a suitable plumbing fixture or, in cases where it is impractical to discharge into a fixture, so as to discharge onto a floor area in an unobjectionable manner. In no case should the relief outlet or relief pipe discharge so as to be a hazard, a potential cause of damage, or a nuisance.

Hot water storage tanks should be installed so that their pressure markings, showing the maximum allowable water working pressure, are in an accessible location for inspection. Such markings are provided in the interest of safety so that by inspection it may be determined that the tank is safe to use under the maximum pressure of operation.

WATER-HEATING METHODS AND HEATER TYPES

The direct heating method consists of heating water by direct contact with surfaces exposed to the high temperatures of fire and flue gases generated by the combustion of fuel, or by direct contact with electrically heated surfaces or immersed electric heating elements. Where direct heating is applied in this way, the temperature to which heating surfaces are subjected is relatively high.

The indirect heating method consists of heating water by contact with copper pipe coils which serve as a means of transferring or exchanging heat from high-temperature hot water or live steam to water in the hot water supply system. With the indirect heating method, heating surfaces are subjected to much lower temperature conditions than prevail generally with the direct method.

All water heaters, whether applying direct or indirect methods of heating, may be classified as either tankless or storage tank heaters. Tankless heaters are designed to heat cold water to standard hot water supply temperature in a single pass through the heater so that hot water may be piped directly from the heater to fixtures. By contrast, storage tank heaters require the use of tanks for storing hot water. Where water is heated in passing through a heater coil and then circulates into a storage tank, the heater is termed a *circulating storage water heater.*

In selecting the particular heating method and heater type to use for any given installation, consideration should be given to such important factors as (1) the availability of an economical source of fuel or heat, (2) the comparative cost of equipment, (3) capacity limitations of available equipment, (4) the temperature to which water will be heated, and (5) the hardness of the water supply. Each of these points should be kept in mind in deciding upon the most appropriate heating method and heater type to use.

Water temperature and hardness are especially important considerations in hard water areas. This is due to the fact that at temperatures above 140°F (60°C), lime and magnesia salts in hard water tend to precipitate from solution and deposit as scale on heating surfaces. Such deposits reduce efficiency in transferring heat to the water, accelerate clogging of water passes in heater coils, and in time may result in cracking or burning of heating surfaces. To prevent or minimize development of these conditions, the temperature to which water is heated may be limited by adjusting operating controls so that water temperature is maintained at not more than 140°F (60°C), or the water supply to the heater may be softened by passage through a water softener, or the water may be heated by the indirect method utilizing an appropriate type of heater.

Tankless heaters, whether heated directly or indirectly, are not recommended for use in areas where the hardness of the water supply exceeds 170 ppm or 10 grains per gal. Under such conditions, clogging of the heater coil occurs rapidly unless the water is adequately softened before passing through the heater. The rate at which clogging develops is much more rapid where tankless heaters are heated directly than where heated indirectly.

As a general rule, the direct method is applied principally in relatively small installations. This is due mainly to the availability of small automatic storage water heaters, produced as economical package units in gas, oil, and electric heater models. However, some large capacity units are available.

The indirect method of heating is applied in the type of heating equipment known as hot water or steam heat exchangers. These are commonly referred to as indirect water heaters. They may be used for any given installation regardless of size.

Indirect heaters are of three general types. One type is specifically designed for installation as a submerged coil unit in hot water storage tanks. With this type, live steam or high-temperature hot water flows through the coil, exchanging its heat to water in the storage tank.

A second type is designed for installation as a submerged coil unit in steam or hot water heating system boilers. In this case, the coil is submerged in high-temperature hot water in the heating boiler, transferring heat therefrom to water flowing through the coil.

A third type is designed for use as an individual external unit, having a casing which houses the coil. Live steam or high-temperature hot water is supplied to the casing of the unit so as to heat water flowing through the coil. This type is frequently installed directly alongside heating boilers for convenience in making connections between the boiler and the heater casing.

DIRECT HEATING EQUIPMENT INSTALLATIONS

The outstanding feature of direct heating equipment installations is that they are designed to burn fuel or to use electrical energy as a source of heat. Thus, they are produced as independent units, and there is no need to rely upon a supply of steam or high-temperature hot water as an available source of heat for domestic hot water heating.

These independent units are available as packages in gas, oil, and electric heater models. They include gas-fired and oil-fired water heaters installed in conjunction with storage tanks, gas-fired water heaters of the tankless type, and electric water heaters of both the tankless and storage tank types. Many such units are produced with completely automatic control features built into them so as to regulate the rate of heating water and to maintain a predetermined range of hot water temperature. Some are also provided with hot water tempering valves so as to furnish hot water at 180°F (82°C) for dishwashing or 165°F (74°C) for domestic dishwashing or laundry machine purposes, in addition to 140°F (60°C) supply for ordinary plumbing fixtures. All package units containing storage tanks are provided with effective insulation so as to minimize heat loss to the environment.

Most of the available direct heating units are of relatively small heating capacity suitable for small installations. Some large capacity units are available and may be used in multiple for meeting the capacity requirements of very large buildings. Consequently, size is not the limiting factor in this regard, but rather the comparative costs of heating domestic water by direct as compared with indirect methods poses limitations. These may be easily determined by cost comparisons for each individual case.

Flues or chimneys must be provided to convey and dispose of the products of combustion where the units are designed to burn fuel, such as coal-fired, oil-fired, and gas-fired water heaters. Electric water heaters do not need such provisions.

Tankless-type direct heaters have copper tube coils and possess a high degree of durability comparable with that of copper water tube of the same type used for potable water service. For the same reason, they provide rust-free hot water service and may be deemed suitable for high temperature water supply service, such as for dual temperature systems where 180°F (82°C) hot water is required for machine equipment and 140°F (60°C) for ordinary hot water supplied fixtures.

Storage-tank-type direct heaters are subject to considerable variation as to durability. This depends principally upon the particular type of tank material used, such as copper, monel, galvanized steel, or a special alloy; and it also depends upon the application of protective linings inside the tanks, or the installation of sacrificial anodes in the tanks to protect them against rapid corrosion. The corrosivity of the water supply should be considered in the selection of tank material. The temperature at which storage-tank-type heaters are operated is also of special significance where the tanks are not of highly durable material for high-temperature service. Corrosion effects in tanks are appreciably increased at temperatures above 140°F (60°C) owing to two facts: an increase in water temperature causes an increase in the electrical conductance of the water and thereby increases the rate of galvanic corrosion, and at about 140°F (60°C) carbonates in water break down releasing carbon dioxide which results in an increase in the rate of general or overall corrosion.

Where original cost of a building is a prime consideration, the use of independent, packaged direct heating equipment, complete with automatic controls, may be a source of appreciable savings. However, the cost of fuel or electrical energy generally is the most important single factor to be considered, and in areas where one type of fuel is especially low, it may be the determining factor with regard to the use of a particular type of direct heating equipment over all others. Similarly, where such cost is high, it may weigh against that type of equipment. The cost of providing flues or chimneys where fuel-burning equipment is installed must be considered in cost comparisons with other water-heating methods. In small installations, this cost may be minor, but it may assume considerable proportions in large installations.

Maintenance costs of both the tankless and circulating types of direct heaters are relatively minor except in areas where the water supply is comparatively hard, that is, where water hardness exceeds 6 grains. In such cases, unless the water is adequately softened, maintenance costs can become excessive as the result of hard water scale deposits accumulating in the heater coils, reducing their heat transfer efficiency and restricting flow.

The cost of maintaining storage tank heaters varies greatly depending upon the durability of the tanks and their susceptibility to corrosion. Mud and scale deposits in the bottom of such tanks must be removed regularly so as not to impair the performance of thermostats located in the lower section of the tanks, and so as to eliminate rumbling noises due to the development of flash steam in such deposits. The latter condition is prone to occur where high recovery heaters are installed in hard water areas unless the water supply is adequately softened.

Heating capacities of direct water heaters are generally given in terms of gallons per hour or gallons per minute, based upon a 60, 80, or 100°F (33.3, 44.4, or 55.5°C) temperature rise, or in terms of Btu per hour input and Btu per hour output. As one gallon of water weighs 8.3 lb, to raise it from 40 to 140°F (4.4 to 60°C) requires 830 Btu of heat. If a heater is assumed to be 83 percent efficient, an output of 830 Btu requires an input of 1000 Btu. Such efficiency assumptions are made by most manufacturers of direct water heaters. Hence, a recommended standard method for determining the normal rated heating capacity of direct water heaters is to divide the Btu input rating, stated on the heater, by 1000. This will give the normal rated capacity of the heater in terms of gallons of water heated through a 100°F (55°C) rise per hour.

The pressure drop occurring in tankless-type heaters is an important factor to be considered. These heaters are designed to heat water to normal service temperature in a single pass through the heater coils. To accomplish this, the coils are comparatively small in internal diameter and long in developed length, features which result in a relatively high pressure drop through the tankless heater at rated flow capacity. Where water pressure is low, this high pressure drop is very important and must be considered in designing the hot water supply system so as to ensure adequate volume and pressure at all hot water supply outlets on the system. But where water pressure is high, the pressure drop through a tankless heater is of less significance in providing for adequate supply conditions. For data on pressure drop through tankless heaters, it is recommended that this be obtained directly from the manufacturers of such equipment.

Pressure drop through storage heaters is much less than through tankless heaters. Only a small drop occurs when water is drawn at peak demand rate from a tank supplied through a bottom tapping, the drop being no more than that equivalent to three 90° standard elbows of the same size as the tank inlet and outlet tappings. Where the tank is supplied through a tapping in the top, having a long, smaller diameter dip tube extending down into the lower section of the tank, the pressure drop through the tank at peak demand rate is appreciably greater and may be assumed to be equal to that of at least ten 90° standard elbows of the same size as the tank tappings. In the case of a circulating type of direct water heater, the pressure drop is of concern insofar as it affects the rate of circulation of heated water from the heater to the storage tank.

Water piping connections to tankless-type heaters and automatic storage water heaters consist of connecting a branch from the cold water supply system to the heater inlet and connecting the hot water supply system to the heater outlet. Size of both cold and hot water piping connections should be the same as the heater tappings. Fuel supply piping connections of adequate size should be made to appropriate supply tappings of the equipment.

A gate, ball, or butterfly valve should be installed in the cold water supply connection to each heater installation so that flow thereto may be shut off when required. Means must be provided for emptying water from the lowest part of the piping of the heating equipment, and for emptying hot water storage tanks when necessary. For small installations, the means may be a drawoff faucet of adequate size, but for large installations a gate, ball, or butterfly valve of adequate size is recommended so that draining and flushing may be performed quickly and efficiently.

Heaters equipped with coils should be provided with suitable means for flushing sediment and loose scale out of heater coils, because all water supplies contain some sediment or solids content which tends to precipitate as the water temperature rises sharply in heater coils. Accumulated loose solid matter must be flushed out of coils periodically in order to maintain their heat transfer efficiency and to prevent flow restriction.

Piping connections for pressure and temperature relief valves should be made in accordance with recommendations given under the heading "Safety Devices" and with regulations given in Standard Code Regulations.

A hot water tempering valve should be provided as standard equipment in the hot water supply line from the outlet of a tankless heater. This is necessary in order to maintain the domestic hot water supply at normal hot water service temperature. Water temperature at the outlet of tankless heaters varies considerably depending upon the rate of flow through the heater and the water temperature at the heater inlet.

In storage heater installations, the control of tank temperature may be accomplished by means of thermostats installed at suitable locations in the tank so as to operate the heater automatically, heating when tank temperature drops below the low-limit setting of the thermostat and shutting off heating when tank temperature rises to the high-limit setting. Where circulating-type heaters are installed in conjunction with storage tanks and circulation pumps are used to circulate water between heaters and tanks, tank thermostats may be employed similarly to control the operation of circulation pumps.

Where storage tanks and storage tank heaters are operated at higher than normal service water temperature so as to supply high temperature water for machines and 140°F (60°C) water to ordinary plumbing fixtures, a hot water tempering valve should be provided in the hot water supply line from the tank. High temperature water may be supplied from a connection between the tank and the tempering valve. Location of the tempering valve should be at least 6 in (150 mm) below the top of the tank if the hot water supply system is not of the return circulation type. This location is recommended so that high-temperature water contained in the uppermost part of the tank and piping directly above, during no-flow conditions, remains trapped therein and does not subject the tempering valve to excessively high temperature conditions which might otherwise impair its normal function in time. Manufacturer's instructions regarding tempering valve location should be observed.

INDIRECT TANKLESS HEATER INSTALLATIONS

Features of an indirect tankless heater installation are inferred in the terms *indirect* and *tankless.* Domestic water is heated by passage through long copper tube coils having sufficient heat transfer surface to absorb heat rapidly from steam or high-temperature hot water surrounding the coils so that the heater may serve as a pipeline heater, heating water instantaneously as it flows through in a single pass and supplying hot water directly to hot water outlets at fixtures. No hot water storage tank is provided with this type of heater. Such installations are especially suitable for service conditions where hot water demand is heavy with long periods between demand, such as in small residential buildings, and also where hot water demand is heavy and relatively uniform over an extended period, such as in residential buildings, apartment hotels, laundries, restaurants, and similar building occupancies. For satisfactory, instantaneous performance, indirect tankless heaters must have available at all times an adequate supply of steam or high-temperature hot water from which the required amount of heat may be absorbed by the domestic water in flowing through the heater.

Indirect tankless heaters are available in an ample range of capacities suitable for most buildings. Several models are marketed. One model is a submerged coil unit designed for installation in steam or hot water heating system boilers. Other models are designed for installation as individual heat exchangers, separate from heating system boilers, and are provided with steam or high-temperature water jackets or casings surrounding the domestic hot water coil unit. These heaters do not come equipped with all the necessary automatic control features built into them, and such controls must be provided additionally in each case.

The durability of copper tube coil units in indirect tankless heaters is high, being comparable with that of copper water tube of the same type used for potable water supply service, and a rust-free supply of hot water may be assured. In addition, indirect tankless heaters are suitable for high-temperature water supply service, such as for dual temperature systems where 180 or 165°F (88 or 74°C) hot water is required for machine equipment and 140°F (60°C) hot water is required for ordinary hot water supplied fixtures.

Original cost of an indirect tankless heater installation usually is appreciably less than that of other types of heaters and storage tank installations of comparable capacity. Some original cost savings may be realized in the construction of boiler rooms in new buildings. As this type of heater is relatively small in size, it does not require much boiler room floor space. When installed as a submerged coil unit in a steam or hot water heating system boiler, no space is required other than the clearance needed for removing the unit from the boiler when necessary. Hence, in designing the building, no additional floor space need be provided in the boiler room for such units, no flues or chimneys are required for them, and corresponding savings may be realized.

Savings may accrue in water heating costs when the heater is submerged in a building steam or hot water heating system boiler. Most heating systems are designed on the basis of high factors of safety and allowances for pickup and piping losses. Very often this results in appreciable unused heating capacity which may be utilized as an available and economical source of heat.

Maintenance costs, in areas where the water supply is not excessively hard, are relatively minor, because indirect tankless heaters are comparatively small in size and essentially just a long copper tube coil. However, in hard water areas, maintenance costs can be prohibitive unless the water supply to the heater is adequately softened. Where hard water is unsoftened, the tendency for scale deposits to form in the heater coil results in severe reduction in its heat transfer capacity, inadequate hot water temperature at the heater outlet, increased pressure drop through the heater, and eventual clogging of the heater coil.

Heating capacities of indirect tankless heaters are rated in terms of the number of gallons per minute of water flow that can be heated from 40°F (4.4°C) at the inlet to 140°F (60°C) at the outlet when the heater coil is immersed in boiler water at 180°F (82°C). In selecting heaters, the standard ratings shown in manufacturers' catalogs should be carefully noted so that heater performance in service will match hot water demand.

The pressure drop occurring in tankless-type heaters is an important factor to be considered. These heaters are designed to produce a 100° temperature rise in water flowing through at rated capacity in a single pass, and to accomplish this, the coils are comparatively small in internal diameter and long in developed length. These features result in a relatively high pressure drop through tankless-type heaters at rated flow capacity. All tankless heaters having the same flow capacity do not necessarily have the same pressure drop. This factor should be taken into consideration in the selection of such heaters, especially where an installation is to be made in an area where water pressure is low, for the hot water supply system must be designed so as to ensure adequate volume and pressure conditions at all hot water supply outlets on the system. For data on pressure drop through tankless heaters, it is recommended that this be obtained directly from the manufacturers of such equipment.

Boiler firing rate and heating capacity, where tankless heaters are installed in heating system boilers, should be adequate to satisfy the total load including that needed for domestic hot water heating. This additional load may require just a small increase in the rate of firing in some instances. In others, it may require a larger and more expensive boiler installation than would be needed just for building heating, and thereby may add appreciably to equipment cost.

In all cases, the firing rate should be at least adequate to maintain boiler water temperature 180°F (82°C) continuously for hot water demand at rated flow capacity of the tankless heater. To heat 1 gal (3.8 L) of water from 40 to 140°F (4.4 to 60°C) requires 830 Btu, and if the boiler efficiency is assumed to be 75 percent, the heat input per hour required to heat water by means of a tankless heater may be calculated as 66,400 Btu/h for each gallon per minute of rated flow capacity. This may be used as a basis for determining the minimum firing rate for boilers equipped with tankless heaters. For example, with fuel oil burners, a firing rate of 1 gph of oil is required for heating 2 gpm (7.57 L/h) or 120 gph (454 L/h) of 140°F (60°C) domestic hot water.

Standard recommendations regarding the allowances which should be made as additional load for domestic hot water, based upon a 100°F (55.5°C) water temperature rise, are as follows: (1) Where the hot water demand is relatively intermittent, allow 1½ ft² of 240 Btu radiation for each gallon per hour of rated heater capacity; (2) where the hot water demand is relatively uniform over an extended period, allow 3⅓ ft² of 240 Btu radiation for each gallon per hour (3.79 L/h) of rated heater capacity; and (3) where a heating boiler is provided exclusively for heating domestic hot water, provide boiler capacity equal to 3⅓ ft² of 240 Btu radiation for each gallon per hour (3.79 L/h) of rated heater capacity.

Piping connections to an indirect tankless heater consist of connecting a branch from the cold water supply system to the heater domestic water inlet and connecting the heater domestic water outlet to the hot water supply connection. Other piping connections are required where separate steam or high-temperature hot water heat exchangers are installed. These units must have piped to them an adequate supply of steam or hot water, and the steam condensate or return water must be piped from them to an appropriate system of returns or to a suitable disposal terminal. Sizes of domestic hot and cold water piping connections should be the same as the heater tappings. Pressure and temperature relief valves should be connected to the piping at such heaters in accordance with recommendations stated under the heading "Safety Devices" and with regulations given in Standard Code Regulations.

In the cold water supply branch to the heater, a gate valve should be installed so that flow thereto may be shut off when required. Means should be provided for emptying water from the lowest part of the piping when necessary. Tankless heaters should have suitable means for flushing sediment and loose scale out of heater coils, because all water supplies contain some sediment and solids content which tends to precipitate as the water temperature rises sharply in heater coils. Accumulated loose solid matter must be flushed out of coils periodically in order to maintain their heat transfer efficiency and to prevent flow restriction. The standard means recommended for this purpose is the installation of a blowoff valve in the cold water supply piping to the heater and a cold water bypass connection with a gate valve installed between the main cold water line and the hot water piping connection at the heater outlet. With this means, sediment and loose scale may be flushed out of the coil by strong reverse flow at full service pressure by shutting off the gate valve controlling cold water supply to the heater inlet and then opening the bypass connection gate valve and the blowoff valve.

A hot water tempering valve should be provided as standard equipment in the hot water supply line from the outlet of a tankless heater. This is necessary in order to maintain the domestic hot water supply at normal hot water service temperature. Water temperature at the outlet of tankless heaters varies considerably depending upon the rate of flow through the heater and the water temperature at the heater inlet. At no flow, the water temperature in the heater coil is approximately the same as that of the steam or high-temperature hot water surrounding the coil. When flow through the coil is at less than rated capacity, and also when the inlet temperature exceeds 40°F (4.4°C), the water temperature at the heater outlet normally exceeds 140°F (60°C). Hence, the installation of a hot water tempering valve is a practical necessity with this type of heater.

Where the hot water supply system is not of the return circulation type, the tempering valve should be installed in the hot water supply line at least 6 in (150 mm) below the bottom of the heater. This is recommended so that the very high temperature water contained in the heater coil during no flow conditions remains trapped therein and does not subject the tempering valve to extremely high temperature conditions which might otherwise impair its normal function. Manufacturers' instructions regarding tempering valve location should be observed.

Water hammer conditions in supply systems may cause damage, leakage, or noise in the coils of tankless heaters. Where such conditions may occur as a result of the use of quick-closing faucets or valves, protection should be provided by means of a shock absorbing device installed in the hot water piping connection at the outlet of the heater. Mechanical shock absorbers are available and are recommended because they provide more constant and effective performance on hot water supply systems than ordinary air chambers such as are usually provided in water supply piping at fixtures.

INDIRECT STORAGE TANK HEATER INSTALLATIONS

The principal features of an indirect storage tank heater installation are inferred in the terms *indirect* and *storage tank.* In this type of installation, a hot water storage tank is provided in addition to an indirect storage tank heater. During periods of low demand, hot water may be heated and stored in the tank to satisfy peak demand requirements. The storage capacity of the tank and the heating capacity of the heater may be selected to match the hot water demand conditions of any building, regardless of occupancy or size, so long as there is an available source of steam or high-temperature hot water which may be used for heating domestic hot water.

Storage tanks of any capacity may be obtained. Standard storage tanks are available as manufactured stock items for a great range of capacities and variety of dimensions. Tanks of larger capacity or special dimensions may be fabricated on the job site in conformity with standards for the construction of pressure vessels. Indirect storage tank heaters are available in a wide range of capacities and several different types. One type is specifically designed for installation as a submerged coil unit in a storage tank of mated design. Where desired, separate small capacity heaters may be arranged in multiple and connected to a single storage tank so as to provide the required capacity to satisfy demand. In this way, the size of a domestic hot water heating job poses no problem.

Indirect storage tank heaters are equipped with copper tube coils and have a high degree of durability comparable with that of copper water tube of the same type used for potable water service. Thus, they do not produce rusty water conditions and may be used very efficiently for high-temperature water supply service.

However, storage tank durability is subject to very wide variation depending upon the particular type of tank material used, such as copper, monel, galvanized steel, or black steel; the thickness of the tank material; and the application of protective linings to the inside of the tank, such as copper linings or special cement linings; or the installation of sacrificial anodes inside the tank to protect it against rapid corrosion. The corrosivity of the water supply should be considered in the selection of tank material. The temperature at which storage tanks are operated is of special significance with regard to tank durability. Water temperatures above 140°F (60°C) cause a sharp rise in the rate of corrosion due to two facts: an increase in water temperature causes an increase in electrical conductance of the water and thereby increases the rate of galvanic corrosion, and at about 140°F (60°C) carbonates in the water break down releasing carbon dioxide which results in an increase in the rate of general or overall corrosion. In order to provide a durable tank installation, careful consideration should be given to the selection of tank materials, protective linings, and sacrificial anodes which will be suitable for the water temperature at which the tank is to operate and for the corrosivity of the water to be stored in the tank.

The original cost of an indirect storage tank heater installation generally is considerably higher than that for an indirect tankless heater of comparable capacity. Comparisons with other types of heater installations vary from less costly for large capacity installations to more costly for small ones. Each case should be separately considered especially where original cost of the building is a prime factor. An item of original cost which should be taken into account is that of boiler room construction in new buildings. For an indirect storage tank heater installation, adequate floor space must be provided for the storage tank and its supports and for external storage tank heaters where used.

Heaters which are installed submerged in the heating system boiler or in the hot water storage tank do not take up boiler room floor space, and so no space is required for them other than the clearance needed for removing the submerged coils when necessary.

Savings generally may be made in water heating costs with this type of installation as compared with other types. It is one which is adaptable for reclaiming heat from low-pressure steam and steam condensate. Indirect storage tank heaters may be used to extract heat from any convenient source of steam or high-temperature water. Heat may be economically absorbed by heaters built into steam or hot water heating system boilers which generally are selected on the basis of high factors of safety and allowance for pickup and piping losses. Very often this results in appreciable unused boiler heating capacity which may be utilized as an available and economical source of heat.

Maintenance costs of storage tank heaters of the circulating type and of the steam tube type are relatively low except in areas where the water supply is excessively hard and unsoftened. This is attributed to the fact that in hard water areas, the tendency for scale deposits to form in the water passes of circulating heater coils and on the outer surface of steam tube heaters poses a greater problem in maintaining heater efficiency and results in higher maintenance costs. The scale deposits cause a reduction in the heater transfer capacity of the heater coil and in the rate of circulation of tank water through the coil of a circulating-type heater as well as eventual clogging of its water passes.

Maintenance costs of storage tanks vary greatly depending upon durability of the tanks, temperature at which they are operated, corrosivity of the water supply stored in the tanks, amount of sediment and solid matter which may be accumulated in the tank without affecting heater efficiency, and other factors. In any event, maintenance costs are directly related to the original selection of storage tank material, the characteristics of the water supply, and the provisions for protection against corrosion so as to avoid short tank life and the need for tank replacement.

There are three different types of indirect storage tank heater installations. One type is designed for built-in installation in a horizontal storage tank. The heater consists of a long bundle of copper tube coils, inserted into the tank through a flanged opening in the lower part of one end of a mated tank, i.e., a tank specially designed for such coil unit installation. Steam or high-temperature hot water is supplied to the coils so as to transfer heat to water in the tank. This type of installation provides the most satisfactory performance of all types where the water supply is excessively hard and unsoftened because hard water does not flow through the coils but instead surrounds them. Hence, scale deposits cannot occur inside the coils to affect heater performance seriously and rapidly.

The second type is designed for built-in installation in a steam or hot water heating system boiler. The heater consists of a long bundle of copper tube coils, inserted into the boiler through a flanged opening in the upper part. In steam boilers, the heater coils are located just a few inches below the normal water line of the boiler so that they are submerged in the hottest boiler water at all times during normal operation. This type of heating unit is of the circulating type, for tank water circulates through the coils and is heated thereby. It is generally satisfactory in performance except where the water supply is excessively hard and unsoftened, in which case gradual accumulations of scale deposits in the coils cause reduced circulation through the heater and impaired performance.

The third type is also a circulating heater designed for installation as an external, separate unit. It consists of a metallic casing housing a copper tube coil. Inlet and outlet connections to the casing are provided for connecting to an available supply of steam or high-temperature hot water, and inlet and outlet connections to the coil are provided for connecting it to the storage tank. Tank water circulates through the coil and is heated thereby. This type is generally satisfactory in performance except where the water supply is excessively hard and unsoftened in which case gradual accumulations of scale deposits in the coil cause reduced circulation through the heater and impaired performance.

Indirect storage tank heater installations require fewer square feet of copper tube heater coil surface to accomplish any given domestic hot water heating job than do tankless heaters. This is due to the fact that the storage tank provides a large reservoir of hot water available for periods of peak demand, thus making it possible to heat water continuously at a lower rate than would be required if it were heated instantaneously to satisfy peak demand. Continuous circulation of heated water from the heater permits recovery of maximum tank water temperature during periods of small or no demand.

How well an indirect storage tank heater installation can satisfy hot water demand depends upon a combination of the tank's capacity and the heater's heating capacity, or what may also be termed the heater's recovery capacity. Only three-fourths of the tank capacity normally should be assumed as the amount of hot water available from the tank at suitable temperature for satisfying demand. The remaining one-fourth is considered to be of inadequate temperature as a result of mixing with cold water entering the tank when hot water is drawn. Therefore, for the duration of the period of demand, the difference between the total amount of hot water required and that available from the tank must be provided by the heating capacity of the heater in that period so as to recover or maintain tank water temperature at a suitably high degree for satisfactory performance. In this way the tank capacity and the heater's heating capacity may be matched for any given hot water demand condition.

Heating capacities of indirect storage tank heaters are rated in terms of the number of gallons per hour of water that can be raised 100°F (55.5°C), based upon 180°F (82.2°C) boiler water temperature and adequate circulation conditions. Where higher boiler water temperatures are used or where the heater is supplied with steam, the actual heating capacity will exceed the rated capacity of the heater. Nevertheless, heaters which are connected to boilers operated at 180°F (82.2°C) boiler water temperature between heating seasons should be selected on the basis of their standard rated heating capacities.

The pressure drop through an indirect storage tank heater installation is much less than through a tankless heater installation. When hot water is drawn at peak demand rate, only a very small drop in available water supply pressure occurs between the cold water supply inlet and the hot water supply outlet of the storage tank. This pressure drop may be assumed to be equal to that produced by the same rate of flow through three 90° standard elbows of the same size as the tank inlet and outlet tappings. A similar assumption is made for the pressure drop through heating boilers in calculations for forced hot water heating systems. For each 90° standard elbow, it is generally accepted that the pressure drop is the same as for a pipe of the same nominal diameter and of a length equal to 25 times the nominal size of the elbow. As an illustration of the pressure drop through a hot water storage tank with 2-in (50-mm) inlet and outlet tappings, the pressure drop at maximum demand may be assumed to be the same as for three 2-in 90° standard elbows, or 3 × 25 × 2 in (50 mm) = 150 in (3.81 m) of 2-in (50-mm) pipe. In an indirect storage tank heater installation, the pressure drop through a circulating-type heater is of concern only insofar as it affects the rate of circulation of heated water from the heater to the storage tank.

Consequently, the indirect storage tank heater installation should be recognized as providing a minimum pressure drop due to flow through water-heating equipment. This is an especially important feature in areas where water pressure is low and the hot water supply system must be carefully designed so as to ensure adequate supply at all fixtures on the system.

Where indirect storage tank heaters are installed in conjunction with heating system boilers, boiler heating capacity should be adequate to satisfy the total load including that required for domestic water heating. In many cases where the domestic hot water load is not very great and boilers are selected on the basis of standard recommendations which include high factors of safety and allowance for pickup and piping losses, no additional boiler load need be provided.

Boilers having net ratings equal to or greater than the heat loss of the building usually are adequate in capacity to handle the additional domestic water-heating load of the typical small one-family residence having not more than two bedrooms. For buildings with greater hot water demand, an allowance should be made in the form of additional boiler load for the domestic hot water demand.

Standard recommendations regarding the allowances which should be made as additional load for domestic hot water, based upon a 100°F (55.5°C) water temperature rise, is as follows: allow ½ ft^2 of 240-Btu radiation for each gallon per hour of the heater's rated hourly capacity.

Hot water storage tanks should be installed above the tops of indirect storage tank heaters so as to permit continuous, active circulation of heated water when water is to be circulated by gravity. The higher the tank is raised above the heater, the greater is the gravity circulation head and the resultant rate at which heated water circulates from the heater into the tank.

For gravity circulation to be continuous at a relatively uniform rate and heat all the water in the tank, the entire tank should be located above the top of the heater. Where part of the tank is below the top of the heater, as frequently is the case with most vertical storage tank installations, the rate of circulation slows down as the lower level of hot water in the tank approaches the level of the top of the heater, and thereafter circulation becomes sluggish.

Horizontal hot water storage tanks are usually preferred because of such practical considerations as (1) the limited headroom generally available for raising the entire tank above heaters in boiler rooms and (2) the matter of providing proper supports for elevated hot water storage tanks.

Regardless of the style of tank or its position with respect to the heater, positive controlled circulation of water between the heater and the tank may be accomplished by installing a circulation pump. A pump installation has advantages which often warrant its use even where gravity circulation may be employed. Pumps should be selected on the basis of delivering, against a head of at least 4 ft (1.22 m) of water column, the number of gallons per hour corresponding to the heater's rated heating capacity. Manufacturers' pump capacity charts should be consulted in the matter of pump selection.

Submerged copper coils of circulating-type heaters should be connected to hot water storage tanks by means of proper circulation piping. Cool water is piped from a tapping in the bottom of the tank to the lower connection of the heater coil. Heated water is piped from the upper connection of the heater coil to a tapping in the upper part of the tank or to an appropriate fitting in the hot water outlet piping from the top of the tank.

When a circulation pump is provided to circulate water between the tank and heater, the pump should be placed in the lower circulation piping. Both upper and lower circulation piping should be installed in such a manner as to preclude the formation of air pockets which might otherwise impair circulation of water. The size of such piping should be no smaller than the size of the heater tappings.

Heaters which are built into heating system boilers or attached externally thereto should be located so as to be heated by the hottest water in the boiler. In steam boilers, built-in type heaters should be submerged in the boiler water just a few inches below the normal waterline.

External heaters, through which boiler water is circulated by gravity, should be located so that the tops of such heaters are just below the normal waterline. For such heaters boiler water should be piped from a boiler tapping, just a few inches below the waterline, to the upper casing, boiler water should be piped to circulate back to a lower tapping in the boiler or to a fitting in a boiler water return line. Connections between the heater casing and the boiler should be installed in such a manner as to preclude the formation of air or steam pockets which might otherwise impair gravity circulation of boiler water through the heater casing.

Regardless of the heater location below the boiler waterline, positive controlled circulation of boiler water through the heater casing may be accomplished by installing a circulation pump in the lower circulation piping. A pump installation has advantages which permit flexibility in locating the tank and the heater and provide positive control of the rate of water heating. Pumps should be selected on the same basis as for circulating water between heater coils and tanks.

Cold water is supplied directly to a tapping in the bottom of the hot water storage tank. Hot water is supplied to the hot water supply piping system from a tapping in the top of the tank. Cold and hot water supply connections to the tank should be remote from each other and near opposite ends of the tank so as to prevent any tendency for cold water entering the tank to flow directly to the hot water outlet and thereby short circuit the supply of hot water from the tank. The cold water piping connection to the tank, for at least several feet from the tank tapping, should be of the same size as the tank tapping so as to minimize the velocity of cold water entering the tank and thereby reduce the mixing of cold with hot water. This is of special importance where tank diameters are less than 16 in (400 mm).

In the cold water piping connection to the tank, a gate, ball, or butterfly valve should be installed so that the supply of water to the system may be shut off when necessary. Between the gate, ball, or butterfly valve and the tank, there should be a pressure relief valve. A temperature relief valve should be installed and located so that its temperature-sensitive element is in contact with the hottest water in the tank, that is, the water in the upper 6 in (150 mm) of the tank. No valve of any type should be installed so as to intervene between either the pressure or temperature relief valves and the tank.

Means must be provided for emptying hot water storage tanks when necessary and for flushing sediment and loose scale accumulations from within the tanks when they are to be maintained. Where steam coil type built-in heaters are installed in hot water storage tanks, sediment and loose scale should be flushed out from around the heater coils which are located close to the bottom of the tanks. Such tanks should be serviced in this way at regular intervals so as to maintain high heat transfer efficiency of the heater coil. The means provided for draining and flushing tanks should be a gate valve of adequate size so as to perform such operations quickly and efficiently.

Some means should also be provided so that sediment and loose scale can be flushed out of the coils of circulating-type heaters. All water supplies contain some sediment or solids content which tends to precipitate as the water temperature rises sharply in heater coils. Accumulated loose solid matter should be flushed out of coils periodically in order to maintain heat transfer efficiency.

The standard means recommended for this purpose is the installation of a blowoff valve, gate valve type, connected to the branch of a T fitting in the lower circulation piping, and a gate valve installed between the T and the connection to the bottom of the tank. By shutting off this gate valve and opening the blowoff valve, sediment and loose scale may be flushed out of the heater coil by strong reverse flow at full service pressure.

In hot water storage tanks heated by built-in heater coils, through which steam or high-temperature hot water flows, the control of tank water temperature may be accomplished by controlling the flow of steam or high-temperature hot water to the heater coils. An automatic thermostat may be installed in a tapping at or above the center of one end of the tank so as to sense tank water temperature at that level. This thermostat should be connected so as to operate a supply control valve in the steam or high-temperature hot water supply line to the heater coils.

Where a circulation pump is provided to circulate either tank or boiler water through a heater, control of tank water temperature may be accomplished by controlling the operation of the circulation pump. An automatic thermostat may be installed in a tapping at or above the center of one end of the tank so as to sense tank water temperature at that level. This thermostat should be connected so as to operate the circulation pump, starting pump operation when tank water temperature drops below the low-limit setting of the thermostat and stopping it when the temperature rises to the high-limit setting.

Wide variations of tank water temperature occur in indirect storage tank heater installations where both boiler and tank water circulate through heaters by gravity. During long periods of little or no demand, tank water temperature may rise to a degree much above the normal service temperature for hot water supply. With continuous gravity circulation of heated water into the tank over long periods of no demand, tank water temperature may approach that of the boiler water to which the heater coils are exposed. Hence, some provision should be made to avoid the hazard of supplying excessively hot water to plumbing fixtures.

Consequently, a hot water tempering valve should be provided as standard equipment in the hot water supply line from the storage tank of an indirect storage tank heater installation where both boiler and tank water circulate through the heater by gravity. This valve is a practical necessity in order to maintain the supply of hot water at normal hot water service temperature at all times.

The location of the hot water tempering valve should be at least 6 in (150 mm) below the top of the tank if the hot water supply system is not of the return circulation type. This location is recommended so that high-temperature water contained in the uppermost part of the tank and piping directly above during no-flow conditions remains trapped therein and thereby does not subject the tempering valve to excessively high temperature conditions which might otherwise impair its normal function in time. Manufacturers' instructions regarding tempering valve location should be observed.

In large indirect hot water storage tank heater installations the amount of heat lost by radiation to the environment can be appreciable when the tanks, heaters, and circulation piping are not insulated. This may raise the boiler room temperature to an excessive and objectionable degree. Such heat loss should be considered as wasted heat which can be conserved by proper insulation. It is recommended that, in large installations, suitable coverings of insulation be provided for hot water storage tanks, external storage tank heaters, and circulation piping so as to minimize the rate of heat loss from such equipment. The thickness and kinds of insulating materials to be used may be selected on the basis of the estimated savings which would result from the reduction in radiant heat loss and the estimated costs of the various thicknesses and kinds of insulating materials.

For small installations, insulation of water-heating equipment is also advisable in view of the cumulative savings in water-heating costs over many years of service. This feature alone may be found sufficient reason to warrant insulation for even the smallest items of water-heating equipment.

DEMAND AND HEATER CAPACITIES

The demand for water at 180°F (82°C) temperature is posed almost exclusively by the high-temperature water requirements for dishwashing machines and dishwashing fixtures in commercial kitchens serving employee dining spaces or public restaurants. Standard regulations require that, where food or drink is served to the public or to employees and the dishes, glasses, or cutlery are to be reused, such articles shall be washed effectively and then sanitized before reuse. The water temperature required for sanitizing purposes in this type of equipment is generally recognized as being 180°F (82°C). As a result of being sanitized by water at this high temperature, washed articles become air-dried in just a few moments after removal from the machine. This eliminates any need for drying articles by other methods.

Where glasses and cutlery are washed and sanitized in individual glass washing and silver washing machines or fixtures, such equipment should be understood to require 180°F (82°C) water for proper performance. In most cases where dishes also are to be washed, there is no special need for separate glass washing and silver washing equipment, for dishwashing machines generally may be used efficiently for washing such items.

The only plumbing fixtures which may be considered suitable for using hot water at 165°F (74°C) safely are those of the machine type having enclosures which protect persons against being scalded during use of the fixture. These fixtures are known as automatic domestic dishwashing machines and automatic domestic laundry washing machines. Although they may perform satisfactorily in most cases with hot water at 140°F (60°C), many manufacturers of such machines have recommended that they be supplied with hot water at 165°F (74°C) for best performance.

To accurately estimate the demand for 180°F (82°C) water for any given installation, the amount of dishwashing to be performed should be determined first, and then suitable equipment and machines should be selected to match the dishwashing needs. Thereafter, the demand may be estimated directly from the data and recommendations contained in specification sheets available from dishwashing machine manufacturers regarding the performance of, and the water requirements for, their products. The rated 180°F (82°C) water requirements for the machines may be used to estimate the maximum instantaneous demand and also the maximum total daily demand.

The demand for 165°F (74°C) water supply to automatic domestic dishwashing machines and laundry washing machines should be estimated from data and recommendations contained in the specification sheets available from the manufacturer in each case. For different sizes and models of machines, there may be considerable variation in the water supply requirements.

Except for the high-temperature water requirements posed by dishwashing machines and dishwashing fixtures in commercial kitchens and for the slightly lower temperature hot water recommendations of manufacturers for automatic domestic dishwashing and laundry washing machines, the demand for hot water supply at ordinary plumbing fixtures in buildings is for water at 140°F (60°C). This is recognized as being the most desirable temperature for supplying hot water to plumbing fixtures as it is sufficiently high so that normal washing tasks may be accomplished and sufficiently low so that scalding of persons may be prevented and excessive corrosion and expansion effects in the system may be avoided.

The demand for 140°F (60°C) water in buildings depends upon numerous factors and conditions. They include (1) the occupancy classification of the building or portion supplied, (2) the number of persons accommodated, (3) the number and kind of plumbing fixtures, (4) any special equipment installed for occupancies, (5) the time of the day,

(6) the season of the year, (7) the installation of water conservation devices, and (8) whether hot water is supplied to occupancies at no extra charge.

Demand is never absolutely constant throughout the day. It is subject to extreme limits of variation. An objective for design of water-heating installations is to satisfy efficiently the peak demand for hot water. This involves two essential considerations: the peak demand rate and the duration of the peak draw period. They form the basis upon which to determine the most suitable capacities of heaters and storage tanks, the arrangement of water-heating equipment, and the sizes of piping between such equipment.

The peak demand rate is the maximum rate at which hot water may be drawn in normal service. This may be stated in terms of gallons per minute or gallons per hour of water to be supplied at a temperature of 140°F (60°C). Where water is to be heated instantaneously, such as by a tankless heater, the rated capacity of the heater should be equal to the maximum rate at which hot water is drawn, i.e., the peak demand rate.

A standard method for estimating the peak demand rate for most types of installations has been developed as the result of the experience of heater manufacturers over the years. This method consists of two parts: (1) a base demand rate established for the particular type of installation or occupancy and (2) additional demand rate allowances assigned for certain individual fixtures, groups of fixtures, or occupancy conditions.

For dwellings, a base demand rate of 300 gph (1137 L/h) is recommended currently for tankless systems, and a rate of 180 gph (682 L/h) is recommended for storage systems. These base demand rates may be applied as the peak demand rate per hour in the case of one-family dwellings. No additional demand rate allowances need be added in this case for provision has been made for all the fixtures normally used including an automatic dishwashing machine and a laundry washing machine in a one-family dwelling.

For multiple dwellings, additional demand rate allowances should be assigned as follows: 12 gph (45.4 L/h) for each dwelling unit, 6 gph (22.7 L/h) for each bathroom group of fixtures exceeding the minimum of one group in a dwelling unit, and 3 gph (11.4 L/h) for each automatic domestic dishwashing or laundry washing machine in a dwelling unit. Where laundry washing machines are grouped in a general laundry room in a residential building, an allowance of 30 gph (114 L/h) for each machine should be assigned.

For all other types of building occupancies, a base demand rate of 180 gph (682 L/h) is recommended. Additional demand rate allowances should be assigned as follows:

Bathtubs: 60 gph (227 L/h) each

Laundry trays: 120 gph (454 L/h) for each

Lavatories or shampoo basins: 10 gph (37.9 L/h) for each

Showers: 120 gph (454 L/h) for each in factories, golf clubs, gymnasiums, and schools; 60 gph (227 L/h) for each in business or residence clubs and hospitals; and 12 gph (45.4 L/h) for each in hotels

Sinks: in public restaurants, allow 1½ gph (5.68 L/h) per meal served during the peak period for all sink facilities; in small luncheonettes and soda fountains, allow 1 gph (3.79 L/h) per meal served during the peak period for all sink facilities

Washfountains, multiple type: 260 gph (985 L/h) for each 54-in (1.37 m) circular unit; 180 gph (682 L/h) for each 54-in (1.37 m) semicircular or 36-in (0.91 m) circular unit and 125 gph (474 L/h) for each 36-in (0.91 m) semicircular unit

For tankless heater systems, the duration of the peak draw period need not be considered. This is due to the fact that tankless heaters must be selected of adequate, rated heating capacity to match the peak demand rate regardless of the length of time over which it lasts.

However, for storage tank systems, smaller heaters may be used than in tankless systems, for a storage heater need only provide the heating capacity necessary to maintain proper tank water temperature for the duration of the peak draw period. Over the period of peak draw, water may be drawn from the tank's available supply which should be considered as being 75 percent of the capacity of the tank. For example, if a system has a peak demand rate extending over a 2-h period, 37 percent of the tank's capacity may be drawn per hour. If the period lasts 3 h, 25 percent of the tank's capacity may be drawn per hour. The difference between the peak demand rate per hour and the amount of water which may be drawn per hour from the tank is the heating capacity per hour required for a storage heater.

In large multiple dwellings, the peak demand rate is generally found to be relatively uniform over a maximum period of about 3 h, extending from 6 to 9 A.M. and 5 to 8 P.M. Hence, for such buildings the duration of the peak draw period may be taken as being 3 h.

In smaller buildings occupied for residental use, the period of peak draw is shorter than 3 h. For a one-family dwelling, the period may be assumed to be no more than 20 min in most cases. The period for larger dwellings may be determined by assuming an extra 10 min per each additional dwelling unit in the building until the 3-h period is reached.

For most large buildings having extensive plumbing systems, the period of peak draw may be safely considered to be 3 h. In factory, office, and mercantile buildings, a shorter period may exist in most instances, such as at lunch time and at closing time. The period may not exceed ½ h in some cases, 1 h in others, and 2 h in a few. But, in any case the 3-h period may be assumed when there is no assurance that a shorter period is clearly definable.

Tankless heaters should be selected on the basis of providing rated heating capacity equal to the peak demand rate for the hot water supply system. For example, the tankless heater to select for a modern one family dwelling, with an automatic dishwashing and an automatic laundry washing machine, is one having a rated heating capacity of 300 gph (1137 L/h) or 5 gpm (18.9 L/min).

To select a tankless heater for a 50-family cooperative apartment building, in which 25 dwelling units have an extra bathroom group of fixtures and also dishwashing machines or laundry machines, the procedure is as follows:

	gph	L/h
Base demand rate for multiple dwellings (tankless)	300	1137
Allowance for 50 dwelling units at 12 gph (45.4 L/h)	600	2274
Allowance for 25 extra baths at 6 gph (22.7 L/h)	150	568
Allowance for 25 dishwashing or laundry washing machines at 3 gph (11.4 L/h)	75	284
Total peak demand rate, or required rated heating capacity of tankless heater	1125	4263

Storage heaters should be selected on the basis of providing rated heating capacity equal to the difference between the peak demand rate per hour and the amount of water which may be drawn per hour from the tank. Heater capacities may be determined after appropriate tank sizes have been selected.

For an indirect storage tank heater installation, the minimum storage tank size recommended for a modern one-family dwelling, with an automatic dishwashing and an automatic laundry washing machine, is 66 gal (250 L) capacity. The minimum recommended tank capacities for larger dwellings are as follows: 30 gal (114 L) per dwelling unit for up to 10 units, 25 gal (94.7 L) per unit for up to 25 units, 20 gal (75.8 L) per unit for up to 50 units, and 15 gal (56.8 L) per unit for buildings having more than 50 dwelling units.

To select a storage heater for an indirect storage tank heater installation in a modern one-family dwelling, the heater size should be based upon the use of a 66-gal (250 L) hot water storage tank and a 20-min peak draw period. The recommended peak demand rate for a storage system in a one-family dwelling is 180 gph (682 L/h). So, for the 20-min peak draw period, the peak demand rate is 60 gal (227 L) per 20 min. The amount of hot water available from the tank is 75 percent of its capacity, 66 gal × 0.75 equals 49.5 gal (188 L). The difference between the peak demand rate of 60 gal (227 L) per 20 min and the 49.5 gal (188 L) available from the tank leaves 10.5 gal (39.8 L) of water to be heated in the 20-min period. The heating capacity required on the hourly basis at which heaters are rated is 10.5 × $^{60}/_{20}$ or 31.5 gal/h (119 L/h). This is the rated heating capacity per hour of the storage heater recommended for the installation.

Selection of a storage heater for the 50-family cooperative apartment building used as an example in the foregoing may be described. The size of the storage tank recommended for 50 dwelling units is one providing 20 gal (75.8 L) capacity per dwelling unit, or a total capacity of 1000 gal (3790 L). Of this capacity, 75 percent or 750 gal (2842 L) is available for being drawn over a period of 3 h. Hence, the rate of drawing tank water per hour is $^{750}/_{3}$ or 250 gph (947 L/h).

For this installation, the total peak demand rate may be established as follows:

	gph	L/h
Base demand rate for multiple dwellings (storage tank)	180	682
Allowance for 50 dwelling units at 12 gph (45.4 L/h)	600	2274
Allowance for 25 extra baths at 6 gph (22.7 L/h)	150	568
Allowance for 25 dishwashing or laundry washing machines at 3 gph (11.4 L/h)	75	284
Total peak demand rate	1005	3808

The rated heating capacity per hour required for the indirect storage heater is the difference between the peak demand rate per hour and the rate at which tank water may be drawn. Thus, the required heating capacity is the difference between 1005 gph (3808 L/h) and 250 gph (947 L/h), or 755 gph (2861 L/h) of rated storage heater capacity.

RETURN CIRCULATION SYSTEMS

Hot water supply piping transmits heat to lower temperature surroundings by convection, radiation, and conduction. This heat is lost by the hot water in the piping. If the water is not reheated and remains stationary in the piping for an extended period, its temperature drops to a degree that the water becomes relatively cold and unsuitable for hot water service. Such water must be drawn from the piping before hot water at normal service temperature is available at outlets. This is objectionable where an excessive length of time is required to purge cold water from the piping before hot water is available, and where an excessive amount of water is wasted in purging cold water from hot water supply systems.

Rarely is there any compelling necessity for circulating hot water continuously through a small system in a one- or two-family dwelling where the length of hot water supply piping is relatively short and pipe sizes are small. In such cases, although hot water may cool off in the piping when no water is drawn, water at normal service temperature is available at outlets within a reasonably short period of time and without excessive drawoff of cool water. For example, if just 1 gal (3.79 L) were drawn, it would be equivalent to purging the contents of 63 ft (19.2 m) of ½-in (12.7-mm) pipe, 36 ft (11.0 m) of ¾-in (19.1-mm) pipe, or 22 ft (6.71 m) of 1-in (25.4-mm) pipe of standard pipe size. Thus, the additional costs involved in installing a return circulation type system and in heating water to compensate for the heat lost by circulation piping are not warranted in most small systems. Of course, where systems supply equipment requiring instantaneous hot water, a return circulation type system may be necessary for satisfactory equipment function.

A return circulation type system definitely is necessary for large extensive hot water supply systems. If they were not designed to circulate hot water continuously, there would be delays in obtaining hot water at normal service temperature, and there would be excessive waste of water of unsuitable temperature. In many instances the length of waiting time could be so great as to cause complaints from building occupants regarding unsatisfactory hot water service where adequate service is paid for in the form of rent. Public water system authorities also object to excessive waste of water because of the general need for conserving the potable water supply, and where buildings are supplied from the public system and are not equipped with water meters, waste may occur without proper charge and cause a corresponding financial loss to the public system.

Consequently, standard code regulations establish a dividing line beyond which return circulation-type hot water supply systems shall be installed. They are required generally in buildings more than four stories in height and in buildings where the developed length of hot water supply piping, measured from the source of hot water supply to the farthest fixture supplied, exceeds 100 ft (30.5 m).

There are three general types of hot water supply-return circulation systems, namely, the upfeed system, the downfeed system, and the combination upfeed and downfeed system. These names are derived from the direction of flow in hot water risers supplying branches to fixtures and are unrelated to the direction of flow in return piping. In each type, return piping is connected to the end or near the end of hot water supply risers so as to circulate water back to the hot water source.

Conventionally, hot water heaters and tanks have been located in cellars and basements for various reasons, such as economy in building design, convenience and economy in locating such equipment in proximity to heating boilers and source of fuel supply, and availability of convenient means for ash disposal. In conventional systems, heaters and tanks are located at the lowest part of the hot water supply system, and under such circumstances, circulation of hot water may be achieved by gravity due to the head induced by the difference in water temperatures in hot water supply and return pipes located above the hot water source. Where systems are extensive or have little effective circulation height, pumps may be used to provide greater circulation.

Inverted design of hot water supply-return circulation systems has been applied advantageously in many tall buildings. In the inverted system, as contrasted with conventional systems, hot water heaters and tanks are located at the highest part of the hot water supply system, and supply and return risers are below the level of the hot water source. Under these circumstances, hot water circulation cannot be achieved by gravity because cooled hot water would settle into the lowest part of the circulation system and remain there. Consequently, in inverted systems circulation of hot water can be achieved only by means of pumps.

The conventional upfeed system is illustrated in Fig. 7-1. In this system, the main hot water supply line is extended from the source of hot water supply and is located in the lowest part of the building. From that location, hot water is supplied to the bottom of all the hot water supply risers. Flow is upward in all the risers supplying branches to fixtures. A hot water return riser is provided for each of the hot water supply risers. The top of the return riser is connected to the supply riser just below the topmost supply branch to fixtures. Return risers extend down to the lowest part of the building where they connect to a main hot water return line through which water circulates back to the source of hot water supply. In this system, air accumulated in the uppermost part of each riser is drawn off when a hot water faucet is opened at a fixture supplied from the top of the supply riser, thereby eliminating air accumulation which might otherwise restrict circulation.

The conventional downfeed system is illustrated in Fig. 7-2. In this system the main hot water supply line is extended from the source of hot water supply to the highest part of the building. From that location,

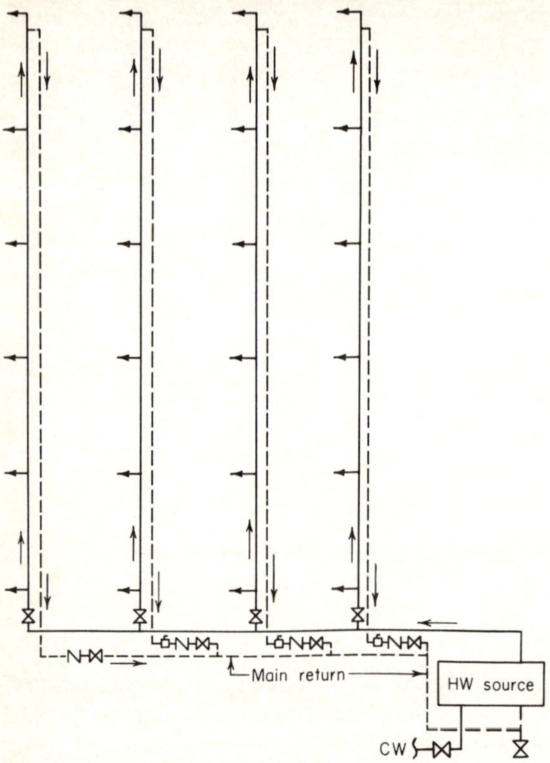

Fig. 7-1 Conventional upfeed system.

hot water is supplied to the tops of all hot water supply risers. Flow is downward in all the risers supplying branches to fixtures. The base of each downfeed riser is connected to a main hot water return line so as to circulate cooled water back to the source of hot water supply. At the top of the main hot water supply line, the highest point in the system, provision should be made to eliminate air so that air pockets do not develop and restrict circulation of hot water. This may be accomplished by connecting a fixture supply branch to the highest point of the system so that air may be drawn off when the hot water faucet is opened at the fixture. Another recommended method is to install an approved type of air relief vent valve at the highest point, with a relief pipe discharging into an approved fixture at which the discharge of air and water from the valve will not be objectionable and where valve leakage will be evident to indicate a need for valve maintenance.

The conventional combination upfeed and downfeed system is illustrated in Fig. 7-3. This system is a combination of the two systems previously discussed. In this case, some hot water supply risers have

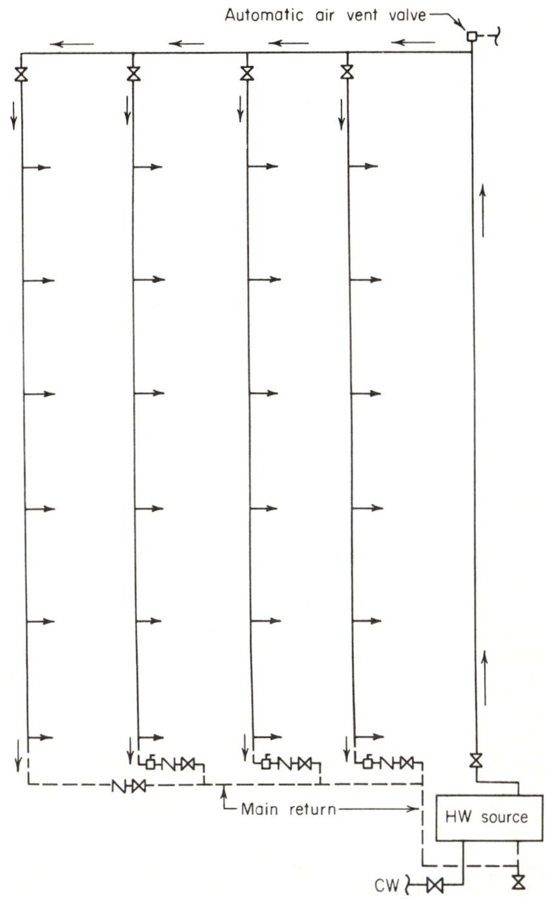

Fig. 7-2 Conventional downfeed system.

upward flow while others have downward flow. Each downfeed riser is supplied from the top of an upfeed riser, and the base of each downfeed riser is connected to a main hot water return line through which water circulates back to the source of hot water supply. Means for preventing air accumulation at high points of the system are necessary. An appropriate method may be selected from those cited for upfeed systems and downfeed systems.

The inverted downfeed system is illustrated in Fig. 7-4. In this system, the main hot water supply line is extended from the source of hot water supply and is located at the highest part of the system. From that location, hot water is supplied to the tops of all the hot water supply risers. Flow is downward in all the risers supplying branches to fixtures. A hot water return riser is provided for each of the hot water supply risers.

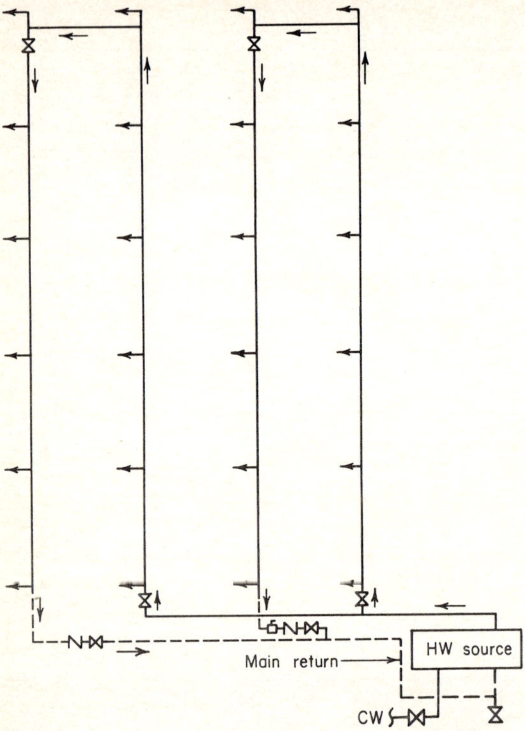

Fig. 7-3 Conventional combination upfeed and downfeed system.

The bottom of the return riser is connected to the supply riser just above the lowest supply branch to fixtures. Return risers extend upward to the highest part of the system where they connect to a main hot water return line through which water is pumped back to the source of hot water supply. Air accumulated in the uppermost part of this system is discharged by means of an automatic air vent valve located at the highest point of the system, thereby eliminating air accumulation which might otherwise tend to restrict circulation, produce noise in the piping, or be objectionable when discharged at a fixture.

Where hot water storage tanks are installed at the highest part of the system, such as is the case in an inverted hot water supply system, it is advisable to provide a tank vacuum breaker, or vacuum relief valve, at the top of the tank so as to permit air to enter the tank whenever a vacuum occurs and thereby prevent collapse of the tank. In such locations, tanks may be subjected to very severe vacuums in the event of cold water supply failure, or when the supply to the tank is shut off and hot water is drawn at a lower floor outlet.

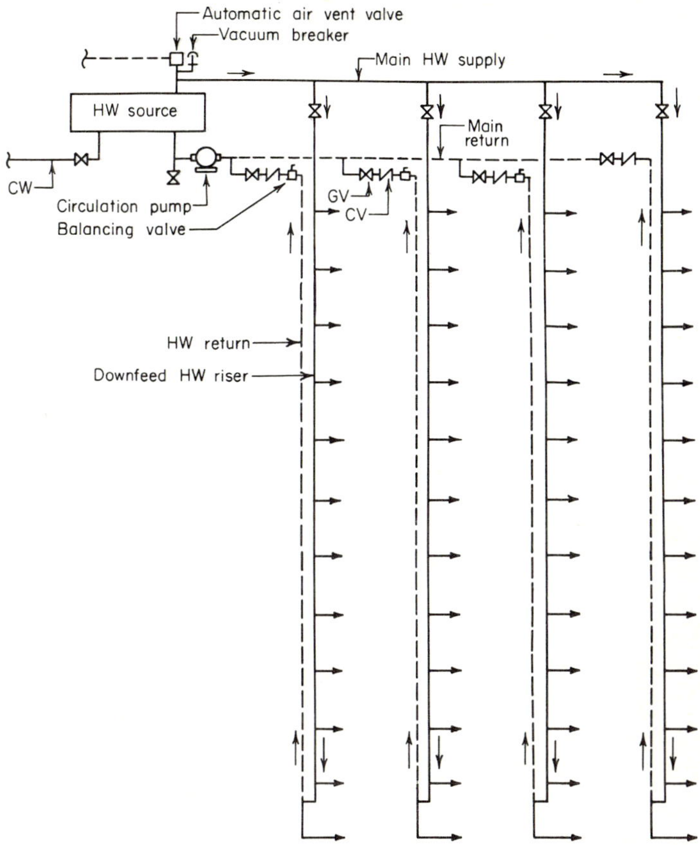

Fig. 7-4 Inverted downfeed system.

The inverted upfeed system is illustrated in Fig. 7-5. In this system, the main hot water supply line is extended from the source of hot water supply to the lowest part of the system. From that location, hot water is supplied to the bottoms of all hot water supply risers. Flow is upward in all the risers supplying branches to fixtures. The top of each upfeed riser is connected to a main hot water return line through which water is pumped back to the source of hot water supply. In this system, air which has accumulated in the uppermost part is automatically discharged by means of an air vent valve located at the highest point of the system, for the same reasons as were given in the case of the inverted downfeed system.

The inverted combination downfeed and upfeed system is illustrated in Fig. 7-6. This system is a combination of the two inverted systems previously discussed. In this case, some hot water supply risers have

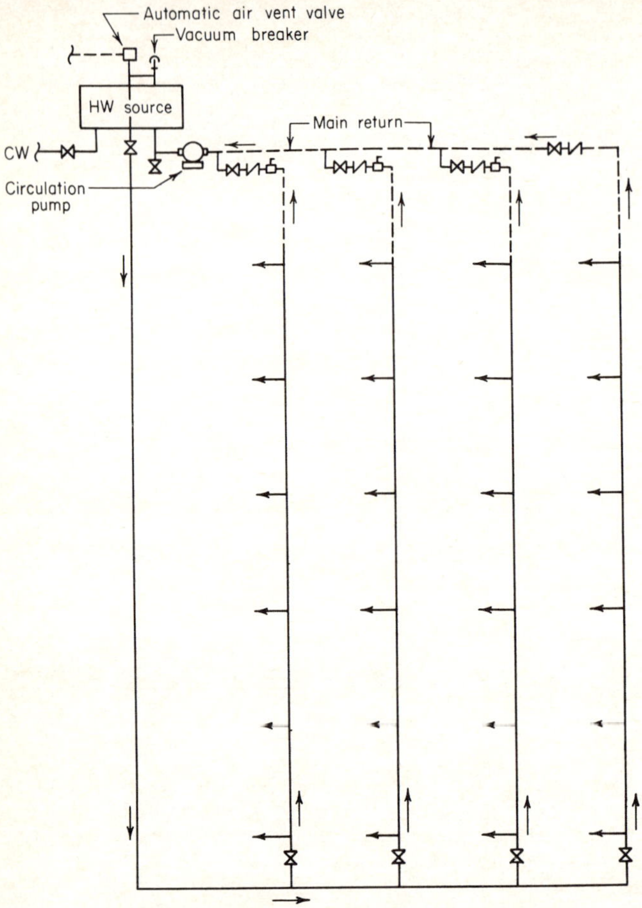

Fig. 7-5 Inverted upfeed system.

downward flow while others have upward flow. Each upfeed riser is supplied from the bottom of a downfeed riser, and the top of the upfeed riser is connected to a main hot water return line through which water is pumped back to the source of hot water supply. Means for eliminating air accumulation in the highest part of the system should be provided for the same reasons as cited for inverted downfeed and inverted upfeed systems.

It is not advisable to connect the main hot water return line to a vertical section of the lower circulation piping between the bottom of a tank and a heater. Water in the main return line generally is hotter than water in the lower circulation piping between the tank and heater, and the hotter return line water tends to rise into the tank while tank

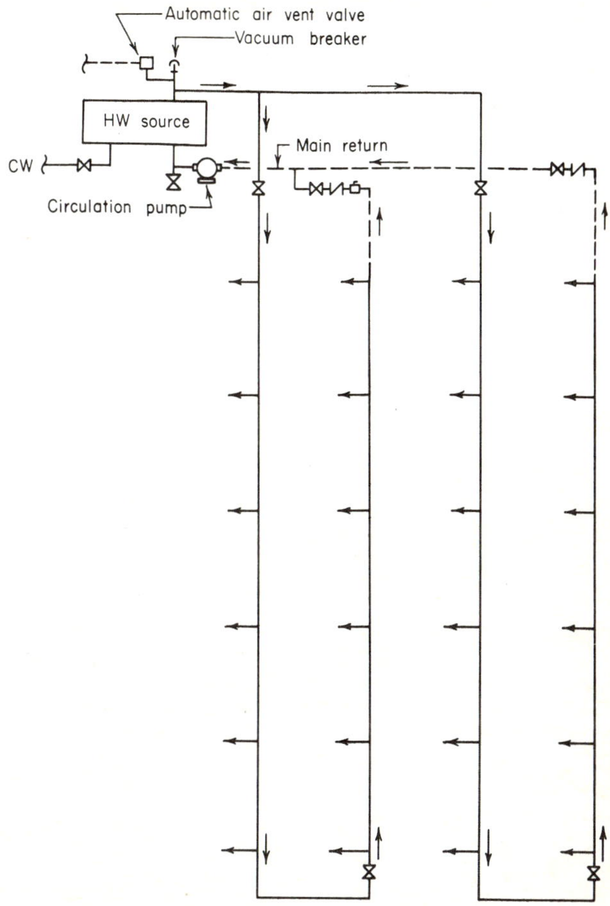

Fig. 7-6 Inverted combination downfeed and upfeed system.

water descends to the heater. These two opposing flows cause circulation to become retarded in both the main return line and in the piping between tank and heater.

Preferably, the main hot water return line of the circulation system should return cooled hot water to the source of hot water supply for reheating by means of a connection to horizontal piping at the heater inlet. Near this connection, a check valve should be provided in the main return line to prevent reverse flow. At the check valve inlet, a gate, ball, or butterfly valve should be installed in the return line and a second gate, ball, or butterfly valve, for blowoff purposes, should be connected to a branch fitting upstream therefrom. By shutting off the gate valve in the main return line and opening the blowoff gate valve, accumulated mud and sediment may be blown out of horizontal sections of the main return line, which should be maintained periodically for satisfactory operation of the system.

Proper valving of all risers is a practical necessity. Gate valves should be provided at the top of all downfeed supply risers and at the bottom of all upfeed supply risers in order to permit such risers to be shut off when necessary. For the same reason, gate valves or plug-type stopcocks should be provided in the horizontal piping at the base of each hot water return riser upstream from its connection to a main hot water return line. In addition, a check valve should be installed adjacent to and upstream from such shutoff valves at the base of each return riser so as to prevent drawing of return line water at fixture outlets.

The force causing circulation in a return circulation type hot water supply system is the difference in head existing in the hot water supply and return lines at the source of hot water supply. In a gravity circulation system, the difference in head is produced by the difference in temperatures, and corresponding densities, of water in the hot water supply and return lines. Water in the return line is cooler and more dense than water in the supply line, and therefore produces a greater hydrostatic head or pressure at the point where they connect to the source of hot water supply. The head induced by the difference in water temperatures in supply and return lines is commonly termed *gravity circulation head,* or at times called *thermal head.* This head varies directly with the heights that the supply and return lines have in common above the heater level.

The recommended design basis for gravity circulation systems is to assume supply and return line temperatures of 140 and 100°F (60 and 37.8°C), respectively. Water density at 100°F (37.8°C) is 61.998 lb/ft^3, and at 140°F (60°C) is 61.386 lb/ft^3. For this 40°F (22.2°C) differential, the density difference is 61.998 − 61.386 = 0.612 lb/ft^3, and the induced gravity circulation head may be calculated as being 0.612/61.386 = 0.01 ft (0.003 m) of 140°F (60 °C) water column or head for each foot of effective circulation system height. Thus, if the highest point of the circulation piping were 100 ft (30.5 m) above the hot water tank or heater, the available induced gravity circulation head would be 0.01 × 100 = 1.00 ft (0.3 m) of head, based on a 40°F (22.2°C) differential in hot water supply and return line temperatures. This illustrates the fact that the head induced in gravity circulation systems is relatively small, yet sufficiently significant to permit adequate design of such systems where a reasonable amount of effective circulation height exists. Most instances of difficulty in achieving adequate gravity circulation occur in systems which are long in horizontal run and short in effective circulation height, a condition where it is generally advisable to provide positive circulation head by means of a circulation pump.

In pump-equipped return circulation type systems, the pump provides circulation at a rate and head corresponding to the performance characteristics of the pump and the frictional resistance of the circulation piping. Charts showing performance characteristics of various sizes and models of circulation pumps may be obtained from pump manufacturers. From the chart for a given pump, the head available at the pump outlet and the efficiency of the pump may be determined for any rate of discharge within the pump's capacity range. These factors should be considered in selecting pumps suitable for the circulation rates and frictional resistance of systems.

The recommended design basis for systems equipped with circulation pumps is to assume hot water supply and return water temperatures of 140 and 120°F (60 and 49°C), respectively. This 20°F (11°C) differential, when compared with the 40°F (22°C) differential in gravity systems, means that the pump-equipped system is designed to circulate hot water at double the rate of the gravity system and, consequently, provides a higher level of performance in maintaining constancy of hot water supply temperature. This feature should be considered in any comparison of operating and maintenance costs of pump-equipped systems and gravity circulation systems. When comparing the original costs of such systems, the additional amount spent in pump-equipped systems for pumps and their automatic controls should be weighed against the savings resulting from the permissible use of smaller return piping sizes. For example, if a pump-equipped system circulates water at double the rate and at 14.4 times as much circulation head as in a gravity system, the return piping sizes may be calculated as being three-fourths that required for a gravity system.

To balance the temperature of water returning from two or more return risers, so that each returns water to the main hot water return line at the same temperature and at a suitable temperature differential between supply and return piping, is a design objective. However, rarely is it possible to design perfect balance into a system, because exactly correct return pipe diameters cannot be selected from the limited number of standard piping diameters that are available. Hence, means for adjusting circulation rates in risers must be provided in most systems, except those which are relatively small and have a high ratio of circulation height as compared with return piping length. The means for adjusting circulation may be either special balancing valves or plug-type stopcocks.

Figure 7-7 is presented herewith for discussion and reference purposes. In systems having two or more risers, circulation is divided among

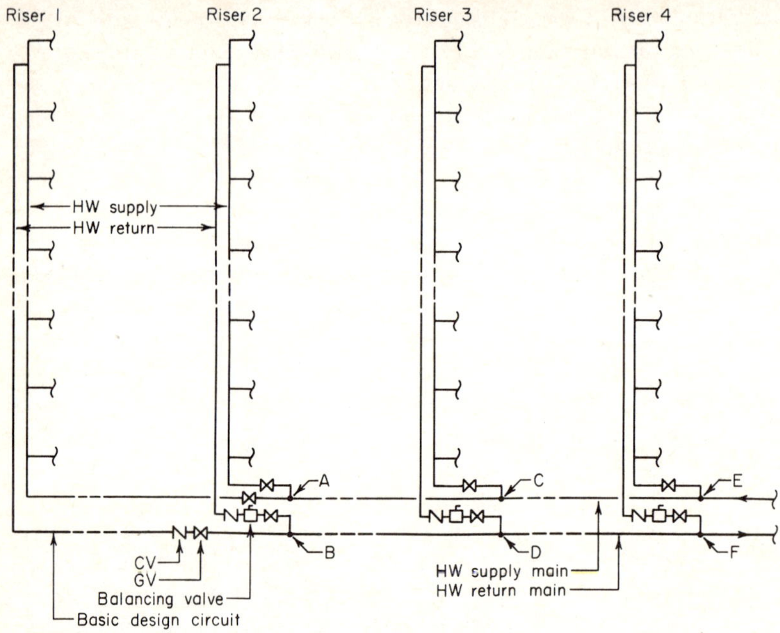

Fig. 7-7 Balancing valves on hot water return risers.

the various risers, which form bypass or parallel circuits through which flow proceeds in accordance with the frictional resistances of the circuits. The riser from which water returns at the lowest temperature is usually at the end of the particular run of return piping having the maximum developed length, and as such, constitutes the basic design circuit of the system of returns.

In the sketch, it may be seen that the distance flow must travel in circulating from point *A*, through riser 1, and back to point *B* is much greater than the distance flow must travel from point *A*, through riser 2, and back to point *B*. In view of the greater distance through riser 1, more frictional resistance occurs in that flow path, and return water temperature will be lower than occurs through riser 2. Similarly, for flow from point *C* to point *D*, more resistance occurs in the flow path through riser 2 than through riser 3; and for flow from point *E* to point *F*, more resistance occurs in the flow path through riser 3 than through riser 4. It may be anticipated that return water temperature at the base of riser 4 will be the highest and at the base of riser 1 will be the lowest.

In order to balance return water temperatures, additional appropriate amounts of frictional resistance must be provided in the returns from risers 2, 3, and 4. This may be done by means of proper adjustment of balancing valves or plug-type stopcocks installed in the horizontal run at the base of those risers. No balancing valve should be provided anywhere in the return piping from riser 1, for this flow path has the greatest frictional resistance in the system; and all other risers in the system must have their circulation rates adjusted so as to achieve return water temperature balance with riser 1.

Before a return circulation system is put into service, it should be adjusted for return line temperature balance under conditions whereby no water is drawn at fixture outlets. The adjustment procedure is simple and may be accurately done using thermometers. First, close all balancing valves so that all circulation goes through riser 1. Then, open the balancing valve on riser 2 to the degree necessary so that it returns water constantly at the same temperature as riser 1. Next, open the balancing valve on riser 3 to the degree necessary so that it returns water constantly at the same temperature as risers 1 and 2. Similarly, adjustment may be made with the balancing valve on riser 4 and with balancing valves on any other risers so that all return water at the same temperature. In this way, relatively perfect return water temperature balance may be achieved.

When a plug-type stopcock is provided as a single device, instead of a gate valve for shutting off the return riser and a balancing valve for adjusting the rate of circulation in the return riser, the stopcock serves a dual purpose. In the event that the stopcock must be shut off during emergencies, it must be properly adjusted when it is open again. Otherwise, the return water temperature balance in the system may be upset sufficiently to cause complaints of inadequate hot water service.

PROCEDURE FOR DETERMINING CIRCULATION RATES AND SIZING RETURN PIPING

The objective of domestic hot water return circulation systems is to maintain hot water temperature within suitable limits in the supply mains and risers. This is done by recirculating hot water through the mains and risers from a storage tank or heater, so that hot water is immediately available at fixtures far removed from the hot water source. To do this economically and accurately, the rates of circulation in, and the sizes of, return piping should be designed in accordance with (1) the heat loss rate of the piping in which circulation occurs, (2) the temperature differential at which the system is to operate, and (3) the pressure or head available for circulation.

A simple, rational method has been developed whereby engineering principles may be applied for determining proper circulation rates and sizes of return piping in any return circulation type domestic hot water supply system to meet particular performance conditions in much the same manner as in hot water heating systems. This method is a complete procedural package, consisting of 11 simple steps that can be applied after hot water supply mains and risers have been sized in accordance with hot water demand requirements of fixtures. The procedure is as follows:

Step 1 Calculate the heat loss rates of the entire hot water supply piping through which hot water is to circulate. This may be done using Table 7-1. It gives heat loss values for various kinds and sizes of piping, bare and insulated, based upon a 140°F (60°C) water temperature and a 70°F (21.1°C) air temperature.

Values for insulated piping assume a covering equivalent to 1 in thick glass fiber. Data compiled in this table are the result of evaluating and converting many items of research information on heat loss from piping gathered bit by bit from scattered reports, charts, manufacturers' literature, and research publications.

Step 2 Determine the heat loss rates of the return mains and risers. As their sizes are unknown at this stage, their heat loss rates cannot be calculated directly. A tentative heat loss rate must be assumed, based upon known facts, experience, and judgment. In this regard, it should be recognized that the length of return piping is approximately the same as that of the supply mains and supply risers in the circulation

Table 7-1
HEAT LOSS RATES FOR HOT WATER SUPPLY PIPING
In Btu/h/linear ft, 140°F Piping to 70°F Air

	Bare			Insulated
Nominal pipe size, in	Galvanized iron or steel pipe, SPS	Brass or copper pipe, SPS or threadless	Copper water tube, type L	All types of piping
½	35	26	19	15
¾	43	32	26	17
1	53	38	32	19
1¼	65	46	39	21
1½	73	53	46	24
2	91	65	58	29
2½	108	75	68	32
3	130	90	81	38
4	163	113	103	46

system. Experience indicates that when final sizes of return piping are established by design, they usually are only about one-half the sizes of comparable sections of supply mains and about three-eights the maximum sizes of supply risers. Table 7-1 may be consulted for the relative heat loss values of the different pipe sizes for such proportions. Based upon these considerations, a reasonable assumption may be made for the heat loss rates of return mains and risers. It is recommended that the assumption be as follows: two-thirds the heat loss rate of the supply piping, when both supply and return piping are insulated, or when both are bare; and four-thirds the heat loss rate of the supply piping, when the supply piping is insulated and the return piping is bare. In this way, heat loss rates for unsized return piping may be tentatively established, and the respective heat loss rates for the various parts of the system may be brought into fairly clear focus.

Step 3 Sum up calculated and assumed heat losses for the entire supply and return piping through which hot water circulates in the system so as to establish its tentative total heat loss rate. Then assign the losses pertaining to individual parts of the system to appropriate main and branch circuits of the circulation piping, in order to establish the proportionate amount of circulation required to compensate for their heat loss loads.

Step 4 Calculate the circulation rates required for the main and branch circuits in accordance with their assigned heat loss loads and with the temperature differential at which the system is to operate. A 20°F (11.1°C) differential is generally recommended for use in designing systems equipped with circulation pumps, while a 40°F (22.2°C) differential is recommended for systems in which circulation is induced by gravity head.

The rate of heat supplied by a circulation rate of 1 gal or 8.3 lb of water per min, in losing 20°F (11.1°C) in temperature, is 166 Btu/min, or 9960 Btu/h. As a matter of convenience, the latter value may be taken as 10,000 Btu/h. The same circulation rate, 1 gpm, in dropping 40°F (22.2°C) supplies about 20,000 Btu/h. These factors may be applied to establish the required circulation rates for all parts of the system. For example, if a pump-equipped system has a total heat loss rate of 61,145 Btu/h, the required pump discharge capacity is equal to 61,145/10,000, or 6.1 gpm; and if the heat loss rate assigned to a given return riser or section of a return main is 8410 Btu/h, the required circulation rate for it is equal to 8410/10,000, or 0.84 gpm. Similar application of the 20,000 Btu/h factor may be made in the case of gravity circulation systems operated at a 40°F (22.2°C) differential. As a result, the circulation rates in gravity systems are just one-half the rates that would be applied if the system were pump-equipped.

Step 5 Determine the pressure or head available for establishing circulation. For a pump-equipped system, this may be determined directly from the performance charts for circulation pumps. Such charts may be obtained from manufacturers. Select the smallest size of pump that has reasonably high efficiency at the required delivery rate. Note particularly the discharge head at which the pump delivers the required rate of flow. For example, if the performance curve for a selected pump shows that the required circulation rate occurs at a discharge pressure of 7.3 ft (2.2 m) of water column or head, this is the head available for circulation.

For a gravity circulation system the available head must be calculated. As a 40° differential is used in designing such systems, the available head may be calculated on the following basis: 0.01 ft (0.003 m) of water column or head multiplied by the vertical height in feet between the elevation of the hot water tank or heater and the elevation of the highest point of the circulation piping. For example, if the highest point of the circulation piping were 109 ft (33. m) above the hot water tank or heater, the available induced gravity head would be calculated as $0.01 \times 109 = 1.09$ ft (0.33 m) of head.

Step 6 Determine which particular run of return piping has the maximum developed length in the system. Measure from the tank or heater to the most extreme point at which the return piping connects to a hot water supply riser. This maximum run of return piping, being the longest in the system, will have the most frictional resistance to flow and require the largest pipe sizes; therefore, it can be called the basic circuit for design purposes. The basic design circuit is illustrated in Fig. 7-8. No balancing valves should be included anywhere in the basic circuit. If they were, someone at sometime certainly would adjust them in a way that would restrict flow, increase the resistance of the basic circuit, and thereby upset the performance of the system. Balancing valves should be provided on all return branches connecting directly to the basic circuit or to branch mains connecting to the basic circuit. Such valves are provided as a means of adjusting and adding resistance to flow in the short return branches so that circulation or flow through them can be reduced to establish temperature balance with the basic circuit.

Step 7 Calculate the pressure drop due to frictional resistance of the

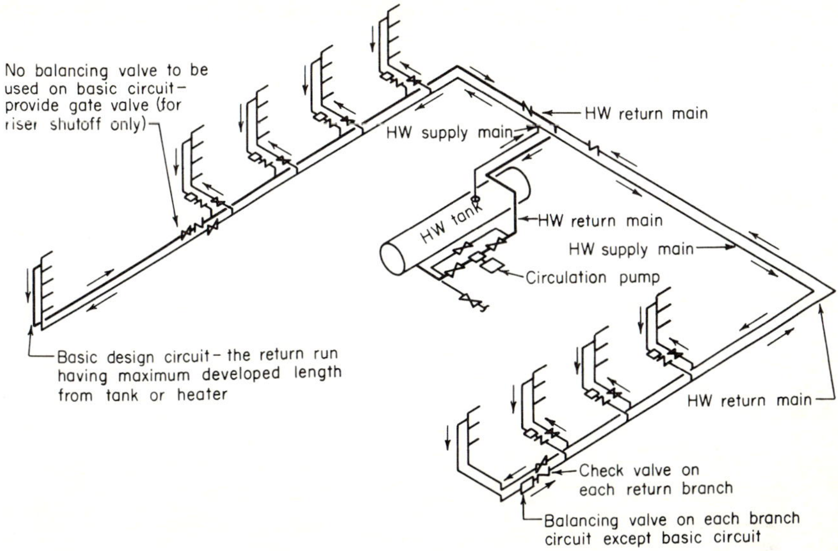

Fig. 7-8 Basic design circuit of hot water recirculation system.

water flowing at required circulation rates in the hot water supply piping extending from the hot water tank or heater, along the supply main, and up the supply riser to the point at which the basic return circuit connects thereto. When this pressure drop is determined (usually found to be so small as to be negligible), it should be deducted from the available head. The difference is the amount of pressure or head which may be dissipated as frictional loss due to flow at the required rates in the basic circuit.

Calculations for pressure drop in the supply piping run may be made directly by applying appropriately one of the following two formulas:

For brass or copper piping:

$$h = 0.000623q^2\frac{L}{d^5} \qquad [7\text{-}1]$$

For galvanized iron or steel piping:

$$h = 0.001246q^2\frac{L}{d^5} \qquad [7\text{-}2]$$

where h = head or pressure loss due to friction, ft of water column
q = quantity rate of flow, gpm
L = length of pipe, ft
d = diameter of pipe, in

In the above formulas, the length of pipe to be used is not just the developed length of the pipe. Rather it is the sum of the developed length plus the equivalent length calculated for fittings and valves. To determine the equivalent length of fittings and valves, use Tables 5-6*a*, 5-6*b*, 5-7*a*, and 5-7*b*.

Step 8 Determine the maximum permissible uniform pressure drop for the basic circuit. The amount of head that may be dissipated as frictional loss in the basic circuit, as established by step 7, should be divided by the total equivalent length of the circuit. As return pipe sizes are unknown at this stage, the equivalent length for fittings and valves must be assumed tentatively. A reasonable assumption in this case would be 10 percent of the developed length of the basic circuit.

For example, suppose a pump-equipped system is being designed and the performance chart for the pump selected indicates that the pump has a 7.3 ft (2.2 m) discharge head at the required circulation rate; and suppose that the friction loss through the supply piping, as calculated by step 7, is 0.2 ft (0.06 m) of head loss, thereby establishing that 7.1 ft (2.1 m) of head remains for dissipation as frictional loss in the basic circuit. Now, if the basic circuit has a developed length of 645 ft (197 m) and if 10 percent or 65 ft (19.7 m) is assumed as the equivalent length for fittings and valves, the total equivalent length of the circuit may be tentatively established as 710 ft (216.7 m). The maximum uniform pressure drop h/L may then be determined as 7.1 ft (2.17 m) head divided by 710 ft (216.7 m) length, or 0.01 ft (0.003 m) head loss per 1 ft (0.3 m) length.

Similarly, the maximum uniform pressure drop for a system circulated by gravity may be calculated. For example, if the head remaining at the top of the basic circuit was 1.09 ft (0.33 m) and if the total equivalent length of the circuit was 297 ft (90.5 m), the maximum uniform pressure drop h/L would be determined as 1.09 ft head/297 ft length, or 0.00368 ft (0.00112 m) head loss per 1 ft (0.3 m) length.

Step 9 Calculate and tabulate the rates of flow for various pipe sizes, of the type of piping selected for the system, that will produce friction losses corresponding to the maximum permissible uniform pressure drop. Actual internal diameters should be applied. For convenience in making direct flow rate calculations, the two formulas given in step 7 have been rearranged as follows:

For brass or copper piping:

$$q = 40.1 d^{2\ 1/2} \left(\frac{h}{L}\right)^{1/2} \qquad [7\text{-}3]$$

For galvanized iron or steel piping:

$$q = 28.3d^{2\ 1/2}\left(\frac{h}{L}\right)^{1/2} \qquad [7\text{-}4]$$

Using these formulas, a table of flow rates, in gallons per minute, has been developed for various pipe sizes and kinds of piping, based upon a uniform pressure drop h/L of 0.01 ft (0.003 m) head loss per 1 ft (0.3 m) pipe length. Table 7-2 is presented herewith.

Similar tables may be developed for other pressure drops simply by multiplying the flow rates shown in the aforementioned table by an appropriate factor. This may be done because the flow rate is proportional to the square root of the uniform pressure drop h/L. For example, if a table were to be developed to show flow rates corresponding to a uniform pressure drop of 0.00368 ft (0.00112 m) head per ft (0.3 m) length, the appropriate factor would be $(0.00368/0.01)^{1/2} = (0.368)^{1/2} = 0.607$. This factor may be used to multiply all the values given in the table of flow rates based on a pressure drop of 0.01 ft (0.003 m) head per ft (0.3 m) length.

Step 10 Size all parts of the basic circuit. Use the tabulated values of flow rates producing a pressure drop corresponding to the maximum permissible uniform pressure drop for the circuit. The same values may be used for sizing all other parts of the return piping, for such other parts are branch circuits of shorter length than the basic circuit. Thus, the sizes established for branches should be adequate.

Step 11 Now that all return piping sizes have been established, apply these sizes in checking the assumptions and calculations made in steps 2 to 9. Determine the exact heat losses for the system, as there no

Table 7-2
FLOW RATES
Producing a Pressure Drop of 0.01 Ft (0.003 m) Head per 1 Ft (0.3 m) Pipe Length

Nominal pipe size, in	Brass or copper pipe, threadless		Brass or copper pipe, SPS		Copper water tube, type L		Galvanized iron or steel pipe, SPS	
	ID in	Flow, gpm	ID, in	Flow, gpm	ID, in	Flow, gpm	ID, in	Flow, gpm
½	0.710	1.71	0.626	1.24	0.545	.89	0.623	.88
¾	0.920	3.22	0.822	2.44	0.785	2.19	0.824	1.71
1	1.185	6.1	1.06	4.6	1.025	4.26	1.048	3.22
1¼	1.53	11.6	1.37	8.82	1.265	7.22	1.380	6.2
1½	1.77	16.8	1.6	13.1	1.505	11.1	1.61	9.15

longer is any need to rely upon assumptions. Recheck the required circulation rates for all parts of the system, the available circulation head, the total equivalent length of the basic circuit, and the maximum permissible uniform pressure drop. If reasonable assumptions were originally made in this procedure, there should be no reason generally to change sizes any further, except for the few branches that were borderline cases according to the table of flow rates. Such branches may be treated individually in accordance with their loads, length, and the pressure differential between their circuit connections.

STANDARD CODE REGULATIONS

Hot Water Supply System Required Buildings and portions thereof shall be provided with plumbing systems designed to dispose of the sewage from all fixtures and to furnish cold water to every water closet and urinal, and hot and cold water to every sink, laundry tray, automatic laundry washing machine, lavatory, bathtub, and shower required therein, except as otherwise provided herein.

Hot Water for Commercial Dishwashing Machines and Fixtures Commercial dishwashing machines and dishwashing fixtures using hot water shall be provided with water at 140 to 160°F for washing, and at 180 to 190°F for sanitizing.

Pressure Relief Valve An approved pressure relief valve shall be provided on plumbing equipment used for heating or storing hot water. Such valve shall be of adequate relief rating for the equipment served.

Temperature Relief Valve or Energy Shutoff Device An approved temperature relief valve or energy shutoff device shall be provided on equipment used for heating or storing hot water. Temperature relief valves shall be of adequate relief rating for the equipment served. Energy shutoff devices shall be of adequate performance rating for the equipment served.

Combination Pressure-Temperature Relief Valve An approved combination pressure-temperature relief valve of adequate rated relief capacity may be provided in lieu of separate pressure and temperature relief valves.

Location of Relief Valves and Energy Shutoff Devices Pressure relief valves shall be installed in the cold water supply line to the heater or tank served, except that, in areas where scale formation due to water hardness is appreciable, such valve may be installed in an approved location in the hot water supply line from the heater or tank served.

Temperature and combination pressure-temperature relief valves shall be installed so that the temperature-sensitive element is immersed in the hottest water, such as: within the top 6 in of the tank of an underfired hot water storage heater; above the hot water inlet to a tank equipped with a side-arm-type water heater; and above the topmost heating element of an electric water heater.

Relief valves shall be installed so that no check valve or shutoff valve intervenes between a relief valve and the heater or tank served.

Immersion-type energy shutoff devices shall be installed so that the temperature-sensitive element is immersed as required for temperature relief valves. Strap-on-type energy shutoff devices shall be installed so that the temperature-sensitive element is mounted on the tank wall and is responsive to the highest water temperature within the tank.

Relief Valve Outlet Connections No relief outlets of relief valves shall be directly connected to drainage or vent piping. Where pipes from such relief outlets discharge into plumbing fixtures, an air gap shall be provided in conformity with the minimum required air gaps for potable water outlets at plumbing fixtures. No relief outlet or relief pipe shall discharge so as to be a hazard, be a potential cause of damage, or otherwise be a nuisance.

Vacuum Relief Valves Where a hot water storage tank or an indirect water heater is located at an elevation above the fixture outlets in the hot water system, a vacuum relief valve shall be installed on the storage tank or heater.

Location of Pressure Markings on Hot Water Storage Tanks Hot water storage tanks shall be installed so that their pressure markings, showing the maximum allowable water working pressure, are in an accessible location for inspection.

Drain Cocks or Valves for Hot Water Storage Tanks Hot water storage tanks shall be provided with cocks or valves for emptying.

Return Circulation Type System—Where Required Hot water supply systems in other than one- and two-family dwellings shall be of the return circulation type in buildings more than four stories high and in buildings where the developed length of hot water piping, from the source of hot water supply to the farthest fixture supplied, exceeds 100 ft.

8

DRAINAGE AND VENT SYSTEMS

GENERAL DESIGN CONSIDERATIONS

This chapter deals with general design considerations for building storm water drainage systems, and for building sanitary drainage and vent systems. These are essential elements in the planning of systems to provide adequate, unobjectionable means for disposal of storm water from the premises, and for safe, sanitary disposal of sewage from all fixtures.

Of primary consideration is the availability of adequate means for unobjectionable disposal of storm water, and safe, sanitary disposal of sewage. Where public sewer systems are available, building systems should be connected thereto for reliability of service and protection of health and safety. Where public sewer systems are not available, adequate approved private treatment and disposal systems must be designed and constructed in lieu thereof. Suitable provisions shall be made to prevent backflow of sewage into buildings.

The sanitary drainage system shall be designed, constructed, and maintained to guard against fouling, deposit of solids, and clogging, and with adequate cleanouts so arranged that the pipes may be readily cleaned. The system shall be designed to provide an adequate circulation of air in all pipes with no danger of siphonage, aspiration, or forcing of trap seals under conditions of ordinary use. Each vent terminal of the sanitary drainage system shall extend to the outer air and be so installed as to minimize the possibilities of clogging and the return of foul air to the building. No substance which will clog or accentuate clogging of pipes, produce explosive mixtures, destroy the pipes or their joints, or interfere unduly with the sewage-disposal process shall be allowed to enter the sanitary drainage system. Proper protection shall be provided to prevent contamination of food, water, sterile goods, and similar materials by backflow of sewage. Where necessary, the fixture, device, or appliance shall be connected indirectly with the sanitary drainage system.

Systems shall be designed and installed so that they may be readily maintained in a safe and serviceable condition from the standpoint of both mechanics and health. Piping shall be designed and installed with due regard to preservation of structural members and prevention of damage to walls and other surfaces through fixture usage. Plumbing systems shall be subjected to such tests as will effectively disclose all leaks and defects in the work or material.

SEWAGE DISPOSAL

Each building in which plumbing fixtures are installed should be provided with a sanitary drainage system for conveying sewage from the fixtures to an adequate and approved means for sewage disposal, such as a public sanitary or combined sewer where available. Where a public system is not available, an approved private sewage disposal system must be provided. Such private systems must conform to regulations of the health authority having jurisdiction in the area. No sewage should be discharged into sewers intended for storm water only, nor be disposed of onto the ground or into a public waterway. Sewage, or other waste which may be deleterious to surface or subsurface waters, should not be discharged into the ground or into a waterway unless first rendered harmless through subjection to treatment in accordance with standards acceptable to the authority having jurisdiction.

Wherever public sanitary or combined sewers are available for disposal of sewage from a building, it is recommended that the sanitary drainage system of the building be connected to the public system. This is the most satisfactory method of assuring disposal of sewage without health hazard or nuisance. No other known method affords the same convenience, reliability, capacity, and trouble-free service for the life of any given building.

In areas where public sanitary or combined sewers have been installed, it is almost always found that public sewer district authorities have regulations which require that the sanitary drainage systems of buildings be connected to the public systems when such buildings are not too distant therefrom. Availability may be considered to be just 100 ft (30.5 m) in the case of a one- or two-family dwelling, and as much as 500 ft (152 m) for other types of building occupancies. In general, wherever it is practical to do so, it is advisable to connect to public sanitary or combined sewers as the most economical method of disposing of sewage.

The disposal of radioactive wastes should conform to applicable regulations of the health authority having jurisdiction. All such wastes do not pose the same degree of hazard, some requiring little or no special treatment and precautions, while others may be too hazardous for disposal into the sanitary drainage system. State health authorities regulate the use of radioactive materials and permit such materials to be used only by licensed and qualified laboratories and their personnel. The manner of disposal to be applied with each particular radioactive substance is carefully prescribed by state authorities in accordance with federal safety standards.

STORM WATER DISPOSAL

Buildings should be equipped with provisions for draining water from roofs and paved areas, including yards and courts. Storm water should be conveyed to an adequate and unobjectionable system of storm water disposal, such as a public storm or combined sewer where available. No storm water should be discharged into sewers intended for sewage only, nor should it be discharged so that water flows across public sidewalks, drains onto adjacent premises, causes erosion of soil, or forms a pond on the premises. Storm water may be disposed of into an existing stream on, or adjacent to, the premises, or into an adequate system of dry wells constructed underground.

Wherever public storm or combined sewers are available for disposal of storm water from a building, it is recommended that the storm drainage system of the building be connected to the public system. The convenience, reliability, capacity, and trouble-free performance recorded for public systems operated and maintained by public authorities have clearly established their superiority over any other method of storm water disposal.

In most areas where public storm or combined sewers have been installed, it is mandatory that the storm water drainage systems of buildings and premises be connected to the public system if the premises are within a reasonable distance of the system. The limits of what is deemed *availability* of the public system may vary in different areas and for different types of building occupancies. For example, a public system may be deemed available for a one- or two-family dwelling when the system is within 100 ft (30.5 m) of the premises as measured along the street. For any other type of building occupancy, the public system may be deemed available when it is within 500 ft (152 m) of the premises.

BUILDING SEWERS

Building sewers should be designed and installed in accordance with regulations of the authority having jurisdiction in the particular area in which they are to be located. Regulations in different areas may vary greatly because of local conditions, such as capacity limitations of the public sewer system, or the public sewage treatment plant, potential infiltration of groundwater into public sewers, maintenance procedures established by sewer authorities in accordance with local sewer conditions, soil conditions below ground, public roadway construction and maintenance standards, roadway traffic and loadings, underground structures and utility piping systems located beneath public roadways, and numerous other conditions singular to given areas in which building sewers may be installed.

A great variety of regulatory agencies exercise jurisdiction over the design and installation of building sewers, such as a public sewer district authority, public highway and sewer department, local board of health, public safety department, and other governmental agencies appropriately designated by law and charged with such duties and responsibilities.

Local health agencies usually exercise jurisdiction over building sewer installations which convey sewage to private sewage disposal systems located on the premises with the building. In this case, the health agency may appropriately serve as the single agency exercising jurisdiction over private sewage-disposal systems, building sewers, and plumbing systems for buildings.

Existing building sewers are not recommended for use in conjunction with new building plumbing systems or major alterations of existing systems, unless it has first been definitely determined by thorough inspection and test that the existing building sewer is in good condition, watertight, in proper alignment, and of adequate size and capacity to perform satisfactorily under the loads to be conveyed from the building. The costs involved in making a proper determination in the case of existing building sewers usually are high and are difficult to justify especially in view of the fact that sewers tend to settle unevenly and become unevenly aligned after years of service unless originally installed of highly durable piping and bedded in hard packed natural soil. Generally, it is recommended that new building sewers be provided for new building plumbing systems and major alterations of existing systems.

COMBINED SYSTEMS IN BUILDINGS

Storm water drainage systems and sanitary drainage systems should be independent of each other unless they discharge into a combined public sewer system. In that case, it may be economically advantageous to bring the two systems together at some appropriate point inside the building and extend therefrom a single dual purpose drain and building sewer to its branch connection with the combined public sewer.

Sanitary drainage systems and storm drainage systems should be joined only at the lowest levels of such systems. To combine such systems at an upper level could cause extensive flooding during storm periods as the result of fouling of drains from sewage flow. Consequently, the different systems should be combined by joining their building drains or building sewers in a manner that will not produce objectionable conditions.

The design of storm drains is based on their full flow capacity for peak load, while sanitary building drains are based on their half-full flow capacity for peak load. To avoid excessive pneumatic effects and flow interference in combining the two systems, it is recommended that the building storm drain connect to the combined drain in the same horizontal plane by means of a single Y branch located at a distance of at least 10 ft (3 m) from any sanitary drainage branch to the combined drain.

STORM WATER DRAINS

Storm water drains which connect to a combined building sewer or building drain should be designed to prevent escape of sewer gases and objectionable odors from the combined system. This should be avoided either by means of an individual trap installed in the horizontal branch serving each leader and area drain, or by means of a single trap installed on the main storm water drain serving all leaders and area drains prior to its connection with a combined building drain, building sewer, or public sewer. Such traps should be of the same size as the horizontal drain in which they are installed; should be provided with an accessible cleanout on the inlet side; should be located within the building; and in the case of leaders, should be located in the horizontal piping at the base of leaders.

Provisions should be made to exclude from storm drains any solid particles and objects that may otherwise cause stoppage conditions in the drains. Strainers should be installed in the inlets to leaders. They should extend at least 4 in (100 mm) above the roof or gutter surface at the leader inlet and should have clear open area at least one and one-half times that of the leader, except that for roof drains of sun decks, parking decks, and similar areas which are usually serviced and maintained, strainers may be of the flat surface type installed level with the deck, but should have clear open area at least twice that of the leader. All openings through roofs for roof drains must be made watertight.

CONTROLLED FLOW STORM WATER DRAINAGE SYSTEM

There is nothing radically new about this type of system. It is known as a "controlled-flow storm drainage system for flat roofs" and has been used for many years in conjunction with the design of numerous large industrial buildings in various areas of the country.

Generally such systems have been specially designed for each building by qualified engineers, in order to provide an adequate storm drainage system at a cost less than would be the case if the system were designed in accordance with most existing plumbing codes. In some cases, such systems have been provided to limit the rate of storm water runoff from buildings into nearby recharge basins, ponds, or watercourses.

In most codes, the maximum drainage area permitted to be served by a given size of leader or horizontal storm drain is stipulated. The area is based upon the assumption that the piping must convey all the rain which falls on the area at the same quantity rate of flow as corresponds to the maximum rainfall intensity over a 10-min period recorded by the U.S. Weather Bureau for the region.

This basis has been found by experience to be valid for smooth roofs pitched several degrees or more. However, for dead-level flat roofs, and those having no more than one degree of slope, the rate of runoff is much slower than for other roof slopes.

Many industrial buildings are designed today with relatively dead-level flat roofs of watertight construction. They lend themselves to more economical design of storm drainage systems. By installing special roof drains equipped with small orifices or weirs through which storm water must run off, the rate of runoff may be controlled. With such control, much of the rainwater during a storm is impounded on the roof and the water depth builds up to a maximum at the time the storm ends. Then the impounded water runs off at a rate controlled by the characteristics of the weir and the depth of water above the crest of the weir.

To obtain balanced design for flat roof drainage, the roof drains for a given area should have their weir crests at the same elevation. The roof drains also should be distributed over the area so that discharge into the drains is equal to the rainfall on the roof area served, while at the same time the storm water is impounded on the roof.

Rainfall records for any given region may be obtained from the U.S. Weather Bureau. Such records provide data relative to intensity of rainfall, duration of storms, and frequency of intense storms. Applying this information, the depth to which water may be impounded on a flat area for any given rainfall rate and duration and frequency of storms can be estimated with reasonable accuracy.

Assuming a maximum design depth to which water should be impounded on the roof, and knowing the discharge rate of the weir to be used, the rainfall data may be applied to determine the number of weirs required to satisfy the maximum design depth limitation, and to establish the rate of controlled runoff.

Table 8-1
CONTROLLED-FLOW STORM WATER DRAINAGE SYSTEMS, LIQUID FLOW CAPACITY

Pipe diameter		Leaders		Horizontal drains, slope, in/ft (mm/m)			
				⅛	10.4	¼	20.8
in	mm	gpm	L/s	gpm	L/s	gpm	L/s
2	50	23	1.45				
2½	62.5	41	2.59				
3	75	67	4.23	34	2.14	48	3.03
4	100	144	9.08	78	4.92	110	6.94
5	125	261	16.47	139	8.77	196	12.4
6	150	424	26.75	222	14.0	314	19.8
8	200			478	30.2	677	42.7
10	250			860	54.2	1214	76.6

When the rate of flow in the various parts of the piping is established in terms of gallons per minute of flow, the matter of sizing the piping is very simple. For easy and accurate sizing, Table 8-1 gives the recommended liquid flow capacity for controlled-flow storm water drainage systems.

This table differs from the usual sizing tables given in codes; the terms are not square feet of roof area permitted but rather gallons per minute of flow rate permitted. Thus, the table may be used in sizing controlled-flow drainage systems in accordance with the established controlled rates of flow.

Manufacturers of roof drains have developed and marketed various kinds of weirs as part of roof drains for flat roofs. Information regarding the flow ratings and installation recommendations for these products may be obtained from the manufacturer.

Presently, some code authorities permit such systems to be installed. In lieu of sizing the system on the basis of permissible areas for various sizes of piping, controlled-flow systems may be acceptable and sized on the basis of controlled flow when the following conditions are satisfied:

1. Calculations and drawings proving adequacy of design must be submitted to the code authority for approval.
2. Design must be based upon the maximum storm of 25-year frequency at least.
3. Design water depth on flat roofs must not exceed 3 in (75 mm).

4. Maximum roof drainage time after storm must not exceed 24 h.

5. Roof must be dead level; have 45° cants at walls and parapets; have flashings extending at least 6 in (150 mm) above roof level; have scuppers in parapet walls at a level 5 in (125 mm) above the roof; also, roofs must be of watertight construction.

6. At least two drains must be installed in roof areas 10,000 ft^2 (929 m^2) or less, and at least four drains in larger areas.

7. Storm and sanitary drainage systems in the building shall be separate.

8. Control of runoff from flat roofs shall be by proportional weirs; no valves or mechanical devices shall be permitted.

9. Drains not equipped with weirs, such as area drains, may be connected to the controlled-flow system, provided the square feet of area so drained is converted to the appropriate flow rate basis and the storm water drain sized to convey the sum of the loads therein.

FIXTURE AND EQUIPMENT CONNECTIONS TO SANITARY DRAINAGE SYSTEM

The most sanitary method for conveying liquid wastes and sewage from fixtures to a safe disposal terminal is to connect the waste outlets of the fixtures directly to the sanitary drainage system of the building. With this method, once the wastes are discharged into the system they flow by gravity to the disposal terminal through a watertight and gastight system of piping designed to provide safe, rapid disposal of sewage and to prevent the escape of any sewer gases and odors from the piping system into habitable spaces in the building.

This method should be applied for all fixtures and equipment which discharge liquid wastes, except for certain specific fixtures and equipment which require special types of connections for a variety of reasons. For example, regardless of how well designed a sanitary drainage system may be, the interior of the piping will become lined with deposits of solid matter, and a stoppage condition eventually can develop and cause backup of sewage into fixtures. Backup of sewage into certain specific fixtures and equipment cannot be permitted.

Fixtures and equipment used for storage, preparation, or processing of food or drink, sterile goods, or similar materials must be provided with air breaks, that is, physical disconnections, at their waste outlets.

The air breaks should be adequate to prevent contamination of the contents of such fixtures and equipment from any possible backup of sewage from either direct or indirect waste piping and should be located within 2 ft of the waste outlet of the fixture or equipment and on the inlet side of the fixture trap. Hence, the waste outlets of all sterilizers, food refrigerators, and food compartments must be protected against contamination from sewage by means of air breaks at their waste outlets.

Refrigerators, ice boxes, or receptacles wherein food is stored, when provided with waste outlets, should be drained by means of drip pipes which discharge through an air break either into a floor drain or sink approved for such use, or into a safe pan, or receptor, which is equipped with a bell trap, preferably, or ordinary type trap and which discharges into the sanitary drainage system by means of an indirect waste pipe.

Fixtures and equipment which have interior surfaces which are not readily accessible to permit effective cleaning and restoration of sanitary conditions therein, after having been flooded by backup of sewage from the sanitary drainage system, should not be connected directly to the system. Instead, they should be connected to indirect waste pipes, which are permitted to convey liquid waste but exclude any human waste, and discharge therefrom into a fixture approved for such use. This method applies especially to automatic laundry machines equipped with squirrel-cage type tumblers which cannot be removed readily for complete cleaning of the interior of the machine after being flooded by backup.

Drinking fountains discharge wastes at such a low rate of flow as to be little more than drippage. In addition, the wastes contain very little or no organic matter. In view of this, relatively no sanitary hazard exists when such fixtures discharge indirectly into a fixture approved to receive such indirect wastes. Consequently, it is permissible for drinking fountains, including electric water cooler drinking fountain units, to be connected to indirect waste pipes. This is a generally accepted alternative method which may be more economical in some particular installations.

Portable household appliances, such as portable laundry and dishwashing machines which are not intended to be permanently connected to the plumbing system of a building, generally are provided with flexible discharge piping of suitable length at the end of which is a bent section which may be hooked over the rim of a fixture. The discharge of such fixtures into kitchen or laundry fixtures in dwelling units poses no additional sanitary hazard and, hence, should be permissible.

Swimming pools and wading pools having overflow connections located at an elevation below street level would be subject to backup of sewage through such connections in the event that they were connected directly to the sanitary drainage system and backup conditions occurred in the public sewer system. This should not be permitted. For such pools, all drainage outlets including pool drains, scum gutter drains, backwash outlets from pool water filters, and floor drains which serve walks around pools should be arranged so as to discharge by means of an indirect waste pipe; and any existing circulation pump for pool water may be used to pump wastes from the pool to an elevation suitable for gravity discharge into a fixture approved for such use.

Where drains are provided in the pits of hoistways, flooding of the pits during backup conditions in public sewer systems should be prevented. Such drains should not be directly connected to the sanitary drainage system or to the storm drainage system, but instead should discharge through an indirect waste pipe.

Certain fixtures and equipment which may be classified as business equipment rather than as building equipment provided for the needs of building occupants pose legal and practical difficulties which weigh against their being connected directly to the sanitary drainage system of a building. These difficulties resulted from the fact that such fixtures: (1) were *business equipment* rather than *building equipment;* (2) were the property of the business owner rather than the property of the building owner; (3) were to be installed upon occupancy by the business to meet its needs; (4) were to be removed by the business when its occupancy in the building terminated; and (5) such fixtures generally were to be installed in unusual locations, such as under bars or counters remote from walls, where the vent piping required for the direct-connection method would usually have to be exposed and thereby detract from the normal appearance and attractiveness of the business space.

Fixtures and equipment which may be classified as business equipment include bar sinks, soda fountains, counter sinks, wash boxes, spoon troughs, glass rinse sinks, barber shop and beauty parlor lavatories, commercial dishwashing machines, glass washers and silver washers, and other equipment specifically required for the business operations of a particular kind of business occupancy. Most codes contain special provisions which permit such fixtures and equipment to be connected by means of an indirect waste pipe discharging into a fixture approved for such use. This is an alternative method specifically permitted for business equipment connections. Generally, this method is the most practical and economical to apply when installing such equipment.

VENTING OF SANITARY DRAINAGE SYSTEM

The sanitary drainage system of a building should be provided with an attendant system of vent piping designed so as to permit gases and odors in all parts of the drainage piping to circulate up through the system and escape into the atmosphere above the building, and to permit the admission and emission of air in all parts of the system so that siphonage, aspiration, or back-pressure conditions will not cause an excessive loss of trap seal under ordinary conditions of use. The sizing, arrangement, and installation of attendant vent piping should be designed so as to limit air pressure variation in all fixture drains to a differential not exceeding 1 in (25 mm) of water column above or below atmospheric pressure.

Adequate circulation of air by induced head or draft through the entire drainage and vent-piping system is an effective aid in avoiding accelerated corrosion of piping which may otherwise occur from such aggressively corrosive gases, hydrogen sulfide and ammonia, normally present in significant amounts in sewer gases. Where inadequate air circulation occurs in drains, fungi find conditions favorable to their growth, and they produce slime. If the slime is not scoured from the piping by flow at sufficiently high velocity, it may accumulate to the degree that a stoppage condition occurs.

The water seals of fixture traps provide a means of keeping objectionable gases and odors confined to the drainage system of the building and preventing them from escaping into rooms in which fixtures are located. To maintain water seals against being lost as the result of siphonage, aspiration, or back-pressure conditions accompanying excessive variation in pneumatic pressures in fixture drains, vent pipes should be provided so as to supply and to remove air at whatever rates may be required in order to limit air pressure variations to a degree that the water seals can effectively resist. As the minimum required trap seal depth is 2 in (50 mm) the permissible air pressure variation in fixture drains is appropriately limited to 1 in (25 mm) of water column. Consequently, this provides a practical basis upon which to size vent piping.

OBJECTIONABLE WASTES

Many kinds of wastes may be detrimental to the sanitary drainage system or public sewer or sewage treatment plant. Such wastes may be classified as objectionable for one reason or another. Special precautions or methods of handling or treating such wastes may be required in order to prevent detrimental effects which might otherwise occur.

Swimming pool wastes might be considered to be objectionable if they were to be discharged into a sanitary drainage system connected to a septic tank system of sewage disposal. The great volume of waste to be handled when a swimming pool is drained may have a seriously adverse effect on the septic tank system and cause extensive flooding of the subsoil leaching field system. This should not be permitted. Under such circumstances, it is recommended that the swimming pool wastes be discharged through an independent sanitary drainage system just for that purpose and conveyed to an independent disposal system.

Industrial wastes which may be detrimental to the sanitary drainage system, or to the public sewer, or to the public or private sewage treatment plant should not be permitted to be discharged in the usual manner. Notice of the type of industrial waste which is proposed to be discharged into a public system ordinarily is required to be filed with authorities having jurisdiction prior to obtaining permission to connect the sanitary drainage system of an industrial building to the public system. If satisfactory treatment of the waste may make it unobjectionable, installation of the treatment process may render the wastes suitable for discharge into the sanitary drainage system. If proper treatment of such wastes is not provided, they should be disposed of by a method acceptable to the authority having jurisdiction.

Excessively high temperature wastes are objectionable because of the unusually large amount of drainage piping expansion and contraction effects resulting therefrom. They may cause pipe joints to be disturbed or pulled apart, or cause solidly bedded piping to be strained or broken, and leakage of sewage may result from the discharge of excessively high temperature wastes into the sanitary drainage system. This should be prevented. It is generally recommended that no high-pressure steam exhaust, boiler blowoff, or similar drip pipe be directly connected to the building drainage system. Such wastes should be cooled to a reasonable temperature before they may be permitted to be discharged. The recommended method of disposing of such wastes after cooling is to discharge them into a branch of the building sewer where they may mix with the total sanitary sewage flow from the building, and no temperature effects may be transmitted into the piping system in the building in the event that the cooling process may be shut down or inoperative at times. (See Fig. 8-15.)

Corrosive liquids, acids, strong alkalies, or other chemicals which might destroy or injure a drain, soil, waste, or vent pipe, or which might create noxious fumes, should not be discharged into the regular sanitary drainage system. Such chemicals should be discharged through an independent sanitary drainage system directly to a public sewer system provided permission has been received for their discharge into the public system from the authority having jurisdiction; or such wastes should be treated by an acceptable method prior to disposal. Where such wastes are treated by passage through an approved dilution or neutralizing device for which adequate maintenance is assured, they may be discharged into the regular sanitary drainage system of the building. However, all drains which convey untreated chemical wastes and all vents attendant upon such drains should be made of materials resistant to the corrosive action of such wastes and their fumes. (See Fig. 8-13.)

Any wastes which may cause clogging conditions in the sanitary drainage system should not be permitted to be discharged into the drainage system unless appropriate, effective means of treatment are provided to render such wastes unobjectionable. Where intercepting strainers or grease or sediment intercepting fixture traps may provide appropriate and effective treatment and render such wastes unobjectionable, they should be installed at the fixtures, and adequate maintenance should be assured for such equipment.

Fixtures through which volatile oil or other flammables could be introduced or admitted into the regular sanitary drainage system of a building, by accident or otherwise, should not be connected to the regular sanitary drainage system. Instead, such fixtures should be connected to an independent sanitary drainage system discharging through an oil separator of satisfactory design and capacity. Where the oil separator discharges to the building sanitary sewer, or building drain, the branch connection should be located downstream from any building trap installed therein. Where the oil separator must discharge into a private sewage disposal system, the system shall be one specially approved for such use. Each oil separator should be provided with an individual vapor vent pipe, at least 3 in (75 mm) in size, to convey flammable vapor from the top of the separator to a suitable and unobjectionable location outside the building. It is recommended that the vapor vent terminal be located at least 12 ft (3.66 m) above grade so as to be safely distant from any source of ignition, and so as to cause no objectionable odors noticeable to passersby. (See Fig. 8-14.)

Radioactive wastes should not be discharged into the regular sanitary drainage system or to a public or private sewer system or sewage treatment plant unless such wastes are specifically treated and disposed of in accordance with standards prescribed by the health authority having jurisdiction.

DRAINAGE SYSTEMS BELOW SEWER LEVEL

Where a drainage system, or part of a drainage system, is located at an elevation below the building sewer or public sewer, the liquids or sewage conveyed thereby should be disposed of through a subbuilding drainage system into a sump, receiving tank, or ejector equipped with automatic equipment for lifting and discharging such liquids or sewage into the building gravity drainage system. The receiving vessel and lifting equipment should be of adequate capacity and suitable design for the volume and kinds of liquids to be conveyed. In large installations where service interruption may result in flooding, or unsanitary conditions, it is recommended that the lifting equipment should consist of two or more units connected in parallel arrangement.

Sumps, receiving tanks, and ejectors which receive sewage from sanitary subbuilding drainage systems should be of airtight design and should be provided with a vent to permit air flow into and from the receiving vessel. Drainage and vent piping of sanitary subbuilding drainage systems should be installed in the same manner as for gravity systems. Sumps, receiving tanks, and ejectors need not be airtight or vented if they do not receive sewage, but instead receive only clear water wastes such as from boiler room floor drains, machinery drips, storm water drains, and subsoil drains.

SUBSOIL DRAINAGE

Subsoil drains, installed under a cellar or basement floor or surrounding the outer walls of a building so as to drain water from such subsoil regions, should be at least 4 in in size and consist of open-jointed, horizontally split or perforated piping of acceptable material. Where such piping discharges to a public sewer, subsoil drains should be connected to an accessible silt and sand intercepting trap, and the liquids therefrom should be disposed of into the storm drainage system of the building. Where the drain from the intercepting trap is connected directly to a gravity storm drainage system which discharges into a public sewer system, an accessibly located backwater valve of acceptable design should be provided in the drain at the outlet side of the intercepting trap.

BACKWATER VALVES

Drainage systems which connect to public sewer systems are subject to backwater and flooding in the event that the public sewer becomes blocked or insufficient in capacity for the load that may be imposed on it, such as may occur during an abnormally heavy storm. To prevent water from backing up into the building drainage system and flooding into the building under such circumstances, backwater valves should be installed on drainage systems wherever it is known or authorities advise that a particular part of the public sewer system may become overloaded or surcharged at times.

In such cases, the drainage piping for fixtures located at an elevation where they may be subject to flooding should be equipped with a backwater valve. An alternative to this method is to install an accessible backwater valve, or a manually operated gate valve, in the building drain at the point of entry inside the building and downstream from any building trap.

The design of backwater valves should provide a positive mechanical seal against backwater. When fully opened, such valves should have a flow capacity not less than that of the piping in which it is intended to be installed. All bearing parts of such valves should be of corrosion-resistant material.

BUILDING TRAPS AND FRESH AIR INLETS

Local authorities having jurisdiction may require because of local conditions that building traps be installed on, or omitted from, the building drain. Wherever such regulations prevail, installations should conform. But where no regulations on the subject of building traps exist, the right to determine whether or not to equip the building's drainage system with a building trap is vested in the owner of the building, who may consider the matter on the basis of relative costs, advantages, and desirability.

The legal basis for plumbing regulation is the reasonable necessity for such regulations in order to protect the health, safety, and welfare of the people. With this in mind, the history of, and need for, building trap and fresh air inlet installations can be viewed in proper perspective. Prior to the period when plumbing fixtures were first installed inside buildings, building drainage systems were provided exclusively for conveying storm water to storm water disposal systems. Early plumbing fixture installations inside buildings were made by connecting the fixtures to existing storm water drainage systems, thus converting them into combination storm and sewage drainage systems.

The combined systems soon became fouled and odorous from decomposing sewage matter and quickly became infested with rats and other vermin which could travel freely from one building to another through the sewer piping. Health departments recognized that this condition presented a serious menace to health. Originally, building traps were installed on each sanitary or combined building drainage system as a means of eliminating this health hazard. Reports of that period indicate that they were generally effective. While subsequent reports revealed that rodents occasionally have been known to pass through the water seals of building traps, this does not invalidate the need or usefulness of building traps.

Health officials, and the plumbing industry generally, recognized that building traps provided a secondary safeguard against the possibility of public sewer gases penetrating into the habitable spaces of buildings.

While the installation of a fixture trap in the drain of each fixture provided a primary safeguard against the escape of sewer gases and objectionable odors from piping of the drainage system, this secondary safeguard was deemed a necessity during the period preceding the discovery and application of the basic principles of venting to protect the seals of fixture traps.

In those days, fixture trap seals very often were sucked into the drainage system or blown out at fixtures because of the strong, sharp air pressure fluctuations developed within unvented drainage systems. Even now many substandard and unvented drainage systems are still in service in communities that have had plumbing regulations for a number of years. In communities that have had no regulations, substandard and unvented drainage systems are common.

Building traps provide protection in other ways, simply because such traps effectively seal off the sewer atmosphere from building drainage systems. Systems frequently must be opened to make repairs, alterations, and additions. During such work, building traps prevent the sewer atmosphere from entering building spaces where pipes are opened. Such protection has been deemed necessary in many areas because very often sewer gases are not only offensive to smell, but are also flammable or explosive.

Following its early period of development, extending from 1866 to about 1900, the design and regulation of plumbing systems progressed rapidly. It had to keep pace with new problems arising from parallel developments in other fields.

Gas manufacturing plants were established, and utility gas piping systems were laid under the streets of most large communities. Natural gas piping systems were similarly laid more recently. Some of these systems conveyed gas at very high pressures. With the advent of the automobile, gasoline selling stations were established in virtually every community, and large underground gasoline storage tanks were installed.

These underground installations caused new hazards to arise in conjunction with building drainage systems. Leakage from underground gas piping systems very often infiltrates the public sewer system and building sewers where laid under paved streets. Where the leakage is from high-pressure gas piping systems, the hazard is increased. Leakage of gasoline from defective underground storage tanks and the inadvertent discharge of flammable liquids with wastes often result in volatile and explosive gas mixtures being conveyed by public sewer systems.

Danger from explosions in public sewer systems has increased seriously with the passing of time. It is now at an all-time high. At present in some large cities, various sections of the public sewer systems are known to be almost constantly high in explosive gas content. Extensive efforts are made to locate the sources of such infiltration and to eliminate them. However, the matter of tracing such infiltration is very often nearly impossible. Further, such conditions recur from time to time, presenting a constant hazard.

Building traps provide effective protection against shock and fire damage within buildings when explosions occur in public sewer systems. The trap effectively seals off explosive gas mixtures in the public sewer; and in the event of an explosion, it acts as a shock absorber, owing to its water seal and 360° change in direction. Presently, this is considered one of the most important reasons for building trap requirements in codes. The use of building drainage systems for venting explosive gases from public sewer systems has been found to be completely inadequate in preventing gas explosions in public sewers and resultant damage within buildings connected thereto.

Durability of standard piping materials used in building drainage and vent systems is a factor that is in part dependent upon the corrosivity of the atmosphere within the systems. In public sewer systems, the corrosivity of the atmosphere is subject to wide variation. This is influenced by the types of wastes the sewer authority permits to be discharged into the public system. Some systems handle just ordinary household sewage; others take in industrial wastes, including highly corrosive acid wastes. Some sewers are specially designed to handle fuming acid wastes. Municipal sewer services may be specially designed to handle the wastes from important industries around which the economic life of the community is centered.

In some communities, building traps are forbidden. This generally occurs only where the atmospheres of public sewer systems are so highly corrosive that maintenance of the systems would be excessively hazardous unless provided with additional ventilation afforded by using building drainage systems as vent stacks for the public systems. In such cases, several of the standard piping materials may have to be excluded from use in the building drainage systems because of insufficient durability, as proved by experience under service conditions.

Where building traps are provided on building drainage systems, fresh air is normally required to be supplied to the system through a fresh air inlet connected to the building drain directly upstream from the building trap. Entrance of fresh air into the system reduces the corrosivity of the atmosphere within the system and tends to establish normal atmospheric pressure conditions in the building drain. Under these conditions, the durability of standard piping materials can be reasonably established on a standardized basis regardless of the corrosivity of the atmosphere in the public sewer system.

Where private sewage disposal systems are installed, building traps very frequently are required so as to avoid the creation of a stench nuisance in the vicinity of vent terminals. In populated areas of private home developments, especially of low ranch-type homes, the stench from septic tank systems ventilated through building drainage systems has often been found to be very objectionable and has resulted in widespread demands by residents that it be prevented. This condition is most objectionable in low or valley areas during hot, humid weather. Building traps effectively prevent such stench nuisance conditions.

Building traps are not necessary for normal operation of building drainage systems, except in one instance. This occurs in very cold regions where the winter is severe and the drainage system is connected to a public sewer system. In such cases, vent terminals of the building system are subject to frost closure because of the relatively high moisture content of the public sewer atmosphere. When frost closure occurs, building drainage systems cannot operate satisfactorily. Under such conditions, building traps are necessary for normal operation of the systems.

When building traps are installed, they offer a resistance to the flow of sewage equal to that offered by four 90° bends of the same material and radius of curvature. They also offer resistance to the flow of air through the trap equal to that of the trap's effective water seal, normally 3 in (75 mm) of water column for standardized building traps.

Where a building trap is provided, a fresh air inlet must be installed so as to ventilate the building drainage system. The fresh air inlet relieves any air pressure rise at the building trap caused by the trap's resistance to the flow of air. Consequently, codes specify no change of minimum pipe sizes whether building traps are or are not installed.

The installation of building traps and fresh air inlets is an item of building cost. However, it is a relatively insignificant factor amounting to no more than about one-tenth of 1 percent in the total cost of a modern private dwelling, and no more than one-hundredth of 1 percent for a large multistory building. Nevertheless, building costs are important to persons who must pay them. Architects, engineers, and contractors, when called upon to serve as authorized agents for building owners in areas where building traps are not required to be installed or omitted, should consider the cost that may be saved by omitting a building trap against the protective values inherent in such installations. In some areas, there may be little or no need for protection; in others, protection may be vital.

Each building trap should be equipped with two brass cleanout plugs of the same size as the trap, except that they need not exceed 4 in (100 mm) in size for larger traps. The cleanout plugs should be accessible so as to permit cleaning the trap interior and rodding upstream and downstream therefrom. Cleanout plugs should be extended to above floor level where required by the authority having jurisdiction in order to exclude from the public sanitary sewer any wastes drained from floors.

Building traps should be installed within the property line of the premises, inside the building wherever practicable, and located on the building drain within 2 ft (0.61 m) of the exterior wall of the structure. Such traps should be located downstream from all drainage branches to the system, except those provided to receive discharges from a sewage lift, oil separator, steam blowoff and condensing tank, or leader. A masonry or concrete pit, or manhole, of acceptable, approved design should be provided for access to trap cleanout plugs when they are located below ground or below a cellar floor.

A fresh air inlet pipe should be provided on every sanitary or combined building drain equipped with a building trap, sewage sump, ejector, receiving tank, oil separator, or similar equipment. The fresh air inlet pipe should be connected to the building drain immediately upstream from, and within 4 ft (1.22 m) of, such trap or equipment and should be at least one-half the diameter of the building drain at the point of connection, but not less than 3 in (75 mm) in size. The connection to the building drain should be made to the upper half of such drain (air space portion), and the fresh air inlet pipe should be extended therefrom to the atmosphere outside the building and terminated in an open end at least 6 in (150 mm) above grade. In skyscraper construction, it is recommended that the open end be terminated above the roof of a setback of the building rather than at street grade so as to avoid condensation drippage and discharge of odors which might otherwise be objectionable to passersby when the system is subject to constant downward flow of wastes and air during peak load periods. Where fresh air inlet pipes are terminated at grade, a perforated metal plate should be permanently installed over the end of the pipe. The plate should have open ventilating area at least equal to the area of the fresh air inlet pipe. In lieu of a perforated plate over the end of the fresh air inlet pipe, a return bend may be installed with its unprotected open end at least 6 in (150 mm) above grade within the property line in an acceptable location.

CONNECTIONS TO SANITARY BUILDING DRAINS

Sanitary building drains are designed to flow half full at peak load. To avoid backup of flow from the building drain into branches, each branch connection to the building drain should be made to its upper half or air space portion. This may be achieved for 90° branch connections by means of a one-sixth bend and a 45° Y branch, or a long

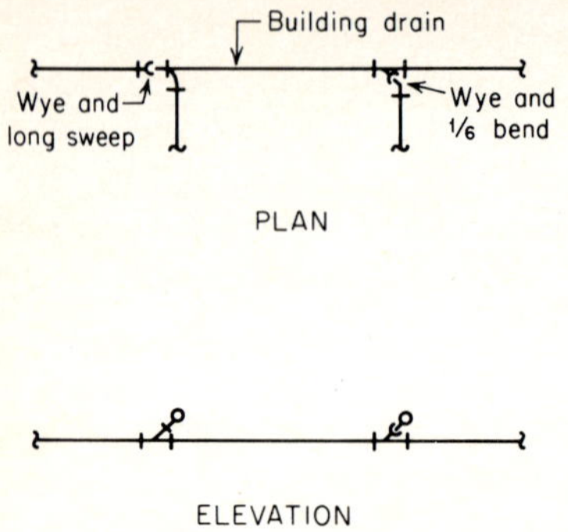

Fig. 8-1 Recommended building drain branch connections.

sweep one-quarter bend and a 45° Y branch. The Y-branch fitting may be rotated so that the branch is at a 45° angle above the horizontal when the one-sixth bend is to be used and at a vertical angle when the long sweep one-quarter bend is to be used. Less invert elevation is lost with the sixth bend and Y combination (see Fig. 8-1).

Branch connections made to the upper half of sanitary building drains are recommended for several additional reasons. First, they reduce the incidence of stoppages occurring in the branch drains; second, they produce less flow interference at the branch connection than would occur were the branch to be made horizontally into the building drain; third, they produce less restriction to air flow in the building drain as the result of reducing flow interference in the drain; and fourth, during periods when one or more branches from drainage stacks are not flowing, their full cross-sectional area is available to relieve pressures occurring in the building drain.

BRANCH CONNECTIONS TO DRAINAGE STACK OFFSETS

Drainage stack offsets are subject to extremely turbulent waste flow conditions and excessively severe pneumatic effects when they convey even as little as one-half the load permitted for a given size of stack. Branches connected to such offsets are subject to conditions prevailing therein. They may be sufficiently severe as to blow out the trap seals of fixtures draining thereto, unless special relief vents of adequate size and number are provided to control effectively any possibility of excessive air pressure development. This is not always practical and is generally uneconomical. Consequently, it is recommended that branch connections to drainage stack offsets should be avoided, except where the portion of the stack above the offset does not serve drains from fixtures at a higher story.

DRAINAGE STACK VENT EXTENSIONS AND VENT STACKS

Each drainage stack connected to the sanitary drainage system should be provided with a vent pipe extending from the top of the highest drainage branch fitting on the stack to the atmosphere above the building roof. This vent extension of the drainage stack is necessary to permit gravity circulation of air up through the stack to the atmosphere at an unobjectionable location and to permit air to enter the top of the stack as rapidly as is required to replace the air which is dragged down by flowing water descending in the stack. Drainage stack vent extensions may be connected together as a vent header above the flood-level rims of the highest fixtures discharging into the drainage stacks and a single vent pipe extended from the vent header to the atmosphere above the roof.

An attendant vent stack should be installed with each soil or waste stack which has drainage branch connections for present or future fixtures on two or more stories. The purpose of the vent stack is to prevent development of excessive air pressure in the lower region of the drainage stack by relieving air therefrom at as rapid a rate as it is carried down the drainage stack by liquids discharged by fixtures into the upper section of the stack. The most effective place for the vent stack base to be connected, consequently, is at the bottom of the drainage stack, below the level of all drainage branch connections and, preferably, to the top of the horizontal drain immediately adjacent to the base fitting, where air pressure rise is maximum and potential stoppage or closure of the vent connection by grease and other deposits is minimum. The recommended vent stack base connection is illustrated in Fig. 8-2.

Each vent stack should extend undiminished in size to the atmosphere above the building roof as an independent vent extension; or it should be connected to the vent extension of the drainage stack it serves at an elevation of at least 6 in above the flood-level rim of the highest fixture discharging into the drainage stack, or to a vent header to which the drainage stack connects.

Offsets in drainage stack vent extensions, offsets in vent stacks, and connections of the bases of vent stacks to the drainage stacks, or the horizontal drains therefrom, should be made at an angle of at least 45° to the horizontal wherever the piping is of a scale- or rust-producing type so as to avoid accumulation of scale or rust therein and resultant loss in venting capacity. Where the entire piping above such offsets is

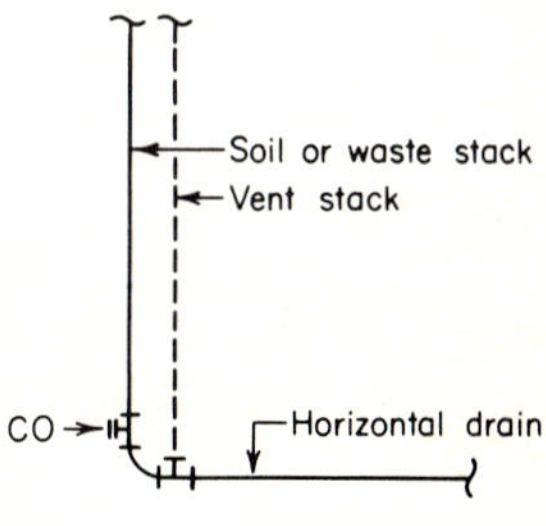

Fig. 8-2 Recommended vent stack base connection.

of a nonscaling type, the offset angle may be reduced provided there is sufficient slope in the vent piping for condensation to drain back to soil or waste pipe connections.

Terminals of all extensions of drainage and vent stacks should be located so as to be unobjectionable. They should not terminate within 10 ft (3 m) of any door, window, or opening for ventilation unless the terminal is located at least 2 ft (0.61 m) above such an opening. Terminals should extend at least 6 in (150 mm) above building roofs, but where a roof is used for other than just incidental access, the terminal should be located at least 5 ft (1.5 m) above the roof. Vent extensions should not be extended through an exterior wall unless it is impracticable to extend the piping through a roof. In such a case, the terminal should open downward and be equipped with a screen so as to prevent birds from nesting therein. In no case should the terminal be located beneath an eave or overhang of the building or within 10 ft (3 m) horizontally from the lot line of the premises.

On several occasions in the past, courts have been called upon to decide cases arising from vent terminals, their effects with regard to adjacent new buildings, and the effects of new buildings erected adjacent to vent terminals. Such decisions may be used as a guide to proper terminal locations. They are as follows: (1) Where a structure is to be built higher than the vent terminal of an adjacent building and thereby adversely affects the vent system of the adjacent building, or when such vent terminal would become a potential nuisance to occupants of the higher structure, the owner of the higher structure shall, at the owner's expense and with the consent of the owner of the adjacent building, cause such vent to be extended or altered to correct the condition; (2) where a vent terminal is to be installed adjacent to an existing higher building, the proposed vent terminal shall be installed by and at the expense of the owner of the lower building, in conformity with plumbing regulations, including any necessary extension of the vent terminal to a location sufficiently remote to prevent the creation of a foul air nuisance to occupants of the higher building.

AIR PRESSURE RELIEF VENTS

Drainage systems, especially those in tall buildings, are frequently found to develop extremely high and objectionable pneumatic effects in several specific portions of such piping. Special air pressure relief vents are recommended to control, within tolerable limits, any air pressure fluctuation that otherwise may occur at these portions of the system. They are generally specific locations where the cross-sectional area of the drain may suddenly become filled with liquid, thereby constricting the area available for passage of air through the drain at the same high quantity rate of flow as it is dragged down into the drain by the flowing liquid. When air is dragged into a zone where airflow constriction or blockage occurs, the pressure may rise sharply and cause excessive back pressure and blowout of fixture trap seals connected to the zone. This must be prevented.

One such zone occurs where a horizontal building drain is offset vertically, or drops at a 45° angle, more than 10 ft (3 m) in invert elevation. Airflow constriction may occur in the horizontal building drain at the base of the vertical drop in the same manner as it occurs in a horizontal drain at the base of a drainage stack. An air pressure relief vent, at least one-half the diameter of the building drain, should be provided at the top of the vertical offset so as to supply such additional air to the drain as may be required by the sudden increase in liquid velocity in the vertical offset. Where a building trap or other sharp change in flow direction is provided in the building drain downstream from the vertical offset, an air pressure relief vent should be provided at the base of, and within 3 ft (0.91 m) of, the vertical offset.

The relief vent connected at the base of the offset should be sized as a vent stack, considering the vertical offset in the building drain as a soil or waste stack. This lower relief vent should be branch-connected to the upper relief vent at a sufficient height so that they cannot serve to bypass sewage flow in the event of a stoppage in the vertical offset.

Another zone occurs in drainage stacks of extended height. Excessive interference with air flowing down a stack, with liquids from upper stories, may occur where additional liquid is simultaneously discharged into the stack at high rates of flow through lower drainage branches. This interference with the flow of air down a tall drainage stack is not unusual. Consequently, provision should be made to prevent occurrence of excessive back-pressure effects. The recommended provision for soil and waste stacks more than ten stories in height is to provide a yoke relief vent at each tenth story of the drainage stack, counting downward from the top story. The lower end of the yoke relief vent should connect to the drainage stack by means of a Y located below the horizontal

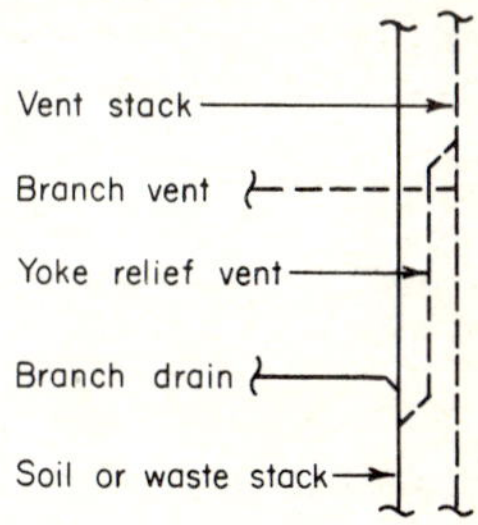

Fig. 8-3 Yoke relief vent.

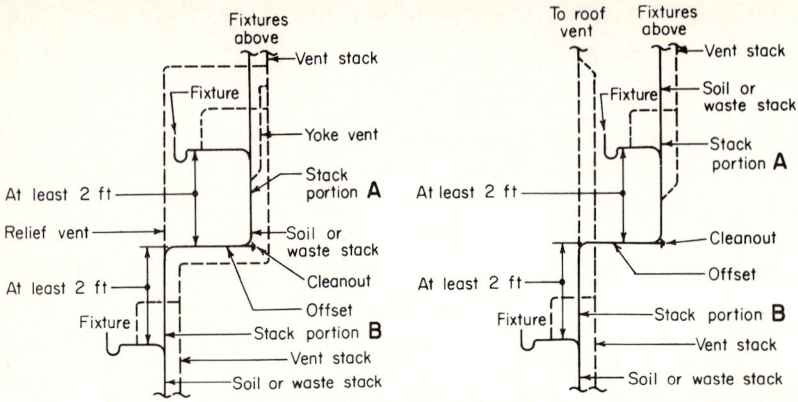

Fig. 8-4 Alternate methods of relief venting for 60 and 90° offsets in drainage stacks.

branch drain serving fixtures in that story, and the upper end should connect to the vent stack by means of a T or inverted Y located at least 3 ft (0.91 m) above the floor level (see Fig. 8-3).

An additional zone occurs in drainage stack offsets which are made at an angle of more than 45° from the vertical and are located more than 40 ft (12.2 m) below the highest drainage branch connection to the stack. Such offsets are subject to extremely turbulent flow conditions and excessively high pneumatic effects.

The recommended relief vent provisions are as follows: either provide for the drainage stack section below the offset and for the drainage stack section above the offset the same venting provisions, such as attendant vent stacks and drainage stack vent extensions, as would be required if they were two separate drainage stacks; or provide a relief vent at the top of the drainage stack section below the offset and a yoke vent at the base of the upper stack section (see Fig. 8-4).

SUDS PRESSURE ZONES AND SUDS PRESSURE RELIEF VENTS

During the past 55 years, the soap and chemical industries have developed and marketed many new products which have replaced the use of bar soap for numerous cleaning tasks. This trend started in the 1920s with the introduction of soap chips, flakes, granules, and powders which did a better job of cleaning and were more convenient to use. Following the close of World War II, the trend accelerated with the introduction of synthetic detergents which produced ample quantities of long-lasting suds in both hard and soft water areas. These detergents have gained wide public acceptance and general use for a large variety of cleaning operations.

Coincident with the extensive use of these new products in place of bar soap, a marked change has occurred in the characteristics of household wastes conveyed by sanitary drainage systems. A tremendous increase in the volume of suds accompanying wastes has resulted. When upper floor fixtures and appliances discharge wastes containing detergents, the suds-producing ingredients mix vigorously with air as the wastes churn down the inner wall of the drainage stack. These suds flow down and settle into the lower sections of the drainage system. It has been found frequently upon investigation that the sanitary building drain and the lower section of soil, waste, and vent stacks were laden with suds and remained in that condition for considerable periods of time.

When upper floor fixtures discharge into a stack, the wastes churn down the inner wall of the stack and drag or force air down into the lower sections of the system. Liquid wastes are heavier than suds and easily displace and flow through the suds-laden lower drainage piping to the sewer. However, the air which is forced down into the body of suds compresses and forces them to move through any available paths of relief.

These relief paths may include the building drain, its branches, the vent stack base connection, branch vents, and individual vents connected to the lower section of the system. These paths of relief may not be available, or they may be cut off or constricted by sudden increases in the cross-sectional area occupied by liquid flow, or they may be inadequate because of arrangement, location, or pipe size. If one or more of these conditions prevail, abnormally high suds pressure may develop in such zones and force trap seals connected to the system.

Cutoff or constriction of the cross-sectional area available for suds flow in drainage piping may occur at sharp changes of direction, such as in the horizontal drain at the base of a stack. Extreme turbulence in liquid flow occurs at such directional changes. The turbulence is accompanied by a *hydraulic jump* in the horizontal drain at a point where velocity reduction is most severe and produces a sudden increase in the cross-sectional area occupied by liquid flow. A hydraulic jump is illustrated in Fig. 9-1.

The cutoff or constriction of suds flow at the hydraulic jump creates a zone where pressure of suds can develop and extend upstream for a considerable distance unless means of relief are provided. Where vent stack base connections, branch vents, or individual vents are required to serve as relief paths from suds pressure zones, very often they are found to be inadequate for such use and suds backup conditions appear at plumbing fixtures. It should be understood that the vent pipe sizing tables given in most current codes, including nationally recognized model codes, are based exclusively upon airflow capacity and give no consideration to suds flow capacity. Hence, sizes determined in applying such tables may prove inadequate for suds pressure relief.

Suds are much heavier than air, do not move with the same ease, and produce considerably more friction loss for the same rate of flow. The density of old or regenerated suds, compared with that of air, varies from a minimum of about 2.7 to a maximum of about 18.7 for various kinds of detergents in common use. On the basis of these values and appropriate assumptions, it may be calculated that for equal rate of flow and pressure drop, the vent pipe diameter for suds relief flow should be from 21.5 to 80 percent larger than for airflow.

Two alternative methods for eliminating suds pressure conditions in drainage systems have been devised on the basis of field tests and investigations of systems in tall residential buildings. The choice of method to apply may be based upon whether a building is in process of being designed or already existing and upon the economies and difficulties involved.

The first method may be applied most advantageously in designing new systems. It is as follows: wherever a soil or waste stack is to receive wastes at an upper floor level from sinks, laundry trays, laundry washing machines, or other fixtures in which sudsy detergents are normally used, the drainage and vent piping for lower floor fixtures should be arranged so as to avoid connection to suds pressure zones in, and adjacent to, the stack.

Suds pressure zones should be considered to exist in the following locations:

1. *In a soil or waste stack offset of* 60 *or* 90°, *serving fixtures on two or more floors and receiving wastes from fixtures wherein sudsy detergents are used:* a zone extending 40 stack diameters upward and 10 stack diameters horizontally from the base fitting for the upper stack section, and a zone extending 40 stack diameters upstream from the top fitting for the lower stack section

2. *At the base of the stack:* a zone extending upward from the base a distance of 40 stack diameters

3. *In the horizontal drain from the base of the stack:* a zone extending horizontally from the base a distance of 10 drain diameters; and, where a 60 or 90° offset fitting is installed in the horizontal drain, a zone extending 40 drain diameters upstream and 10 drain diameters downstream from the fitting

4. *In a vent stack having its base connected to a suds pressure zone in the drainage stack or horizontal piping therefrom:* a zone extending from

the vent stack base connection upward to above the level of the suds pressure zone in the drainage stack

The four suds pressure zone locations listed above are also illustrated in Fig. 8-5.

The second method provides a practical solution which may be applied to existing systems. In such systems, where suds backup conditions develop at lower floor fixtures connected directly to such pressure zones, it is usually found to be excessively costly, impractical, or otherwise undesirable to change the arrangement of drainage and vent piping connections for the fixtures. A more appropriate method under these circumstances is to provide suds pressure relief vents. *A suds pressure relief vent should be connected to each suds pressure zone and installed so as to relieve suds pressure therefrom to a nonpressure zone downstream from and at a lower elevation in the sanitary drainage and vent system.*

At 60 or 90° offsets in soil or waste stacks, the suds pressure relief vent should be connected to the stack at fixture rim level just above the offset and then dropped down to a connection just below the offset so as to permit suds to overflow and run off through the suds pressure relief vent and bypass the zone of constricted flow through the offset.

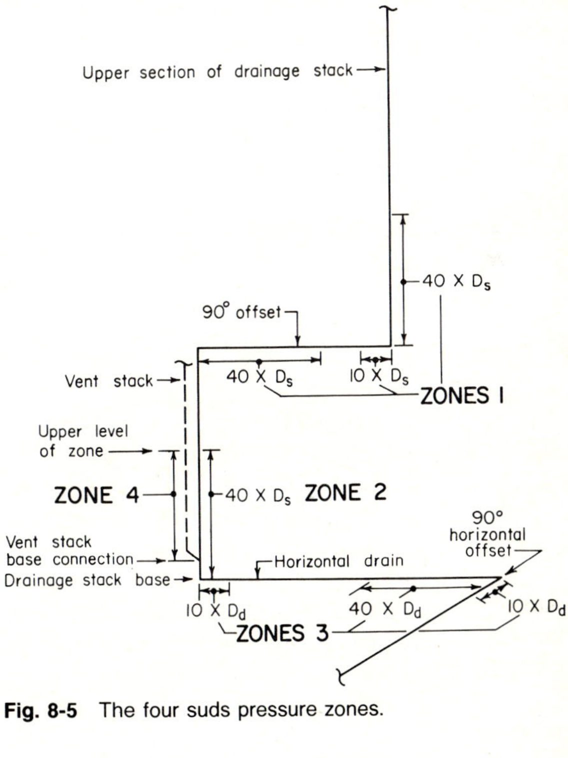

Fig. 8-5 The four suds pressure zones.

At the base of soil or waste stacks, the suds pressure relief vent should be connected to the stack at fixture rim level just above the base of the stack and then dropped down to a connection to a nonpressure zone in the house (building) drain. This permits suds to overflow and run off through the suds pressure relief vent and bypass the zone of constricted flow through the horizontal drain.

Figure 8-6 illustrates recommendations for relief of suds pressure conditions at the base of soil or waste stacks in high-rise residential buildings. Note that the suds pressure relief vent connection is made to the drainage stack at fixture rim level [36 in (0.91 m) maximum] so as to avoid creating a potential sewage bypass condition in the event of a stoppage occurring at the base of the stack. Note also that the size of the suds pressure relief vent is 3 in (75 mm), or three-fourths the diameter of the stack in which the suds pressure zone occurs. The suds pressure relief vent drops down and connects to a nonpressure zone in the upper end of the house (building) drain where pressure conditions usually are minimum, although other nonpressure zones farther downstream in the house (building) drain might be found equally suitable. In most instances, this is the most practical, economical, and desirable method to apply to existing systems where suds backup conditions develop at lower floor fixtures connected directly to suds pressure zones.

Probably, other satisfactory alternative methods may be applied to eliminate the development of excessive suds pressure in the lower section of sanitary drainage systems. Undoubtedly, larger pipe sizes may be applied advantageously for horizontal drains at the base of drainage stacks, and for the lower sections of vent stacks. However, this should be properly tested and developed in the field, or conclusively established by computations or research.

In industrial or loft buildings of the multistory or high-rise type occupied or used for light industry, sudsy detergents are frequently used in processes such as washing, dyeing, and shrinking of cloth and textiles, or washing, separating, and recovery of gold and precious metals. In such cases, the disposal of sudsy industrial wastes has caused serious suds backup problems at lower floor plumbing fixtures when the sudsy wastes were discharged into the regular drainage stacks at upper levels. *For such process wastes, it is recommended that they be disposed by an independent waste system discharging directly into the house (building) drain at its lowest elevation (at the front wall of the building) or into the building sewer. The objectionable effects of process wastes should be avoided, and the industrial occupancy should be prevented from adversely affecting the performance of the regular plumbing system of the building.*

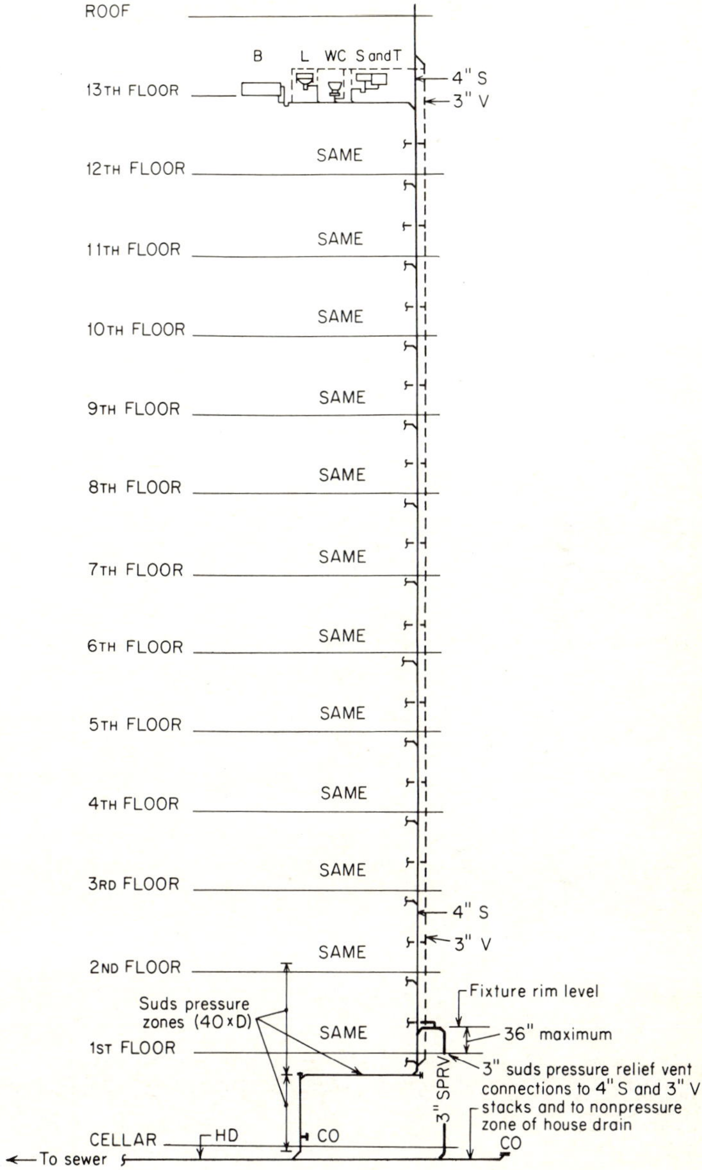

Fig. 8-6 Suds pressure relief vent connections for high-rise stack systems.

AIR PRESSURE RELIEF FOR PNEUMATIC EJECTORS

A pneumatic ejector is a special type of device designed to receive and dispose of liquids and sewage from subbuilding drainage systems. This device is specially designed to apply compressed air as the means of expelling liquids and sewage accumulated in the receiver, forcing them out and up through a discharge pipe to a suitable height, and discharging them into the gravity building sewer. An inlet and an outlet check valve are provided at the pneumatic ejector. Liquids and sewage enter the device at atmospheric pressure and displace air therefrom into an air pressure relief pipe. When the ejector is filled to a predetermined level, the device operates automatically, shutting off the air pressure relief pipe, allowing compressed air to enter the ejector and expel its liquid contents, shutting off the supply of compressed air when the liquid contents have been expelled, and opening the air pressure relief pipe to permit air pressure in the device to drop to atmospheric pressure conditions again.

The minimum air pressure required in the device to expel liquids therefrom is the pressure equivalent to the head of liquid in the discharge pipe plus the pressure loss due to flow in the discharge pipe. After discharging, the device must be relieved from such relatively high air pressure to atmospheric pressure before additional sewage may flow by gravity into the ejector. In no case should the air pressure relief pipe for a pneumatic ejector be connected to the attendant vent piping serving a drainage system. To do so would produce severe pneumatic effects in such a system and result in blowout of fixture trap seals.

It is recommended that the air pressure relief pipe from a pneumatic ejector be extended independently to the atmosphere and terminated at an unobjectionable location, preferably above the building roof. The size of the vertical section of such relief piping should be adequate to avoid frost closure at the roof terminal and to permit adequate venting capacity. It is generally required that such piping be at least 3 in in diameter and that a cleanout be provided at its base so as to permit removal of any scale or deposits that may accumulate at the base of the vertical section. The size of piping from the pneumatic ejector to the vertical section should be adequate to permit pressure in the ejector to be completely relieved to atmospheric conditions within a reasonable period. This has been deemed to be just 10 s, and the minimum size recommended for such piping under usual conditions is 1¼ in (31 mm).

PIPING INSTALLATION

Horizontal drainage piping should be installed in practical alignment at a uniform downstream slope sufficient to yield flow velocity of at least 2 fps (0.61 m/s) as may be determined for uniform flow conditions in piping of various sizes and materials. To attain such minimum velocity for scouring effect in horizontal drains, it is recommended that the pipe slope be at least ¼ in/ft (20.8 mm/m) for piping of 3 in (75 mm) diameter or smaller, and at least ⅛ in/ft (10.4 mm/m) for larger piping. Lesser slope should be used only when computations clearly establish that sufficient velocity will be attained. It should be noted that standard threaded and solder-joint drainage fittings having inlet openings at a nominal 90° angle, outlets of P traps, and both inlets and outlets of running traps are designed to provide a pitch of ¼ in/ft (20.8 mm/m) in the horizontal drain.

Vent piping should be installed in practical alignment and sloped upward continuously from its lowest connection with soil or waste piping to its terminal in the atmosphere above the building roof. This is necessary so as to permit ventilation of all parts of the drainage and vent-piping system by gravity circulation of air. Sags or traps in vent piping should not be permitted because condensation may collect therein and cause restricted air circulation and reduced venting capacity. Vertical drops in upward sloping vent piping should not be permitted because such drops tend to entrap warm moist air in the top of the *bowed* piping section, restrict air circulation, and permit accelerated corrosion of the piping.

There are certain locations in buildings where drainage and vent piping should not be installed, such as in stairways, in a hoistway, or under an elevator or counterweight, or where such piping would interfere with the normal operation of windows, doors, or other building openings. Horizontal drainage piping should not be located directly above nonpressure water supply tanks, access holes of pressure water supply tanks, or floor areas used for the manufacture, preparation, packaging, storage, or display of food unless a watertight barrier is provided to intervene between the piping and such tanks or space immediately below. Leakage from such horizontal drainage piping may contaminate potable water or food under those circumstances, as has been amply evidenced by epidemics which have occurred in the past. (See Fig. 3-1.)

Unless adequate provision is made to protect soil and waste piping against damage from frost conditions, such piping should not be installed outside of buildings or concealed in exterior walls in climate zones where freezing temperatures may occur. Similarly, such piping should not be installed in rooms or spaces of buildings where freezing temperatures may occur normally, such as in food freezer rooms, lockers, refrigerators, cold storage rooms, etc.

Protection against damage from external corrosion should be provided for drainage and vent piping which must be installed in or beneath cinders or other corrosive material. Although adequate protection may be afforded in many instances by application of one or more coats of suitable paints and wrapping of joints, it is recommended that piping installations in highly corrosive beds be avoided wherever possible or that the piping be encased in a special bed of chemically neutralized, noncorrosive material.

Drainage and vent piping should be installed in such a manner as to avoid damage and breakage due to strain accompanying normal expansion and contraction of the piping, and to building settlement. Where piping passes through foundation or bearing walls, protection should be provided by means of iron or steel pipe sleeves two sizes larger than the pipe passing through the wall, or by means of masonry relieving arches directly above the top of the piping. Flexible sealing material should be caulked into the annular space between the pipes and the sleeves or arches. Where outside leaders are installed along alleyways, driveways, or other locations where piping may be exposed to damage, protection may be afforded by the installation of suitable guards, or the piping may be recessed in a wall.

Underground drainage and vent piping should be laid on a firm natural bed of earth for its entire length or on an equally firm means of continuous support. Tunneling is not recommended as a satisfactory method for installing such piping because of the resulting misalignment of the piping when the soil above it settles down upon it in time and a rut forms in the ground surface at grade level. Open trenchwork is generally recommended for such piping installations. Proper compactness of backfill should be assured without damage to piping. Clean earth, sand, or screened gravel should be placed under, around, and above the piping, to at least 1 ft above it, and compacted carefully. Thereafter, backfilling should be completed to grade, compacting the backfill at least every 2 ft (0.61 m). Heavy boulders and corrosive cinder fill should not be allowed in the trench as backfill material.

Drainage and vent piping aboveground should be securely supported and attached to the building construction. Where it is deemed necessary to prevent movement of the piping, it should be attached securely to an anchor rigidly affixed to the building construction. Hangers, piers, and pipe anchors should be of durable materials having adequate strength to perform their respective functions for the anticipated life of the building.

The maximum distances between supports for drainage and vent piping of various types of materials commonly used are as follows:

Vertical piping:

Cast-iron pipe: at base, and at each story

Screwed pipe (standard pipe size): every other story, not to exceed 25 ft (7.6 m), or where fitting is installed within story

Copper tube: every story, but not more than 10-ft (3.0-m) intervals

Lead pipe: 4-ft (1.2-m) intervals

Plastic pipe (DWV, rigid): 4-ft (1.2-m) intervals

Glass pipe: every story, but not more than 15-ft (4.6-m) intervals

Horizontal piping:

Cast-iron pipe: 5-ft (1.5-m) intervals, except that where 10-ft pipe lengths are used, 10-ft (3.0-m) intervals are acceptable

Screwed pipe (standard pipe size): 10-ft (3.0-m) intervals

Copper tube: 10-ft (3.0-m) intervals

Lead pipe: on continuous metal or wood strips for entire length

Plastic pipe (DWV, rigid): 4-ft (1.2-m) intervals

Glass pipe: 8-ft (2.4-m) intervals

Changes in direction of drainage piping should be made by means of fittings which will permit flow to proceed without excessive reduction in velocity or other adverse effects. The use of 45° Ys, long 90° sweeps, sixth, eighth, and sixteenth bends, and combinations of such fittings and equivalents of such fittings, is recommended. In the interest of economy, it is recommended that available combination fittings which are the equivalent of several individual fittings be used in place of them wherever the opportunity presents itself. Short 90° sweeps are not recommended for use in drainage piping except where it is 3 in (75 mm) or larger in size. Single and double sanitary tees should not be used, except in vertical drainage piping.

Running threads, bands, and saddles should not be permitted in drainage or vent piping, nor should the drilling or tapping of such piping be allowed. Experience has vividly shown that connections made by means of running threads, bands, and saddles seldom if ever remain tight and are prone to leakage of waste and objectionable gases and odors. Drilling and tapping of piping are disapproved generally because of the many instances in which the piping has been damaged or split, the inadequate pipe wall thickness available for the minimum number of threads required for joint tightness, and the projection of branch pipe ends into the drilled pipe.

Any method of installation, fitting, device,or connection which retards flow in drainage or vent systems to a greater degree than normal frictional resistance should not be permitted. Hence, double hubs should not be used in drainage piping. Neither should a fitting having a hub facing downstream be so used. Nor should a T branch of a drainage fitting be used as an inlet branch for wastes. Heel- or side-inlet quarter bends should not be used as vent connection fittings in drainage piping when the heel or side inlet is placed in a horizontal position. In that position, the vent connection is prone to becoming blocked by sewage matter which may enter and accumulate in the horizontal vent and eventually stop airflow through it.

Installation methods which may result in damage or material reduction in the durability of piping should be avoided wherever possible. For example, the expanding or swaging of 3-in (75-mm) lead bends and stubs, so as to connect them to 4-in (100-mm) flanges for fixtures, results in a corresponding reduction in pipe wall thickness. Numerous cases of leakage have been known to occur within a short time after installations were completed and found tight under test. Such leakage incidence should be avoided. It is recommended that 3- by 4-in (75- by 100-mm) lead bends and stubs, having uniformly proper wall thickness, be used for connections to 4-in (100-mm) floor flanges and that 4- by 3-in (100- by 75-mm) floor flanges be used for connections to 3-in (75-mm) lead bends and stubs.

For future fixture installations, provisions should consist either of plugged branch fittings in stacks or branch piping or completely installed drainage and vent piping for such fixtures, except for exposed short fixture drain and trap connections which may be required for completion of such installations. Drainage or vent branch pipes which terminate at a distance greater than 2 ft (0.61 m) from a ventilated line of piping are considered to be "dead ends" in which inadequate air circulation conditions exist. They are subject to the development of fungi and accumulations of slime and sludge. Hence, they should not be permitted except where it is necessary in order to extend the piping for a cleanout in an accessible location.

Hazardous piping should always be conspicuously identified so as to warn maintenance personnel of danger. This applies to acid waste piping systems, high-pressure steam blowoff systems, and most particularly to piping and equipment for conveying radioactive wastes. In the latter case, the piping and equipment should be conspicuously identified by adequate labeling with the standard radiation danger symbol as required by state health authorities having jurisdiction.

EXPANSION AND CONTRACTION OF PIPING

Drainage and vent piping should be installed in such manner as to avoid damage and breakage due to strain accompanying normal expansion and contraction of the piping. *The recommended design basis for sanitary drainage and vent piping aboveground in buildings is to provide accommodation for the normal expansion and contraction in length which may occur in the materials corresponding to a temperature change of 50° F (27.8° C), or a pipe temperature range from 40° F (4.4° C) to 90° F (32.2° C).*

Table 4-1 presents coefficients of linear expansion for piping materials in general use, including plastic pipe. It should be noted that ABS-DWV, type I, and PVC-DWV, type II, expand at a rate 9.4 times as much as cast iron, and 5.9 times as much as copper, and necessitate careful attention.

Calculations made applying this information show that a 100-ft (30.5-m) straight run of either ABS-DWV, type I, or PVC-DWV, type II, expands 3.36 in (85.3 mm) in length when subjected to temperature change of 50°F (27.8°C). For a 10-ft (3.1-m) straight run, the corresponding change in length would be approximately ⅜ in (9.5 mm). The calculations apply to piping which is free to expand.

If the piping were restrained completely so that expansion could not take place, a 50°F (27.8°C) temperature change would cause development of an axial compressive stress in the piping. The amount of such stress as calculated would be 588 psi (4054 kPa) in ABS-DWV, type I, and 825 psi (5688 kPa) in PVC-DWV, type II. Under these conditions, the pipe, acting as a column, will tend to bow between points of anchorage. Although it is highly probable that the effect of lateral deflection on the axial compressive stress in the pipe will be relatively small, it remains to be proved that no permanent deformation or failure will result when these plastic materials are subject to complete constraint of expansion over an extended period of service.

Appropriate accommodation for normal expansion and contraction effects in plastic piping systems may be provided by recognized methods in common usage. Four such methods are: (1) packless expansion joints; (2) slip joints; (3) swivel joints; and (4) flexural offsets, bends, or loops.

Methods 3 and 4 may be applied in the field using available materials in appropriate arrangements. To apply them, determine first the amount of expansion to be accommodated for a given straight run of piping, and then determine the developed length of pipe and fittings which should be provided in the swivel joint, or flexural offset, bend, or loop.

The amount of expansion to be accommodated for any given length of straight run of ABS-DWV, type I, or PVC-DWV, type II, may be established as being approximately ⅜ in (9.5 mm) per each 10 ft (3.1 m) of length. Knowing this, the developed length which should be provided in the swivel joint or flexural offset or loop, so as not to exceed an allowable extreme fiber stress in the piping, may be determined from equations which have been developed for this purpose. They are as follows:

For ABS-DWV, type I piping:

$$L = 1.71\sqrt{de} \quad [8\text{-}1]$$

For PVC-DWV, type II piping:

$$L = 1.89\sqrt{de} \quad [8\text{-}2]$$

where L = developed length of piping in expansion loop or swing joint, ft
d = outside diameter of piping, in
e = amount of expansion to be absorbed, in

An example of how this may be applied can be given. A straight run of piping, such as a building drain or a drainage stack, has a T-Y branch connection located 10 ft (3.1 m) from the point at which the straight run is anchored or restrained from movement. For a 50°F (27.8°C) temperature differential in the straight run, the amount of expansion to be accommodated at the branch connection will be approximately ⅜ in (9.5 mm). To accommodate this amount of expansion, the branch pipe must have sufficient developed length to flex or twist without being subjected to excessive strain. The developed lengths calculated in this example for various sizes of branch pipes made of ABS-DWV material are as follows: for 1¼-in (31.8-mm) pipe—1.3 ft (0.4 m) developed length; 1½-in (38.1-mm) pipe—1.4 ft (0.43 m); 2-in (50-mm) pipe—1.6 ft (0.49 m); 3-in (75-mm) pipe—1.9 ft (0.58 m); and 4-in (100-mm) pipe—2.2 ft (0.67 m). For PVC-DWV material, the respective lengths would be 11 percent more.

To accommodate expansion and contraction in building drains and in drainage stacks, 90° offsets may be provided. The developed length which should be provided, so as to absorb safely a given amount of expansion or contraction, may be calculated for a given size of piping offset in the same way as explained above.

At roof flashings for vent extensions through roofs, adequate means to accommodate the normal expansion and contraction of stacks must be provided for a satisfactory installation. Where the amount of stack expansion is relatively small, not more than about ½ in (12.5 mm), this may be accommodated by a flexible type of flashing. Where the amount of expansion is greater, this may be accommodated by means of a double-sleeve-type flashing, the lower section surrounding the pipe and attached to the roofing in a watertight manner, while the upper section is attached to the pipe in a watertight manner and extends down to sleeve or slip over the lower section so as to provide a weathertight flashing. Buckling and leakage at flashings as the result of stack expansion must be avoided for a satisfactory installation.

CLEANOUTS IN DRAINAGE PIPING

All drainage piping is subject to stoppage development at some time, and no system, regardless of how well designed, can be considered to be immune from such conditions. This has been amply demonstrated by experience. In the past when drainage piping was not required to have provisions for clearing stoppages, the piping system had to be broken into and opened for cleaning at various appropriate locations. Damage to the piping, inadequate clearing of stoppages, improper repairs to the piping where openings were made, resulted in leakage, vermin, and other objectionable unsanitary conditions in buildings. Hence, adequate provision of readily accessible cleanout openings of adequate size at appropriate locations in the drainage piping system is now deemed a practical necessity.

Cleanouts should be provided on the building drain near its junction with the building sewer outside the building, or at a Y-branch fitting or building trap immediately inside the building; and cleanouts should be provided at all changes in direction of the building drain greater than 45°. All horizontal drainage piping should be provided with cleanouts spaced not more than 50 ft (15.2 m) apart for piping 4 in (100 mm) or less in diameter, and not more than 100 ft (30.5 m) apart for larger piping; except that, for underground piping over 10 in (254 mm) in diameter, access holes of acceptable design and equipped with suitable covers should be installed at each 90° change in direction and at maximum intervals of 150 ft (45.7 m). An accessible cleanout should be provided at the base of each soil stack, waste stack, and leader. All underground traps should be provided with an accessible cleanout, except for P-traps located directly beneath and serving floor drains which are equipped with removable strainers or gratings.

The size of cleanouts should be the same as the nominal size of the pipe they are intended to serve, but need not be larger than 4 in (100 mm). On concealed or underground piping, cleanouts should be extended so as to be accessible at the grade, floor, or wall, whichever location is the most suitable for maintenance operations. Cleanouts should be installed so that the opening is in a direction opposite to that of flow in the drain, or at right angle thereto. They should also be installed so as to permit adequate clearance for rodding. For piping 3 in (75 mm) and larger, at least 18 in (0.45 m) of clearance should be provided, while at least 12 in (0.3 m) of clearance should be available for smaller piping.

Wherever a fixture trap, or a fixture having an integral trap which is readily removable without disturbing concealed piping, is located at a part of the drainage piping where a cleanout is required, such trap or fixture may be considered to be acceptable as the equivalent of a cleanout, provided there is not more than one 90° bend in the piping to be rodded.

FIXTURE TRAP VENTS

Adequate vent pipes should be provided to protect the water seals of all fixture traps against excessive seal loss or voidance due to siphonage, aspiration, and back pressure under conditions of normal use. Where fixtures discharge directly into drainage stacks, or into branch drains which discharge into drainage stacks, and the drainage stacks are provided with adequate air supply at their tops and adequate attendant vent stacks to relieve air from their bottoms, the only additional vent protection required to prevent trap seal loss for such fixtures is that which is necessary to avoid excessive self-siphonage effects when the fixture discharges and to relieve excessive pneumatic effects occurring in the branch drains when other fixtures discharge through them.

Regulations prevailing in many areas of the nation require that the drain of each individual fixture be provided with a vent pipe, connected in a prescribed manner, to protect the water seal of each fixture trap. But in numerous other areas, alternate special methods of venting fixture traps may be permitted for groups of fixtures in various arrangements. These methods are generally known as *wet venting, stack venting, circuit and loop venting,* and *combination waste-and-vent system.* Recommendations regarding application of these special methods have been included in model plumbing codes, based upon research reports published by the National Bureau of Standards and other research agencies. As a result, these methods may be applied for economy in vent-piping design wherever they are acceptable to the authority having jurisdiction. They are described in the following sections under their respective titles (see Figs. 8-7 to 8-14).

Blowout-type water closets and urinals, during the initial phase of their discharge, produce very sudden and heavy surges with appreciable shock effects in the fixture drain. This necessitates providing an individual vent connected to the fixture drain of each blowout-type fixture in order to avoid transmitting such effects to other fixture traps connected to the same branch drain.

To protect a fixture trap against loss of seal due to siphonage when the fixture discharges, air must be allowed to enter the fixture drain at a level above the dip of the fixture trap so as to disrupt or prevent the development of siphonic action in the fixture drain as the last of the volume of water drains from the fixture. This prevents the sloping fixture drain from acting as the long outlet leg of a gravity siphon which otherwise might occur and siphon out the last amount of water flowing into the trap. An additional factor which tends toward producing siphonage effects is the velocity head, or momentum, of the water during peak rate of flow in the fixture drain. Research has shown that even though siphonic action occurred during fixture discharge, the trap seal might be completely refilled by a sufficient amount of trickling or trailing flow at the last phase of drainage from fixtures having flat bottoms of 1 ft^2 (0.093 m^2) or more, such as sinks, laundry trays, and bathtubs.

Nevertheless it is recommended that vent connections to fixture drains be made above the level of the dip of the fixture trap for all fixtures except floor-outlet-type water closets and urinals, which are specifically designed to produce strong siphonic action for adequate fixture performance, and service or slop sinks, which are equipped with floor-outlet trap standards and sufficiently large flat bottoms to refill the trap seal when the fixture discharges. These exceptions are recommended in view of the practical difficulties and other objections encountered in venting such fixtures in the usual manner.

A single vent pipe which connects at the junction of two fixture drains above the level of the dip of each fixture trap may serve as an individual or *common* vent to protect both traps. This type of vent is shown with other types of individual vents in Fig. 8-7.

Vent connections to fixture drains should be installed so that the developed length of fixture drain between the weir of the fixture trap and the vent connection does not result in excessive self-siphonage effects. This length varies with the type of branch fitting into which the fixture drain discharges, the slope of the fixture drain, and its diameter. In view of the permissible types of branch fittings and drain slopes, the maximum permissible developed length may be based just on the diameters of fixture drains. Recommended maximum distance of a vent connection from the weir of a fixture trap is given in Table 8-2.

Vent connections should not be permitted to be made to the crown of any trap. Such vent connections have been proved to be worthless in a very short period of normal service. When wastes flow through a fixture trap, they are centrifuged against the crown of the trap, whereby they can enter and deposit waste matter in a vent connection attached to the crown. This results in rapid stoppage of the vent connection and failure in its capacity to provide necessary venting.

Vent piping should be installed in such a manner as to minimize the possibility of stoppage therein and especially prevent it from serving as a means of bypassing waste flow to other drains in the event of a stoppage development in the drain to which it is connected. This is

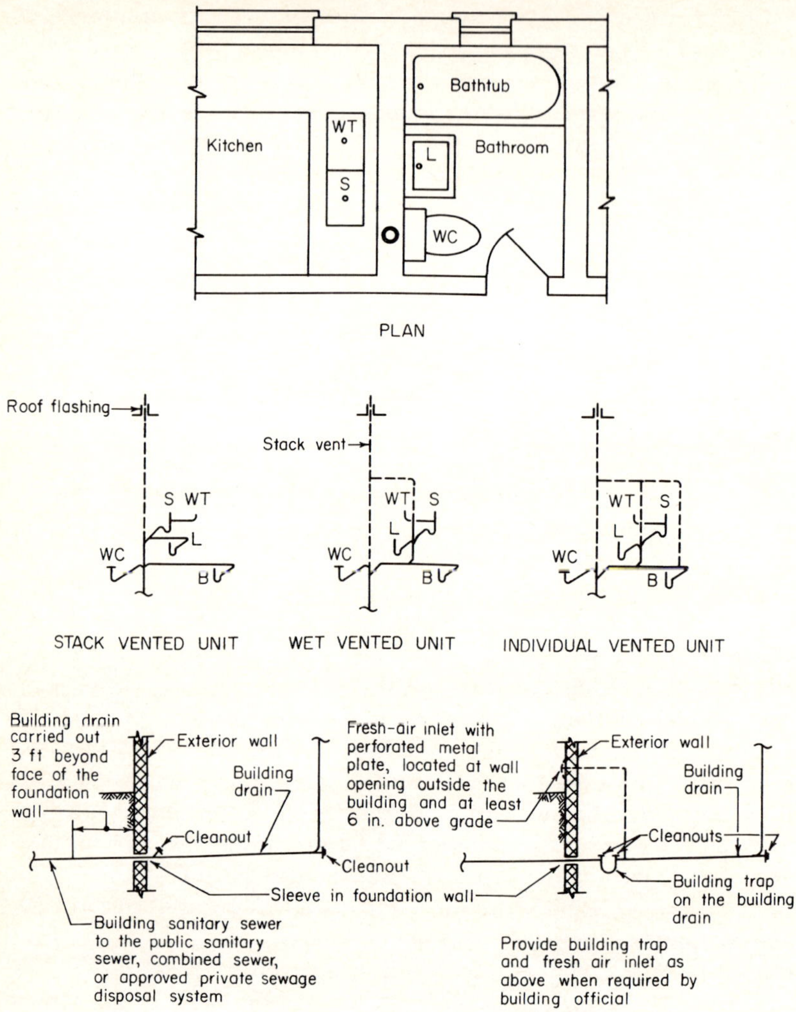

Fig. 8-7 Typical sewage drainage system for one-family dwelling.

necessary in order to maintain the vent piping at its maximum capacity for service.

Vent connections should be made to the top half of fixture drains and should rise at least at a 45° angle above the horizontal before offsetting horizontally. Each vent pipe connected to a fixture drain or branch drain should rise to a level at least above the flood-rim level of the highest fixture discharging into the branch drain before the vent pipe connects to a branch vent, vent stack, or drainage stack vent extension.

Table 8-2
MAXIMUM DISTANCE OF VENT FROM FIXTURE TRAP

Size of fixture drain		Maximum distance of vent to trap	
in	mm	ft	m
1¼	31.8	2½	0.75
1½	38.1	3½	1.05
2	50.0	5	1.50
3	75.0	6	1.80
4	100.0	10	3.0

WET VENTING

This special method of venting is intended as an economical means of providing adequate protection of fixture trap seals for a group of fixtures, such as a bathroom and kitchen group in a dwelling unit, when such fixtures discharge into a main drain or drainage stack in which only minor pneumatic effects may be anticipated. A single vent is used in this case to relieve whatever minor pneumatic effects may occur in the fixture and branch drains, and to prevent excessive self-siphonage of water seals during fixture discharge.

On the top story of a building, an individually vented fixture drain of a lavatory, kitchen sink, or combination fixture may serve as a wet vent to protect the traps of a bathtub, shower stall, and water closet provided:

Not more than one fixture unit of load is served by a 1½-in (38.1-mm) wet vent, nor four fixture units by a 2-in (50-mm) wet vent

The length of each fixture drain does not exceed the maximum distance permitted between a fixture trap and its vent connection

The horizontal branch drain connects to a drainage stack at the same level as, or below, the water closet drain; or, the horizontal branch drain connects directly to the upper half of the horizontal portion of the water closet drain at an angle no greater than 45° from the direction of flow (see Figs. 8-7 and 8-9)

On the top story of a building, a common vent and drain for two lavatories, back-to-back, may serve as a wet vent to protect the traps of two bathtubs or shower stalls installed back-to-back provided:

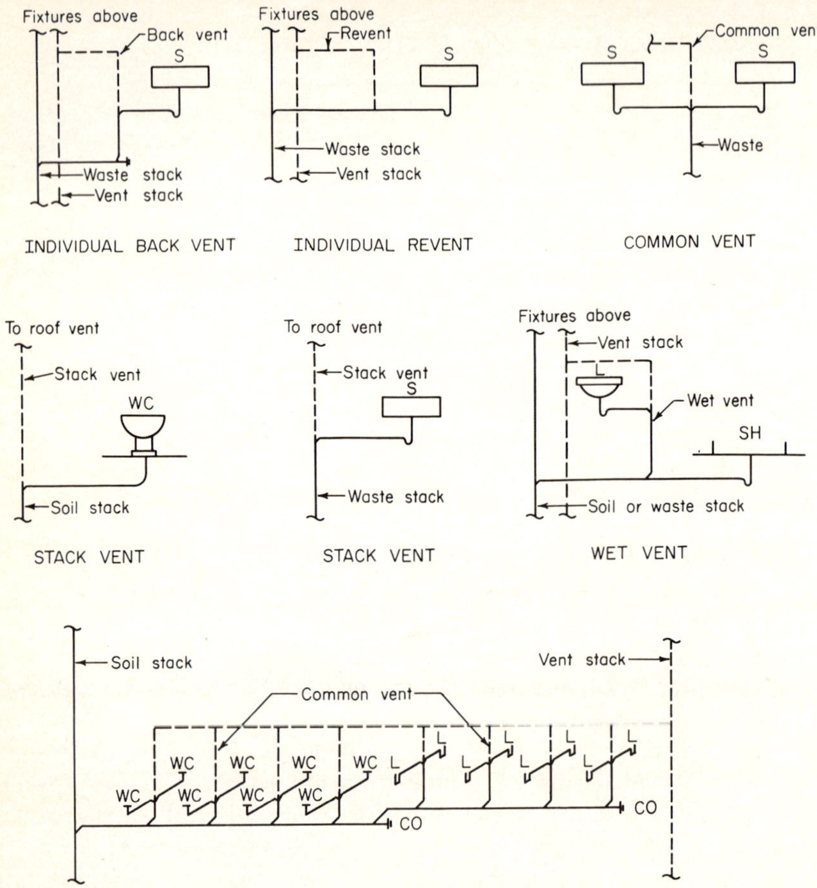

Fig. 8-8 Fixture trap vents.

The fixtures discharge into the same horizontal branch drain

The length of each fixture drain does not exceed the maximum distance permitted between a fixture trap and its vent connection

The wet vent is at least 2 in (50 mm) in size (see Fig. 8-9)

Below the top story of a building, an individually vented fixture drain of a lavatory, or a common vent and drain for two lavatories, back-to-back, may serve as a wet vent to protect the traps of one or two bathtubs or shower stalls provided:

The wet vent and its extension to the vent stack is at least 2-in (50-mm) in diameter

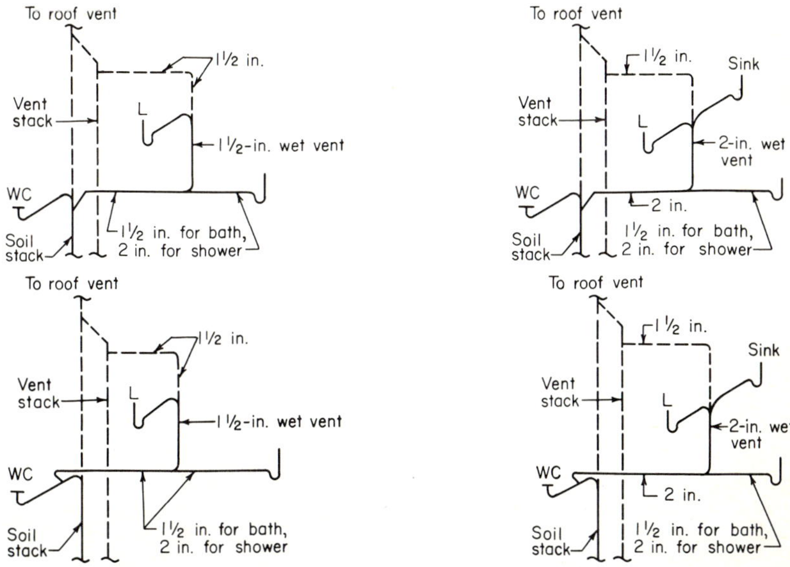

Fig. 8-9 Wet-vented single bathroom and kitchen fixture group on a stack or at top floor of a stack serving multistory bathroom groups.

Each water closet below the top story is protected by an individual vent

The length of each fixture drain does not exceed the maximum distance permitted between a fixture trap and its vent connection

The vent stack is sized in accordance with Table 8-3

Water closets below the top story of a building need not be protected by individual vents provided a 2-in (50.0-mm) wet-vented waste pipe connects directly to the upper half of the horizontal portion of the water closet drain at an angle no greater than 45° from the direction of flow (see Fig. 8-10).

Table 8-3
SIZE OF VENT STACKS FOR WET-VENTING BATHROOM GROUPS

Number of wet-vented fixtures	Diameter of vent stack in	mm
1–2 bathtubs or showers	2	50.0
3–5 bathtubs or showers	2½	63.5
6–9 bathtubs or showers	3	75.0
10–16 bathtubs or showers	4	100.0

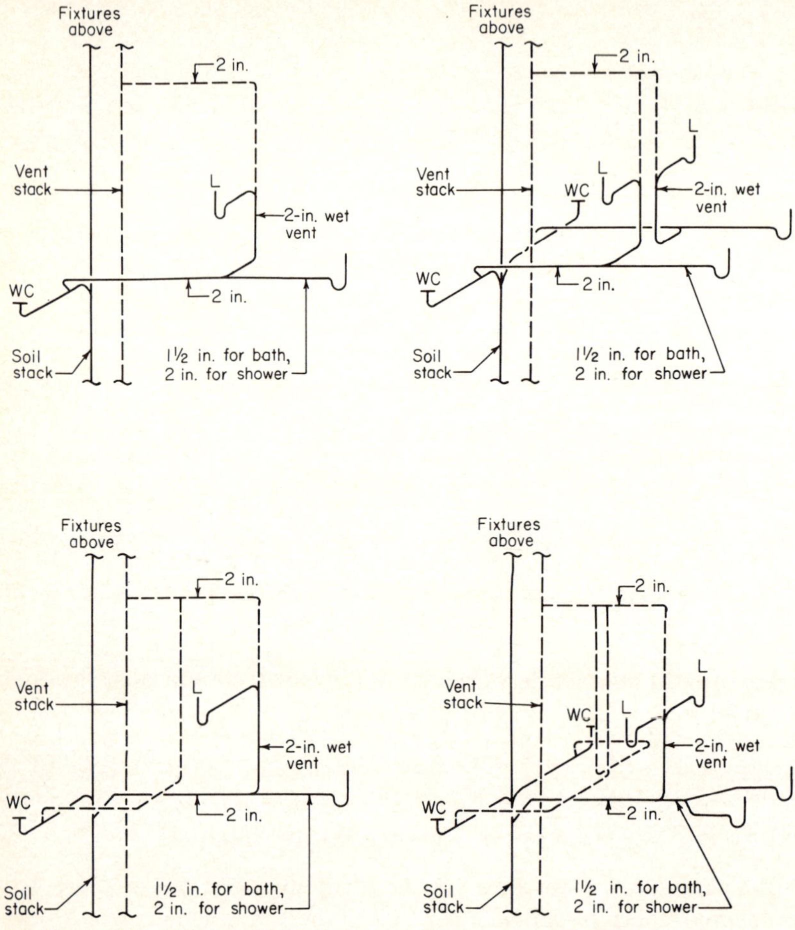

Fig. 8-10 Wet-vented multistory bathroom fixture groups below top floor group.

STACK VENTING

This special method of venting is intended as an economical means of providing adequate protection of fixture trap seals for fixtures which are grouped adjacent to and at the top of a drainage stack, where only minor pneumatic effects may be anticipated in the drainage stack. In this case, the fixtures are individually connected to the drainage stack which serves as a vent connection to prevent excessive self-siphonage of water seals during fixture discharge.

Where a fixture discharges into a soil or waste stack above all other drainage branches, the drainage stack and its vent extension may serve as an individual vent to protect the fixture trap provided that the fixture drain connects to the drainage stack above the level of the dip of the trap, except for the fixture drains of floor-outlet-type water closets and urinals, and of floor-outlet type trap standards for service and slop sinks; and that the connection to the drainage stack is not more distant than the maximum permitted between a fixture trap and its vent connection.

Where the highest two fixture drain connections to a soil or waste stack are for two horizontal fixture drains serving fixtures on the same floor level, the drainage stack and its vent extension may serve as an individual vent to protect the water seals of both fixture traps provided that the soil or waste stack is at least one pipe size larger than the highest fixture drain and not smaller than the lower fixture drain and that both fixture drains do not exceed in length the maximum permitted between a fixture trap and its vent connection.

In a one-story building, or on the top floor of a building, a group of fixtures on the same floor level, consisting of one bathroom group and a kitchen sink or combination fixture, may be deemed to be adequately protected by the drainage stack and its vent extension provided that each fixture drain connects independently to the soil stack, and that the water closet and the bathtub or shower stall drains enter the stack at the same level, and that all fixture drains do not exceed in length the maximum permitted between a fixture trap and its vent connection. In an area where the public sewer may become overloaded sufficiently to cause frequent backwater conditions in the building sewer, a relief vent or an individually vented fixture drain should be connected to the soil stack below the fixture drain connections serving a stack-vented water closet, bathtub, or shower stall (see Fig. 8-7).

CIRCUIT AND LOOP VENTING

This special method of venting is intended as an economical means of providing adequate protection of fixture trap seals of floor-outlet-type fixtures which connect in battery arrangement to a horizontal branch soil or waste pipe wherein only minor pneumatic effects may be anticipated, and the self-siphonage characteristics of the type of fixtures in this arrangement pose no problem. A single vent pipe connected to the most upstream section of the horizontal branch drain serves to relieve whatever pneumatic effects may occur therein.

With this method, a uniformly sized horizontal branch soil or waste pipe, to which two or more, but not exceeding eight, floor-outlet type water closets, shower stalls, or floor drains are connected in battery arrangement, may be deemed adequately vented by means of a circuit or loop vent connected to the horizontal branch drain at a point between the two fixture drain connections most upstream on the drain. Lavatories and similar fixtures may be connected to the circuit or loop-vented branch soil or waste pipe provided individual or common vents are installed to protect the traps of such fixtures (see Figs. 8-11 and 8-12).

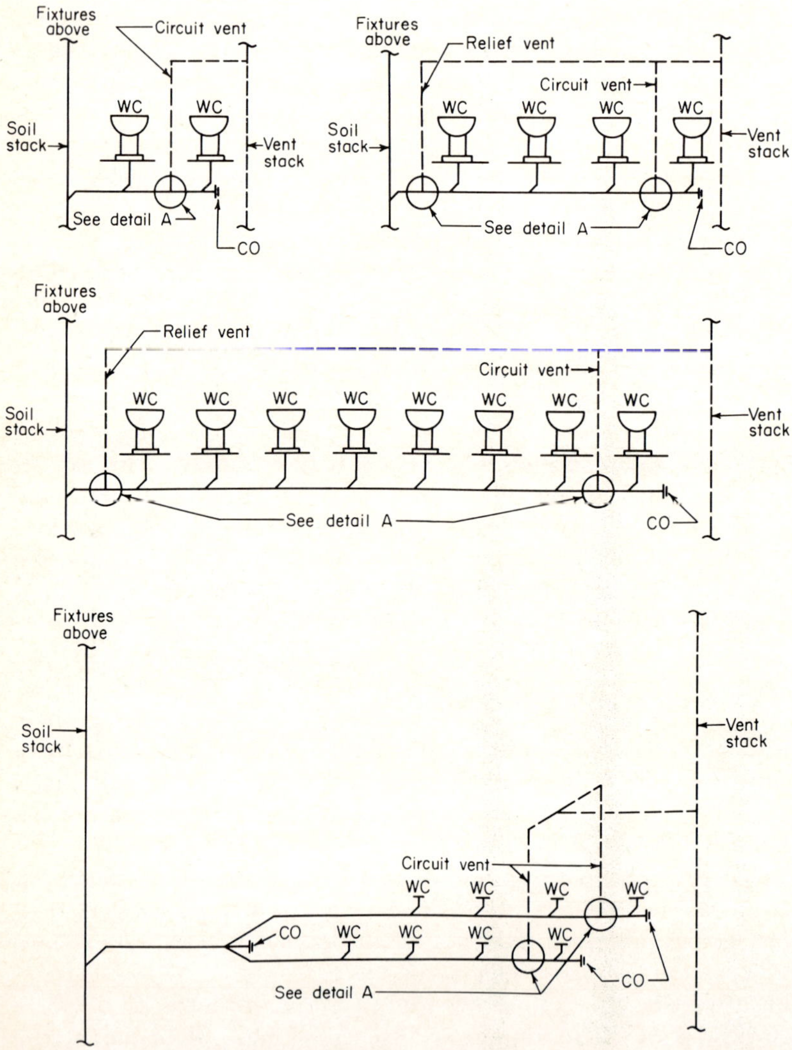

CIRCUIT VENTING OF PARALLEL BRANCH DRAINS

Fig. 8-11 Venting for batteries of fixtures: circuit venting.

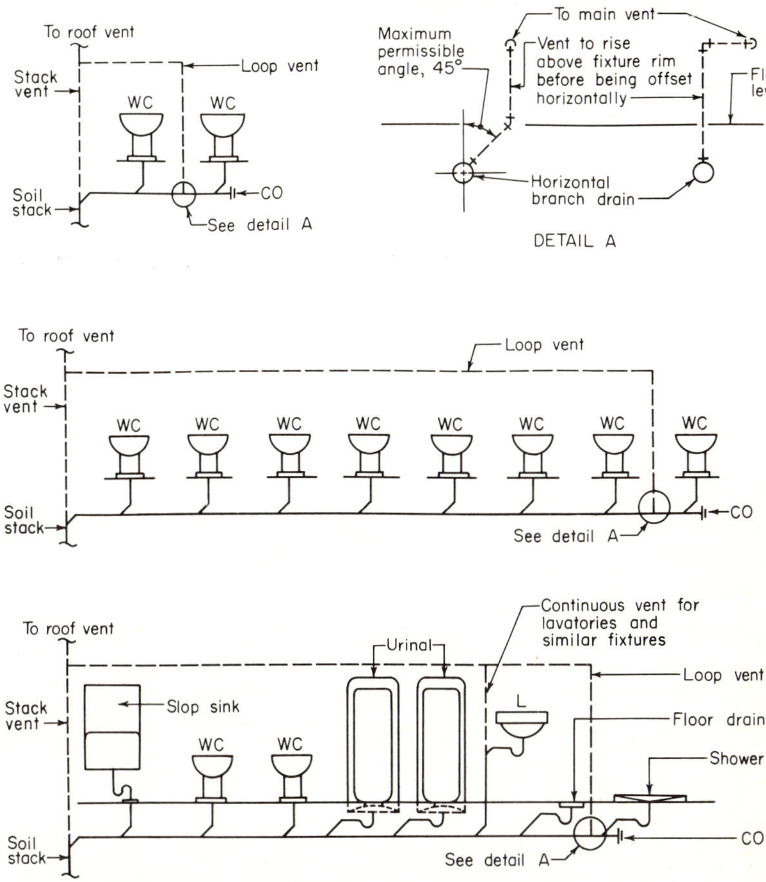

Fig. 8-12 Venting for batteries of fixtures: loop venting.

COMBINATION WASTE AND VENT SYSTEM

This special method of venting is intended as an economical means of providing adequate protection of fixture trap seals for extensive floor and shower drain installations, floor sinks and drains in large markets, laboratory and work tables in school buildings, and similar installations where individual venting of fixture drains to protect trap seals is either impractical or causes undue hardship. With this method, the waste piping is purposely oversized so as to permit it to serve as both a waste and a vent pipe to avoid excessive pneumatic effects at fixture drains.

Combination waste and vent piping systems, limited for use as a means of venting the traps of floor drains and laboratory sinks, should be per-

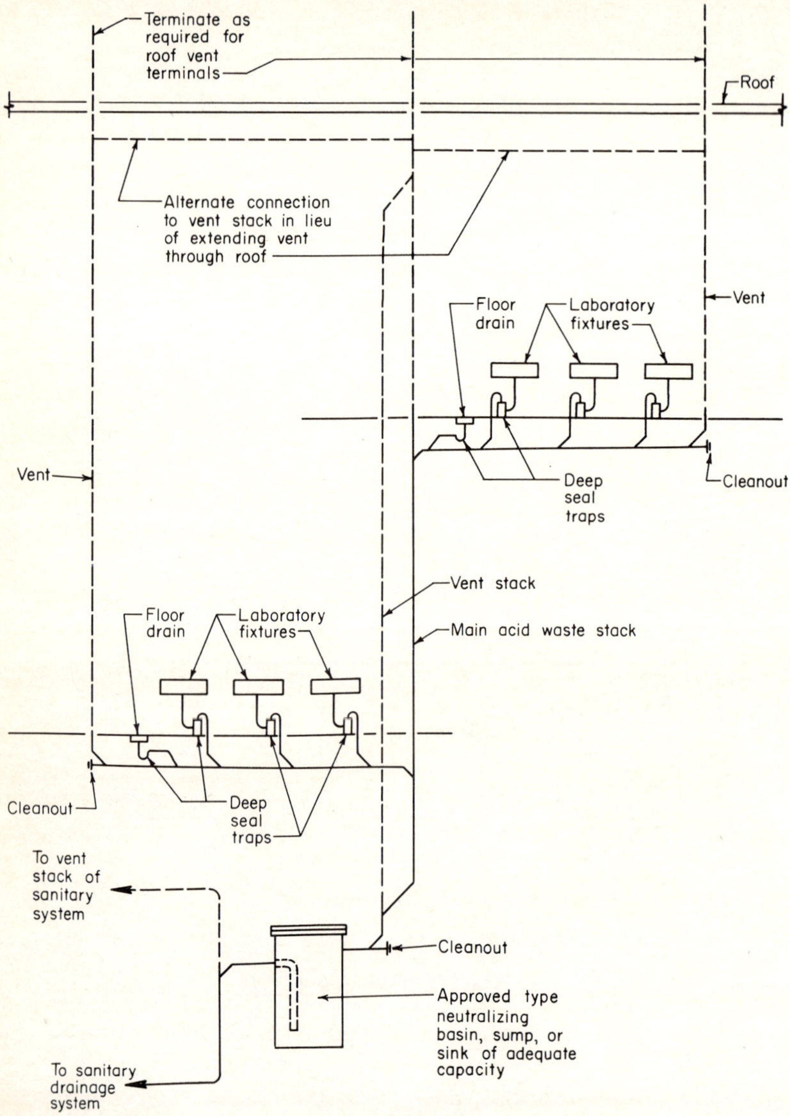

Fig. 8-13 Acid waste system with neutralizing unit.

mitted in conjunction with horizontal branch waste piping of an independent acid waste system, or an independent flammable oil waste system, or where deemed acceptable for other systems. Combination waste and vent piping shall be two sizes larger than otherwise required for drainage only. Horizontal branch waste and vent piping should be provided with

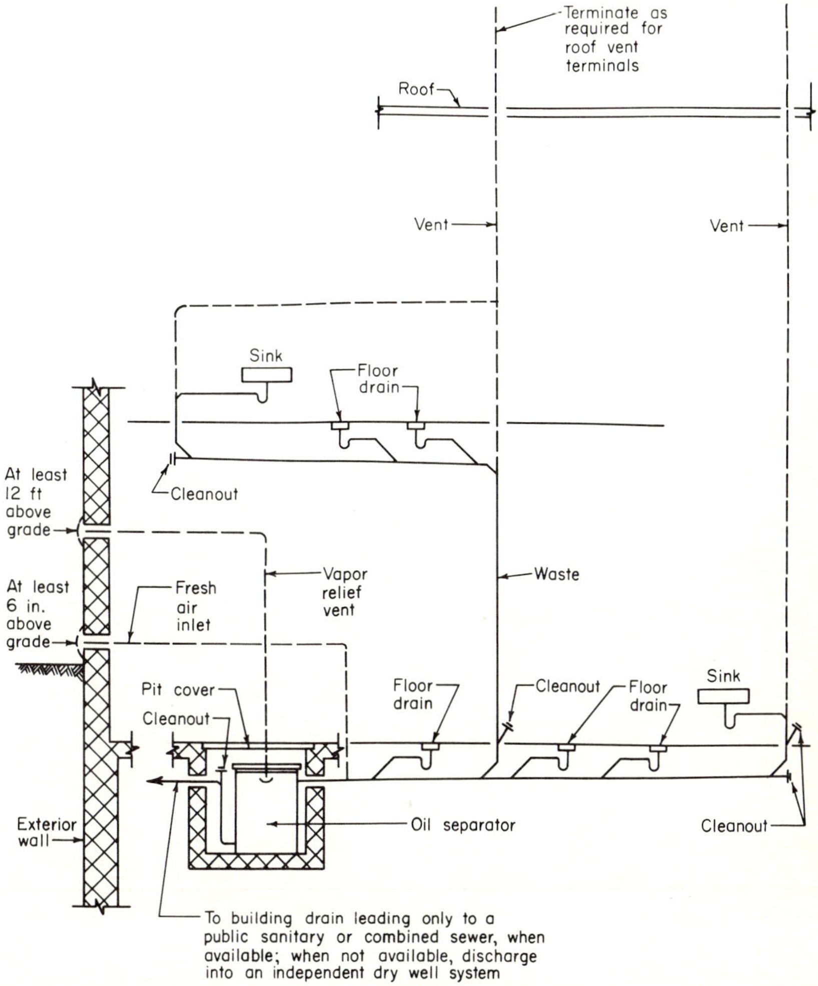

Fig. 8-14 Combination waste and vent arrangement for volatile, flammable oil waste drainage.

vent pipe connections so as to permit air circulation through such horizontal branch drains (see Figs. 8-13 and 8-14).

INDIRECT WASTE PIPING

The installation and sizing of indirect waste piping should in most respects be the same as for direct waste piping. However, indirect waste piping is subject to much greater incidence of stoppage development because of the relatively slow rates at which wastes are discharged into such piping from the various types of fixtures permitted to discharge indirectly. One of the most important requirements for such piping is the provision of an accessible cleanout at each change of direction in horizontal piping, regardless of the angle at which the change is made.

Indirect waste pipes should discharge through an air break into a suitable, water-supplied fixture which is directly connected to the sanitary drainage system. The air break should be provided by terminating the open end of the indirect waste pipe at least 1 in (25 mm) above the flood rim of the receiving fixture. The receiving fixture in each case should be of a type acceptable for such use, such as a sink or floor drain, and should be located in a well-ventilated area where the discharge of waste pipe odors into the atmosphere at the terminal will not become objectionable.

Vent pipes need not be provided to protect the water seals of fixture traps connected to indirect waste piping. Such piping seldom is subject to severe pneumatic effects because of the low rates of flow therein. However, where indirect waste piping exceeds 15 ft (4.6 m) in developed length and conveys just drippage from refrigerators and showcases and where indirect waste piping for other fixture wastes exceeds 100 ft (30.5 m) in developed length, such piping should be extended to the atmosphere, preferably above the building roof, independent of vent piping for the regular sanitary drainage system, and terminated as required for vent extensions through roofs. Such ventilation of indirect waste piping is necessary to prevent rapid fouling of the piping due to the activity of slime molds in the absence of adequate air circulation.

SPECIAL WASTES

Special methods for discharging wastes are recommended for use in conjunction with emptying, overflow, and relief pipes of the water supply system. Such pipes should not be connected directly to the drainage system because of the possibility of contaminating the potable water supply system. They should discharge either through an air break into a fixture acceptable for such use, or onto a roof. The same precautions and methods should be applied with regard to drain pipes of expansion tanks and sprinkler systems and to the discharge pipes from cooling jackets.

Drip and overflow pans, and similar equipment which discharge clear water only, should be discharged through an air break into a fixture acceptable for such use, or onto a roof. Such equipment should not be directly connected to the regular sanitary drainage system, for in the event of a stoppage condition in the system, sewage backup could

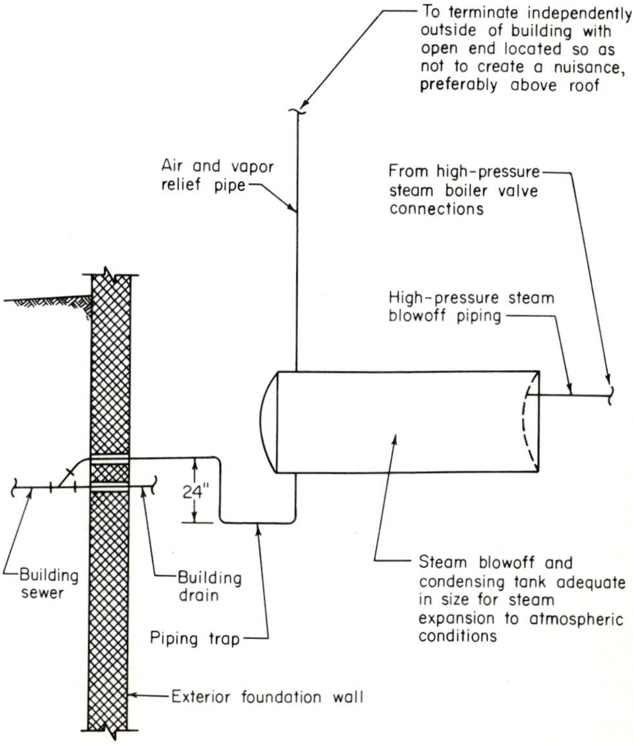

Fig. 8-15 High-pressure steam blowoff tank connection to drainage system.

occur at the equipment and cause an unnecessary, unsanitary, and objectionable condition. This should be avoided.

Steam expansion, blowoff, condenser and cooling tanks, provided to receive and handle discharges from steam exhaust and boiler blowoff pipes, should discharge through an air break into a fixture acceptable for such use, or should discharge by means of a direct connection to the building sewer where permissible (see Fig. 8-15).

TESTING OF DRAINAGE AND VENT SYSTEMS

All drainage and vent piping, except outside leaders, open-jointed or perforated subsoil drainage pipes, and exposed short fixture drain connections, should be tested and proved watertight upon completion of the rough piping installation prior to covering or concealment.

A water flow test should be applied as a practical means of testing drainage piping where the developed length of continuous drainage and vent piping installation is less than 10 ft (3.0 m), such as may be the case for a minor alteration or addition to an existing system. Water flow through the drainage piping should be at a rate similar to the service condition under which the piping is to function.

However, a water pressure test should be applied to drainage and vent piping where the developed length of continuous drainage and vent piping exceeds 10 ft (3.0 m). Test pressure should be equal to at least a 10-ft (3.0-m) column of water at all parts of the piping except the uppermost 10 ft of the system, measured downward from the highest vent terminal above the building roof. Test fittings should be provided at appropriate locations in the piping where it may be desirable or necessary to test the system in sections. As a safety precaution which should be observed, the test pressure at any point in the system should not be allowed to exceed the equivalent of a 100-ft (30.5-m) column of water. Higher test pressures cause practical difficulties, such as the need for blocking lead bends and stubs to prevent them from being blown out of hubs or distended from normal shape, and cause unnecessary and undue hazard to persons in the test area.

In the event that a water pressure test may not be appropriate because of special circumstances prevailing at the building site, an air pressure test may be acceptable to the authority having jurisdiction. The test pressure should be at least 5 psig (34.5 kPa). The air pressure test has the disadvantage of not providing prompt visible evidence to pinpoint sources of leakage. Once it is determined that a system loses air pressure, the source of leakage must still be located. This may require extensive time for application of soap solution to the entire piping system and for detection of leaks by means of soap bubble indications.

Sanitary drainage and vent systems upon completion should be tested and proved gastight. After all fixtures are installed and all traps are filled with water, the system should be subjected to a final test during which all flow of water in the system should be stopped. A smoke pressure test is recommended. It should be applied by introducing into the lowest part of the system a pungent, thick smoke produced by one or more suitable smoke machines. When smoke belches from all root terminals, they should be sealed and the system subjected to smoke pressure equivalent to a 1-in (25-mm) column of water for the duration of testing. The system should be proved smoketight.

A peppermint vapor test may be applied in lieu of a smoke pressure test only under appropriate circumstances and where acceptable to the authority having jurisdiction. In applying this test, at least 2 oz (0.0582 L) of oil of peppermint should be introduced into the roof vent terminal of every line or stack under test and should be vaporized by the immediate introduction of 10 qt (9.46 L) of boiling water.

Roof vent terminals should be sealed promptly for the duration of testing. Oil of peppermint, and all persons having had contact with oil of peppermint, must be excluded from the interior of the building during testing so as to permit the detection of leakage by the presence of peppermint vapor odor in the atmosphere. This test also has the disadvantage of not providing prompt, visible evidence to pinpoint sources of leakage. Consequently, other means for pinpointing leaks must be applied.

EUROPEAN SINGLE STACK DRAINAGE SYSTEMS

Regardless of progress made in the field of venting sanitary drainage systems in the United States, the use of the oldest system of all, the single stack drainage system, has continued in a number of foreign countries as a matter of economy and simplicity. Numerous modifications have had to be developed by research overseas for high-rise building construction. These modifications include use of special stack fittings for branch drain connections, diverter fittings to reduce the rate of flow down the drainage stack, and special stack relief vent fittings to reduce the pressure developed near the base of stacks, especially where suds pressure is a serious concern.

Nevertheless, use of the single stack drainage system in Europe has expanded throughout most of Europe in recent years. European plumbing differs from American plumbing in many significant respects. Measured against American codes and standards for plumbing design, installation, and performance, the codes and standards of some Western European countries are gradually catching up. In a few years, they may surpass those of America in certain ways. One of these ways is the permissible use of single stack drainage systems under specially prescribed conditions.

Much of this progress in Europe may be attributed to plumbing research work conducted by governmental and private agencies regarding the performance of building systems, including plumbing systems, and to prompt application of cost-saving innovations developed in this research. These innovations were first introduced into a great number of relatively small buildings rebuilt following World War II. More recently, they have been applied in many large government owned or sponsored housing projects and high-rise buildings erected there in the past 15 years.

In Europe, governments are financially involved on an extensive scale in new building construction and rehabilitation of existing buildings. This has resulted in expansion of government activity into the field of building research, including plumbing research and development, in order to keep construction costs as low as possible. Government interest in this work has facilitated the rapid introduction of cost-saving innovations, new methods, and new materials in the construction of buildings.

As a cost-saving innovation, the single stack drainage system has been the subject of research in three different countries almost simultaneously. These countries are France, England, and Switzerland.

In France, the Ministry of Reconstruction, following the end of World War II, had the responsibility for rehabilitation of a great number of damaged buildings and much new building construction. In 1947, the ministry created the Centre Scientifique et Technique du Batiment (CSTB), and assigned to it the task of carrying out research on all matters of building construction. At about the same time, the ministry instituted a system called "l'Agrément," or "authorization for use," for application to new materials and methods of construction, or possibly to new ways of using old materials and methods. At first, CSTB acted as evaluator and technical adviser for the ministry, which issued the actual certificate. But, in 1948, CSTB was assigned to perform all these functions. In 1953, the procedures to be followed were clearly established by a decree and a regulation.

Single stack drainage systems are acceptable in France under conditions for which they have been proved adequate. Some of these conditions may be cited. The stack should not be loaded to a greater degree than has been proved satisfactory for a given size. Fixtures should be grouped within prescribed distances from the stack, fixture loads should be similar to those fixtures used in acceptance tests, and the fixture drains should be of appropriate size and slope. The stack should extend vertically from its base without any offsets to at least above the highest fixture drain connection. The base fitting of the stack should be of long radius and located at least 3 ft below the lowest drainage branch. It is recommended that the horizontal drain extending from the base of the stack should be at least one size larger than the stack and be installed at a slope greater than the minimum otherwise acceptable.

In Great Britain, the Building Research Station (BRS) of the Ministry of Housing and Local Government has been of great influence in the modernization of building regulations. It reported that specification-type regulations in effect at the time hindered development of new building materials and methods of construction, and recommended use of regulations based on performance requirements. The model regulations since 1953 have been changed to contain performance requirements so far as practicable.

In 1966, the British Agrement System was instituted. It is similar to the French Agrément System. The British Agrement Board, appointed by the Minister of Public Buildings and Works, has power to issue agrement certificates, or "authorization for use" certificates, for application to new materials and methods of construction, or possibly to new ways of using old materials and methods. The board uses existing laboratory facilities, such as the British Research Station, to perform the technical investigations required for evaluation of applications for certificates.

Compliance with the performance requirements of the National Building Regulations may be achieved by conformity with British standards developed under procedures of the British Standards Institution. There are British standards for plumbing materials, and there are British standard codes of practice for design and installation of plumbing systems. Compliance may also be achieved by conformity with the terms of an agrement certificate.

The British Standard Code of Practice, CP 304, 1968, "Sanitary Pipework Above Ground," contains detailed design and installation requirements for the British single stack drainage system, which may be applied for housing. This design is based upon close grouping of fixtures in dwelling units or apartments around a single vertical drainage stack, with branches connected to the stack by fittings that are covered by British standards, a method that avoids the need for vent stacks and branch vents from fixture drains.

The British single stack drainage system is the result of research by the British Research Station. In 1954, a 4-in single stack system was recommended for use in dwellings up to five stories in height. In 1966, BRS recommended a limit of 10 stories for a 4-in system, 15 stories for a 5-in system, and 25 stories for a 6-in system.

BRS testing is not done just in the laboratory. It is also carried out on installations being made in high-rise buildings in the process of construction. Site testing of systems, using specially installed test connections, is one method employed to check on performance of systems. *Performance tests are required to be made on the British single stack drainage system. The conditions and criteria prescribed for the performance test are carefully detailed in CP 304, 1968.*

In Switzerland, plumbing regulations enacted as law are set forth as performance requirements similar to those in France and Great Britain, rather than as specification-type regulations. Compliance with plumbing performance requirements may be achieved by conformity with published recommendations issued by authorized agencies recognized as foremost experts in their fields. The recommended code of practice is entitled "Basic Principles for Sewage System Installations."

It presents the design details for single stack drainage systems in Switzerland as relatively standard practice. This design may be considered conventional in most of Western Europe, rather than singular to Switzerland, as may readily be seen by inspecting new building sanitary drainage system installations in the various countries.

Another innovation in the design of single stack drainage systems developed in Switzerland is known as the "Sovent system." It was devised by Fritz Sommer, a teacher in the trade school of the city of Berne, Switzerland. Development and testing were done on a 10-story test tower there. A United States patent has been issued on Sommer's invention. The Sovent system differs from the conventional Swiss single stack drainage system in two significant respects. One is the installation of a special Sovent junction mixer fitting in the stack at each floor level so as to divert flow in the stack and to receive branch flow from fixtures there. The second is the installation of a special Sovent deaerator fitting just above the base of the stack. These special fittings have been included in United States standards for cast bronze drainage fittings.

The Sovent system has achieved general acceptability in Western Europe. This has been facilitated to a great degree by the application of on-site performance tests of the system at the appropriate stage of construction for high-rise buildings. The system has been installed in multiple dwellings ranging up to 30 stories in height.

In the United States, the Copper Development Association has been a leading proponent in advocating introduction of Sovent systems made of copper tube and cast bronze drainage fittings. Many such Sovent installations have been made in different areas of the United States. They have in general been installed on an experimental basis. Several have been installed in high-rise multiple dwellings in New York State under its performance code. As nonstandard systems, special performance tests had to be devised and applied both during construction and upon completion to prove adequacy of performance as installed. Since completion, their performance in service has been monitored and recorded by appropriate official agencies.

In 1972, the U.S. Department of Housing and Urban Development (HUD) sponsored a study of innovative plumbing, including the above-mentioned European single stack drainage systems, and subjected them to test at the new 10-story test tower installed at Davidson Laboratory, Stevens Institute of Technology, Hoboken, NJ. The results of the study have been completed and presented to HUD. This has added greatly to the fund of knowledge regarding the performance of European single stack drainage systems.

STANDARD CODE REGULATIONS

Method of Storm Water Disposal Buildings shall have drainage provisions for conveying storm water from roofs and paved areas on the premises to a public storm or combined sewer, except that where such sewer is not available, an approved method of storm water disposal shall be provided.

For a one- or two-family dwelling, a public storm or combined sewer shall be deemed available when such sewer is within 100 ft of the premises on which the dwelling is located, measured along a street, and a connection may be made lawfully thereto.

For buildings of any other occupancies, a public storm or combined sewer shall be deemed available when such sewer is within 500 ft of the premises on which the building is located, measured along a street, and a connection may be made lawfully thereto.

No storm water shall be drained into sewers intended for sewage only, nor be discharged so that water will flow onto public sidewalks.

Method of Sewage Disposal In buildings in which plumbing fixtures are installed, a sanitary drainage system shall be provided for conveying sewage from the fixtures to a public sanitary or combined sewer system. Where a public system is not available, an approved method of sewage disposal shall be provided.

For a one- or two-family dwelling, a public sanitary or combined sewer shall be deemed available when such sewer is within 100 ft of the premises on which the dwelling is located, measured along a street, and a connection may be made lawfully thereto.

For buildings of any other occupancies, a public sanitary or combined sewer shall be deemed available when such sewer is within 500 ft of the premises on which the building is located, measured along a street, and a connection may be made lawfully thereto.

Private sewage disposal systems shall conform to the regulations of the authority having jurisdiction.

No sanitary drainage system or part thereof shall be installed so as to discharge sewage into sewers intended for storm water only.

Disposal of radioactive wastes shall conform to applicable regulations of the State Sanitary Code. The method of disposal of such wastes shall be as specially approved by the authority having jurisdiction.

No sewage from a plumbing system shall be disposed of into public water ways except if specially approved by the authority having jurisdiction.

Combined Storm Water and Sanitary Drainage Systems Sanitary and storm drainage systems of a building shall be entirely separate from each other, except that where a combined public sewer is available for disposal of such drainage, a building storm drain may connect to a combined building drain in the same horizontal plane through a single Y fitting located at least 10 ft from any sanitary drainage branch to the drain.

Fixture and Equipment Connections to Sanitary Drainage System Plumbing fixtures and equipment which discharge liquid wastes or sewage shall be directly connected to the sanitary drainage system, except as otherwise provided in this section.

Fixtures and equipment used for storage, preparation, or processing of food or drink, sterile goods, or similar materials shall have their waste outlets equipped with air breaks, adequate to prevent contamination of such contents from any possible backup of sewage through the direct or indirect waste piping. Such air breaks shall be located within 2 ft of the waste outlet and on the inlet side of the trap. Waste outlets of sterilizers shall be equipped with such air breaks.

Bar sinks, soda fountains, drinking fountains, and dishwashing machines may be connected to indirect waste pipes.

Fixtures and equipment which have interior surfaces not readily accessible to permit effective cleaning shall be connected to the sanitary drainage system by means of indirect waste pipes.

Portable household appliances, such as portable laundry and dishwashing machines, which are not permanently connected to the plumbing system, shall discharge by means of suitable flexible piping into a sink, laundry tray, or other fixture approved for such use.

Refrigerators, ice boxes, or receptacles where food is stored, when provided with waste outlets, shall discharge by means of drip pipes; such drip pipes shall discharge through an air break into a floor drain or sink approved for such use, or into a safe pan or receptor which is equipped with a bell trap or ordinary type trap and discharges to the sanitary drainage system by means of an indirect waste pipe.

Kitchen and similar equipment which is not water supplied but is equipped with waste outlets shall be discharged to the sanitary drainage system by means of an indirect waste pipe.

Swimming pools and wading pools which have overflow connections located at an elevation below street level shall have their drainage outlets connected to the sanitary drainage system by means of an indirect waste pipe. Such drainage outlets shall include pool drains, scum gutter drains, backwash outlets from pool water filters, and floor drains which serve walks around the pools. When such drainage piping is below the grade of the building sewer, any existing circulation pump for pool water may be used for pumping the wastes to an elevation suitable for gravity discharge into a fixture approved for such use.

Drains provided in the pits of hoistways shall be connected by means of indirect waste pipes to the sanitary drainage system, or to the storm drainage system where approved.

Venting of Sanitary Drainage System The sanitary drainage system shall be provided with an attendant system of vent piping designed to permit adequate circulation of air in all pipes, and the admission and emission of air so that the seals of fixture traps are subjected to an air pressure differential of not more than 1 in of water column.

Methods for Handling Objectionable Wastes Where sewage from the sanitary drainage system is disposed of through a private sewage disposal system, swimming pool wastes shall not be discharged into the regular sanitary drainage system, but shall discharge through an independent sanitary drainage and disposal system.

No steam exhaust, boiler blowoff, or similar drip pipe shall be directly connected to a building drainage system. Where such wastes are discharged into the drainage system they shall be at a temperature not higher than 140°F.

Industrial wastes which may be detrimental to the sanitary drainage system or to the public or private sewage treatment plant shall be treated and disposed of by an approved method.

Wastes which may produce clogging conditions in the sanitary drainage system or sewer shall not be discharged thereinto, except where intercepting strainers, or grease or sediment intercepting fixture traps are provided and approved as satisfactory for rendering such wastes unobjectionable.

No corrosive liquids, acids, strong alkalies, or other chemicals which might destroy or injure a drain, soil, waste, or vent pipe, or which might create noxious fumes, shall be discharged into the regular sanitary drainage system. Such chemicals shall be discharged through an independent sanitary drainage system directly to a sewer, or to a dilution or neutralizing device, or to some other means of disposal, approved for such use. Where means are provided for the dilution or neutralization of such chemicals by passage through an approved dilution or neutralizing device for which adequate maintenance is assured, such treated wastes may be discharged into the regular sanitary drainage system. Chemical waste and vent piping shall be of materials resistant to the corrosive action of chemicals and fumes.

Where the authority having jurisdiction determines that a hazard would exist due to oil or other flammables which could be introduced or admitted into the regular sanitary drainage system by accident or otherwise, the fixtures receiving such wastes shall be connected to an independent sanitary drainage system discharging through an approved oil separator.

Where an oil separator discharges into a public sanitary or combined sewer, the oil separator shall be connected to the building sanitary sewer, or to the building sanitary drain, on the outlet side of any building (house) trap. Where the oil separator must discharge into a private disposal system, the system shall be one approved for such use. Oil separators shall be equipped with an individual 3-in vapor vent pipe extending from the top of the separator and terminating in the open air at an approved location at least 12 ft above grade. The depth of liquid retained by oil separators shall be at least 2 ft. The capacity of oil separators shall be specially approved for each installation, but such capacity shall be at least 6 ft^3. Where provision is made for draining rain water from diked or enclosed areas around storage tanks containing flammable liquids, located aboveground and outside of buildings, such area drains shall be provided with suitable and accessible shutoff valves.

No radioactive wastes shall be discharged into the regular sanitary drainage system or to a public or private sewer system or sewage treatment plant, except where such wastes are treated and disposed of by a method specially approved by the authority having jurisdiction.

Methods for Handling Drainage below Sewer Level Drainage from parts of drainage systems which cannot drain by gravity into the sewer shall be disposed of through subbuilding (subhouse) drainage systems and discharged into the building gravity drainage system by automatic equipment or by another approved method. Drainage and vent piping of sanitary subbuilding drainage systems shall be installed in the same manner as for gravity systems, except that the building drains of such systems shall drain into airtight and vented sumps, ejectors, or receiving tanks from which the sewage shall be discharged as required herein. Sumps, ejectors, and receiving tanks which receive only clear water drainage and from which sewage is excluded need not be airtight and vented.

Methods for Handling Subsoil Drainage Where subsoil drainage is discharged to a public sewer, subsoil drains shall discharge into an accessible approved silt and sand intercepting trap, the drainage from which shall be disposed of into the storm drainage system. Where such piping from intercepting traps is directly connected to the gravity storm drainage system, such piping shall be provided with an approved and accessibly located backwater valve.

Prevention of Overflow into Building Where fixtures are subject to overflow as the result of backwater from the public sewer system, accessible backwater valves shall be installed in the fixture drains of such fixtures, or an accessible gate valve shall be installed in the building drain at its point of entry inside the building and downstream from any building trap.

Backwater valves shall be designed so as to provide a positive mechanical seal against backwater, and when fully opened such valves shall have flow capacity not less than that of the piping in which they are installed. All bearing parts of such valves shall be of corrosion resistant material.

Piping Installation No drainage or vent piping shall be located in stairways, nor so as to interfere with normal operation of windows, doors, or other building openings.

No drainage or vent piping shall be located in a hoistway or under an elevator or counterweight.

No horizontal drainage piping shall be located directly above nonpressure water supply tanks, access holes of pressure water supply tanks, or floor areas used for the manufacture, preparation, packaging, storage, or display of food unless a watertight barrier is provided to intervene between the piping and such tanks or space immediately below.

Building sewers shall be installed in accordance with the regulations of the authority having jurisdiction.

No soil or waste piping shall be installed outside of buildings, or concealed in exterior walls or located where it may be subjected to freezing temperatures, unless adequate provision is made to protect such piping against damage from freezing.

Drainage and vent piping passing through or under cinders or other corrosive material shall be provided with approved coating, wrapping, or other means of protection against damage from external corrosion.

Drainage and vent piping shall be installed so as not to be subject to undue strain. Provision shall be made to protect the piping against damage from strain due to normal expansion and contraction, and to building settlement.

Drainage and vent piping passing through foundation or bearing walls shall be protected by means of sleeves or arches, or approved equivalent protection shall be provided. The space between sleeves or arches and the pipes passing through the wall shall be filled with approved sealing material. Sleeves shall be of iron or steel pipe two standard sizes larger than the pipe passing through.

Outside leaders installed along alleyways, driveways, or other locations where they may be exposed to damage shall be protected by guards or recessed in a wall.

Excavation for the installation of underground drainage and vent piping shall be open trenchwork. Such piping shall be supported on a firm bed for its entire length.

Precautions shall be taken to assure proper compactness of backfill without damage to the piping. Trenches shall be backfilled and compacted to at least 12 in above the top of piping with clean earth, sand, or screened gravel, which shall not contain boulders, cinders, or other substances which may damage or break the piping or cause corrosive action. Thereafter, backfilling shall be completed up to grade and be properly compacted.

Drainage and vent piping aboveground shall be securely attached to the building construction at no greater distances between supports than given in the following:

Vertical piping:

- Cast-iron pipe: at base, and at each story
- Screwed pipe (standard pipe size): every other story, not to exceed 25 ft (7.6 m), or where fitting is installed within story
- Copper tube: every story, but not more than 10-ft (3.0-m) intervals
- Lead pipe: 4-ft (1.2-m) intervals
- Plastic pipe (DWV, rigid): 4-ft (1.2-m) intervals
- Glass pipe: every story, but not more than 15-ft (4.6-m) intervals

Horizontal piping:

- Cast-iron pipe: 5-ft (1.5-m) intervals, except that where 10-ft pipe lengths are used, 10-ft (3.0-m) intervals are acceptable
- Screwed pipe (standard pipe size): 10-ft (3.0-m) intervals
- Copper tube: 10-ft (3.0-m) intervals
- Lead pipe: on continuous metal or wood strips for entire length
- Plastic pipe (DWV, rigid): 4-ft (1.2-m) intervals
- Glass pipe: 8-ft (2.4-m) intervals

Hangers, anchors, and piers for the support and attachment of drainage and vent piping shall be of approved material and have sufficient strength to support the piping and its contents.

Horizontal drainage piping shall be installed in practical alignment at a uniform downstream slope of not less than ¼ in/ft for piping of 3-in diameter or smaller, and not less than ⅛ in/ft for larger piping. Lesser slope may be used only when specially approved.

Vent piping shall be sloped upward continuously from its lowest connection with soil or waste piping to its terminal so as to provide ventilation of all parts of the drainage system by gravity circulation of air.

Changes in direction of drainage piping shall be made by the use of 45° Ys; long sweeps, sixth, eighth, or sixteenth bends, or approved combinations of these or equivalent fittings, except as otherwise provided in this section.

Short sweeps may be permitted only in drainage piping which is 3 in or larger in size.

Single and double sanitary Ts may be permitted only in vertical drainage piping.

No running threads, bands, or saddles shall be used in drainage or vent piping. No drainage or vent pipes shall be drilled or tapped.

No fitting, connection, device, or method of installation which retards the flow of water, wastes, sewage, or air in the drainage or vent systems to an extent greater than the normal frictional resistance to flow shall be installed. Double hubs are prohibited for use in drainage piping. No fitting having a hub faced downstream shall be used as a drainage fitting. No T branch of a drainage fitting shall be used as an inlet branch for wastes.

No heel- or side-inlet quarter bend shall be used as a vent connection fitting in drainage piping when the heel- or side-inlet is placed in a horizontal position.

The expanding or swaging of 3-in lead bends or stubs to 4-in size, thereby causing a reduction in pipe wall thickness, is prohibited. Approved 3- by 4-in lead bends and stubs which have uniformly proper wall thickness may be used for connection to 4-in floor flanges, and approved 4- by 3-in floor flanges may be used for connection to 3-in lead bends and stubs.

Dead ends shall be prohibited in the drainage system, except where necessary to extend piping for a cleanout so that it will be accessible.

Drainage and vent piping provisions for future fixture installations shall consist of plugged fittings at the stacks, or of piping installed without dead ends.

Piping and piping equipment conveying radioactive wastes shall be adequately and properly identified as hazards by means of conspicuous warning signs conforming to regulations of the State Sanitary Code.

Cleanouts for Piping There shall be an accessible cleanout on the building drain near its junction with the building sewer outside the building, or at a Y branch fitting or building trap immediately inside the building.

Cleanouts shall be installed at changes in direction of the building drain greater than 45°.

Horizontal drainage piping shall be provided with cleanouts spaced not more than 50 ft apart for piping 4 in or less in diameter, and not more than 100 ft apart for larger piping, except that for underground piping over 10 in in diameter, approved access holes with covers shall be installed at each 90° change in direction and at maximum intervals of 150 ft.

An accessible cleanout shall be provided at the base of each waste stack, soil stack, and leader.

Cleanouts shall be installed in such manner that the cleanout opening is in a direction opposite to the direction of flow in the drain or at a right angle thereto.

Cleanouts on concealed or underground piping shall be extended so as to be accessible at the wall, floor, or grade.

A fixture trap or a fixture with an integral trap, readily removable without disturbing concealed piping, may be accepted as a cleanout equivalent provided there is not more than one 90° bend in the line to be rodded.

Cleanouts shall be of the same nominal size as the pipe, but need not be larger than 4 in.

Cleanouts on piping less than 3 in in size shall be so installed that there is at least 12 in of clearance to permit rodding; for larger piping such clearance shall be at least 18 in.

Underground traps, except P-traps into which floor drains having removable strainers discharge, shall be provided with accessible and removable cleanouts.

Existing Building Sewers and Building Drains Existing building sewers and building drains may be used for new drainage systems only when tested and approved as conforming to the requirements for new building sewers and building drains. The enforcement officer shall have authority to require exposure of part or all of such piping.

Building (House) Traps and Fresh Air Inlets Building (house) traps may be required in accordance with local conditions. Where the local authority having jurisdiction requires a building trap to be installed on, or omitted from, a building drain, the installation shall conform to such local determination.

Building traps shall be provided with two brass cleanout plugs. Such plugs shall be of the same size as the trap, for traps up to 4 in in size, and shall be at least 4 in in size for larger traps. Cleanout plugs shall be located at the trap so as to provide access for cleaning the trap interior and for rodding upstream and downstream from the trap, except that, where the authority having jurisdiction requires the cleanouts of traps to be extended above the floor level because of local conditions, the cleanouts shall be installed in conformity with such local determination.

Where a building trap is installed, it shall be located within the property line, inside the building wherever practicable, on the building drain within 2 ft of the exterior wall of the structure, and on the sewer side of all connections except those provided to receive the discharge from a sewer lift, oil separator, blowoff and condensing tank, or leader. Where the cleanouts of such trap are located underground or below a cellar floor, an approved masonry or concrete pit or manhole shall be provided for access to the cleanouts.

Every sanitary or combined building drain equipped with a building (house) trap, sewage sump, ejector, receiving tank, oil separator, or similar equipment, shall be provided with a fresh air inlet pipe connected to the building drain immediately upstream from and within 4 ft of such trap or equipment. Such connection shall be made in the same manner as prescribed herein for vent connections to horizontal drains, and the fresh air inlet pipe shall be extended to the outer air and shall be terminated in an open end at least 6 in above grade. The open end shall be protected by a perforated metal plate permanently fixed in the mouth of the inlet and having an open ventilating area at least equal to the area of the pipe, or a return bend with its unprotected open end at least 6 in above grade shall be installed within the property line in an approved location. The size of the fresh air inlet pipe shall be at least one-half the diameter of the building drain at the point of connection, but at least 3 in.

Prohibited Drain Connections at Soil and Waste Stack Offsets No drainage branch shall be connected to a soil or waste stack within 2 ft above or below a stack offset, made at an angle of more than 45° from the vertical, except where no other drainage branch is connected to the stack at a higher story.

Soil and Waste Stack Connections to Building Drain Where two or more soil or waste stacks discharge into a building drain or a main branch thereof, they shall connect into the upper half (air space portion) of the horizontal building drain or main branch where practicable.

Vent Stacks and Stack Vents A vent stack shall be installed with a soil or waste stack which has provision for the connection of present or future fixtures in two or more stories.

Vent stacks shall connect full size at their base to the building drain, or to the soil or waste stack at or below the level of the lowest drainage connection to the soil or waste stack.

Vent stacks shall extend undiminished in size to a point at least 1 ft below the roof and connect to an independent vent extension through the roof, or to a vent header, or to the stack vent portion of the soil or waste stack, at least 6 in above the flood level of the highest fixture discharging into the soil or waste stack.

Offsets in the stack vent portion of soil and waste stacks (above the highest fixture drainage connection), offsets in vent stacks, and connections of vent stacks at the bottom to a soil or waste pipe or to the house drain shall be made at an angle of at least 45° to the horizontal. However, where the entire piping above such offsets is of nonscaling type, the offset angle may be reduced, provided there is sufficient slope for condensation to drain back to soil or waste pipe connections.

Where stack vents and vent stacks are connected into a vent header, such connections shall be made at the tops of the stacks. The vent header shall connect to a vent extension through the roof.

Vent Extensions through Roofs Vent extensions through roofs shall terminate at least 6 in above roofs, except that where a roof is used for other than incidental access, such vent extension shall terminate at least 5 ft above the roof.

No vent terminal shall be located within 10 ft directly beneath any door, window, or ventilating opening. Vent terminals shall not be located within 10 ft horizontally of such openings unless the vent terminal is at least 2 ft above the top of the opening.

Where a structure is to be built higher than the vent terminal of an adjacent building and thereby adversely affects the vent system of the adjacent building or when such vent is a potential nuisance to the occupants of the higher structure, the owner of the higher structure shall, at the owner's expense and with the consent of the owner of the adjacent building, cause such vent to be extended or altered to correct the condition.

Where a vent terminal is to be installed adjacent to an existing higher building, the proposed vent terminal shall be installed by and at the expense of the owner of the lower building, in conformity with this section, including any necessary extension of the vent terminal to a location sufficiently remote to prevent the creation of a foul air nuisance to occupants of the higher building.

Each vent extension shall be at least as large as the soil stack, waste stack, vent stack, or vent header served thereby but in no case less than 3 in in size. Where it is necessary to increase the size of a vent pipe at its vent extension, the change in size shall be made by use of a long increaser at the base of the vent extension.

Vent extensions shall not run through an exterior wall unless specially approved. Where installed, the terminals shall open downward and be effectively screened, and shall not be located under an overhang of the building nor within 10 ft measured horizontally from any lot line.

No antenna, flag pole, or similar equipment shall be attached to a vent extension.

The openings through roofs for vent extensions shall be made weathertight.

Fixture Trap Vents Individual vents shall be provided for the traps of blowout type fixtures. The traps of other type fixtures shall be provided with individual vents, except that special methods of venting as prescribed in this chapter under the headings of "Wet Venting," "Stack Venting," "Circuit and Loop Venting," and "Combination Waste and Vent System" may be used in accordance with the special conditions stated for such installations.

A common vent may serve as an individual vent for not more than two fixture traps. Such common vent shall connect at the junction of the two fixture drains and shall rise vertically from the connection before offsetting horizontally. No vent connection shall be made to the crown of a fixture trap nor to a fixture drain within two drain pipe diameters of the trap weir.

The vent connection shall be installed so that the developed length of fixture drain between the vent connection and the weir of the fixture trap does not exceed the distance set forth in Table 8-4.

The vent connection to the fixture drain shall be above the level of the dip of the fixture trap, except in the case of fixture drains of floor-outlet type water closets and urinals, and of floor-outlet type trap standards for service sinks.

The vent pipe connected to a soil or waste pipe shall rise to a level at least 6 in above the flood rim of the highest fixture discharging into such soil or waste pipe before connecting to a branch vent, vent stack, or stack vent. The

Table 8-4
MAXIMUM DISTANCE OF VENT FROM FIXTURE TRAP

Size of fixture drain		Maximum distance of vent to trap	
in	mm	ft	m
1¼	31.8	2½	0.75
1½	38.1	3½	1.05
2	50.0	5	1.50
3	75.0	6	1.80
4	100.0	10	3.0

vent connection to a horizontal soil or waste pipe shall be made to the upper half of such pipe.

Wet Venting On the top story, the drain from a back-vented lavatory, kitchen sink, or combination fixture may serve as a wet vent for the traps of bathtub, shower stall, and water closet provided:

Not more than 1 fixture unit is drained into a 1½-in wet vent or not more than 4 fixture units are drained into a 2-in wet vent

The length of each fixture drain conforms to Table 8-4

The horizontal branch drain connects to the stack at the same level as, or below, the water closet drain; or the horizontal branch drain connects directly to the upper half of the horizontal portion of the water closet drain, at an angle no greater than 45° from the direction of flow

On the top story, the drain from two common vented lavatories may serve as a wet vent for the traps of two bathtubs and shower stalls installed back-to-back provided:

The fixtures discharge into the same horizontal branch drain

The length of each fixture drain conforms to Table 8-4

The wet vent is at least 2 in in size

Below the top story, the drain from one or two individually vented lavatories may serve as a wet vent for the traps of one or two bathtubs or shower stalls provided that:

The wet vent and its extension to the vent stack is at least 2 in in diameter

Each water closet below the top story is individually vented

Table 8-5
SIZE OF VENT STACKS FOR WET VENTING BATHROOM GROUPS

Number of wet-vented fixtures	Diameter of vent stack	
	in	mm
1–2 bathtubs or showers	2	50.0
3–5 bathtubs or showers	2½	63.5
6–9 bathtubs or showers	3	75.0
10–16 bathtubs or showers	4	100.0

The length of each fixture drain conforms to Table 8-4

The vent stack is sized in accordance with a special vent stack sizing table (Table 8-5)

In bathroom groups, vented in accordance with this section, water closets below the top story need not be individually vented if the 2-in wet vented waste pipe connects directly to the upper half of the horizontal portion of the water closet drain, at an angle no greater than 45° from the direction of flow.

Stack Venting Where a fixture discharges directly into a soil or waste stack at a level above all other drain connections thereto, the stack vent may serve as the vent for the fixture trap provided that:

Such vent connection is above the level of the dip of the trap, except for fixture drains of floor-outlet type water closets and urinals, and of floor-outlet-type trap standards for service sinks; and such vent connection is within the permitted distance given in Table 8-4.

Where the highest two drain connections to a soil or waste stack are for two horizontal fixture drains serving fixtures on the same floor level, the traps of both fixtures may be vented by the stack, provided that the soil or waste stack is at least one pipe size larger than the highest fixture drain and not smaller than the lower fixture drain, and that both fixture drains have their fixture traps no farther from the stack than permitted by Table 8-4.

Except as provided above in this section, a group of fixtures on the same floor level, consisting of one bathroom group and a kitchen sink or combination fixture, may be installed without individual vents for fixture traps in a one-story building or on the top floor of a building provided that each fixture drain connects independently to the soil stack and that the water closet and the bathtub or shower stall drains enter the stack at the same level and in accordance with the distances permitted in Table 8-4.

Where a public sewer is overloaded sufficiently to cause frequent backwater conditions in the building sewer, a relief vent or an individually vented fixture drain shall be connected to the soil stack below the fixture drain connections serving a stack-vented water closet, bathtub, or shower stall.

Circuit and Loop Venting A uniformly sized horizontal branch soil or waste pipe to which two or more, but not exceeding eight, floor-outlet-type water closets and urinals, floor-outlet-type trap standards for service sinks, shower stalls, or floor drains are connected in battery arrangement may be vented by means of a circuit or loop vent connected to the horizontal branch soil or waste pipe at a point between the two fixture connections farthest from the stack or main drain. Lavatories or similar fixtures may be connected to a circuit or loop vented branch soil or waste pipe provided the traps of such fixtures are protected by individual or common vents.

Combination Waste and Vent System A combination waste and vent piping system, limited for use as a means of venting the traps of floor drains and laboratory sinks, shall be permitted in conjunction with horizontal branch waste piping of an independent acid waste system, or an independent flammable oil waste system, or where specially approved for other systems.

Relief Vents Where an offset between horizontal portions of the building drain rises vertically more than 10 ft, a relief vent shall be provided at the top of the vertical offset. The size of such relief vent shall be at least one-half the diameter of the building drain at the offset. Where the building drain is equipped with a building trap, a relief vent also shall be provided at the base of, and within 3 ft of, the vertical offset. The relief vent connected to the base of the offset shall be sized as a vent stack, considering the vertical portion of the building drain as a soil or waste stack, and shall be branch connected to the upper relief vent at a sufficient height so that the relief vents cannot serve as soil or waste pipes in the event of a stoppage in the vertical offset.

Soil and waste stacks more than 10 stories high shall be provided with a yoke relief vent at each tenth story, counting from the top story. The lower end of the yoke vent shall connect to the soil or waste stack through a Y located below the horizontal branch drain serving fixtures in the story and the upper end shall connect to the vent stack through a T or inverted Y not less than 3 ft above the floor level.

Soil and waste stack offsets made at an angle of more than 45° from the vertical and located more than 40 ft below the highest drain connection thereto, shall be equipped with relief vents as follows:

> Provide for the stack section below the offset and for the stack section above the offset the same venting provisions as would be required if they were two separate soil or waste stacks.
>
> Provide a relief vent at the top of the stack section below the offset, and a yoke vent at the base of the upper stack section.
>
> Where a drainage branch connects within 2 ft above or below a soil or waste stack offset made at an angle of 30 to 45° from the vertical and located more than 40 ft below the highest drain connection thereto, a

relief vent shall be provided at the top of the stack section below the offset.

Suds Pressure Zones and Suds Relief Vents Where sinks, laundry trays, laundry washing machines, and similar fixtures in which sudsy detergents are normally used discharge at an upper floor level into a soil or waste stack which also serves fixtures in other occupancy units at a lower floor level, the drainage and vent piping for such lower fixtures shall be arranged so as to avoid connection to suds pressure zones in the sanitary drainage and vent systems, or a suds relief vent, relieving to a nonpressure zone, shall be provided at each suds pressure zone where such connections are installed. The size of suds relief vents shall be at least three-quarters the diameter of the piping in which the pressure zone occurs, but not less than 2 in.

Suds pressure zones shall be considered to exist at the following locations in sanitary drainage and vent systems:

> In a soil or waste stack, which serves fixtures on two or more floors and receives wastes from fixtures wherein sudsy detergents are used, a zone shall be considered to exist in the vertical portion within 40 stack diameters of the base fitting.
>
> In the horizontal drain at the base of a soil or waste stack, which serves fixtures on two or more floors and receives wastes from fixtures wherein sudsy detergents are used, a zone shall be considered to exist in the horizontal portion within 10 stack diameters of the base fitting and, where a 60 or 90° fitting is installed in the horizontal drain, a zone shall be considered to exist in the horizontal portion within 40 drain diameters upstream of, and 10 drain diameters downstream of, the fitting.
>
> In a soil or waste stack offset of 60 or 90°, which serves fixtures on two or more floors and receives wastes from fixtures wherein sudsy detergents are used, a zone shall be considered to exist in the vertical portion of the stack within 40 stack diameters of the base fitting for the upper section of the stack, and zones shall be considered to exist in the horizontal offset within 10 stack diameters of such base fitting and within 40 stack diameters of the top fitting for the lower section of the stack.
>
> In a vent stack, which has its base connected to a suds pressure zone in the sanitary drainage system, a zone shall be considered to exist in the portion of the vent stack extending from its base connection up to the lowest branch vent fitting located above the level of the suds pressure zone in the sanitary system.

Air Pressure Relief Pipe for Pneumatic Ejector The air pressure relief pipe from a pneumatic ejector shall not be connected to the regular venting system, but shall be connected to an independent 3-in vent stack terminating as required for vent extensions through roofs. Such relief pipe shall be of sufficient size to relieve air pressure inside the ejector to atmospheric pressure within 10 s but shall be not less than 1¼ in in size.

Indirect Waste Piping Indirect waste pipes shall discharge through an air break into a water-supplied sink or into a water-supplied floor drain directly connected to the sanitary drainage system and approved for such use.

The air break for an indirect waste pipe shall be provided by terminating the open end of the pipe at least 1 in above the flood rim of the receiving fixture.

No vents need be provided for the traps of fixtures which are connected to indirect waste piping.

Indirect waste piping which exceeds 15 ft developed length and is used exclusively to convey drippage from refrigerators and showcases, and all other indirect waste piping which exceeds 100 ft in developed length, shall be extended through the roof, independent of vents for the regular sanitary system, or to an approved location in the outer air, and terminated as required for vent extensions through roofs.

Indirect waste piping shall be installed with cleanouts at each change of direction on horizontal runs.

The size of indirect waste piping shall be the same as required for direct waste piping.

Special Wastes Emptying, overflow, and relief pipes of the water supply system shall discharge through an air break into a fixture approved for such use, or onto a roof.

Expansion tanks, cooling jackets, sprinkler systems, drip or overflow pans, or similar equipment which discharge clear water only shall discharge through an air break into a fixture approved for such use, or onto a roof.

Where steam exhaust and boiler blowoff pipes discharge through properly installed and approved expansion, blowoff, condenser, or cooling tanks, such tanks shall discharge through an air break into a fixture approved for such use or shall discharge through a direct connection to the building sewer.

Roof Drains Strainers shall be provided in the inlets to leaders. Strainers shall extend not less than 4 in above the roof or gutter surface immediately adjacent to the leader inlet, and shall have total open area not less than 1½ times the area of the leader. However, strainers at roof drains for sun decks, parking decks, or similar areas which are normally serviced and maintained may be of the flat surface type installed level with the deck, but such strainers shall have total open area not less than twice the area of the leader.

The openings through roofs for roof drains shall be made watertight.

Traps on Storm Water Drains Where leaders and storm drains connect to a combined building drain or sewer, individual traps shall be installed in the horizontal branch serving each leader and each area drain, or a single trap shall be installed in the main storm drain before its connection with a combined building drain, combined building sewer, or combined public sewer.

Traps installed on storm water drains shall be of the same size as the horizontal drain in which they are installed and shall be provided with an accessible cleanout on the inlet side.

Test Methods Building drains, drainage and vent piping, except outside leaders, open jointed or perforated subsoil drainage pipes, and exposed short fixture connection drain pipes, shall be tested and proved watertight upon completion of the rough piping installation, prior to covering or concealment.

Where the developed length of continuous drainage and vent pipe installation is less than 10 ft, a water flow test shall be applied to the drainage piping. Water flow through the drainage piping shall be provided at rates similar to the service conditions under which the piping is to function. This test method may be applied in testing existing buried building drains when permission is granted by the enforcement officer.

Where the developed length of continuous drainage and vent pipe installation is 10 ft or more, a water pressure test shall be applied to the drainage and vent piping. Test pressure shall be equal to at least a 10-ft column of water at all points, except that the uppermost 10 ft of the system, measured downward from the highest roof vent terminal, need be subjected only to the pressure produced when water overflows from that terminal. The piping may be tested in sections when approved test fittings are provided at appropriate locations. Test pressure at any point in the system shall not be allowed to exceed the equivalent of a 100-ft column of water. This test shall be applied to all building drains except those for which a water flow test is specially permitted.

An air pressure test, at 5 psig pressure, may be applied instead of the water pressure test when special permission is granted by the enforcement officer.

Sanitary drainage and vent systems, upon completion, shall be tested and proved gastight.

After all plumbing fixtures have been installed and all traps have been filled with water, every part of a new sanitary drainage and vent system within building walls shall be subjected to a final test as prescribed herein. For the duration of testing, flow of water in the system shall be halted and the building drain shall be sealed adjacent to its point of entry inside the building. The enforcement officer may require the removal of any cleanout plugs to ascertain that the testing is effective in all parts of the system. Wherever there is reason to believe that the sanitary drainage or vent system of an existing building has become defective, a final test as prescribed herein shall be applied when deemed necessary by the enforcement officer.

A smoke pressure test shall be applied to the system by introducing into its lowest part a pungent, thick smoke produced by one or more approved smoke machines. As smoke belches from all roof vent terminals, they shall be sealed. Then the system shall be subjected to smoke pressure, equivalent to a 1-in column of water, for the duration of testing.

A peppermint vapor test may be applied instead of a smoke pressure test when special permission is granted by the enforcement officer. The peppermint vapor shall be applied to the system by introducing into the roof vent terminal of every line or stack under test, at least 2 oz of oil of peppermint, followed immediately by the introduction of 10 qt of boiling water. Then, the roof vent terminals shall be promptly sealed for the duration of testing. The presence of oil of peppermint or persons in contact with such vapor shall be excluded from the test area inside the building.

9

FLOW OF WATER IN DRAINAGE PIPING

PRINCIPLES OF HYDRAULICS

This chapter deals with the principles of hydraulics that are relevant to flow of water in drainage piping in buildings. Such principles concern physical and mechanical properties of water, and their application in engineering design of drainage piping systems.

The physical and mechanical properties of water include its density, scouring action, uniform and turbulent flow conditions, and frictional resistance.

Energy principles include potential and kinetic energy, static and velocity heads, and pressurized flow from fixture outlets. Friction is an essential energy consideration in limiting velocity of flow through drainage piping flowing partially full.

These properties and principles have specific application with regard to gravity flow in sloping drains and vertical drains, velocity of flow for scouring action, hydraulic jump at base of vertical drains, and quantity rates of flow in fixture drains and branch drains.

GRAVITY FLOW IN SLOPING DRAINS

Gravity flow in long, sloping horizontal drains of a plumbing system is comparable to flow in open channels in which the flowing water is not completely enclosed by walls and has a free surface exposed to atmospheric pressure. Under such circumstances, flow is not dependent upon pressure applied to the water, but instead is caused just by the amount of gravitational force induced by the slope of the drain and of the water surface therein.

In an open channel or sloping horizontal drain of constant shape and size, and of extensive length, flow has a chance to adjust itself and reach a state of equilibrium which is termed *uniform flow.* For this condition, the slope of the water surface in the drain matches the slope of the drain, and the friction head loss (in feet of water column per foot of drain length) is equal to the slope of the drain (in feet per foot).

One of the most widely used formulas for determining the rate of uniform flow in sloping horizontal drains was reported by Robert Manning in 1890. It is as follows:

$$V = \frac{1.486 \times m^{2/3} S^{1/2}}{n} \qquad [9\text{-}1]$$

and

$$Q = \frac{A \times 1.486 \times m^{2/3} S^{1/2}}{n} \qquad [9\text{-}2]$$

where V = velocity of flow, fps
Q = quantity rate of flow, ft^3/s
A = cross-sectional area of flow, ft^2
m = hydraulic mean depth of flow, ft
S = hydraulic slope of the surface of flow, ft/ft
n = a coefficient or factor that depends upon the roughness of the pipe surface, degree of fouling in service, and pipe diameter (sometimes referred to as Kutter's n)

The hydraulic mean depth of flow, in feet, frequently termed *hydraulic radius,* is the ratio of the cross-sectional area of flow, in square feet, to the wetted perimeter of pipe surface, in feet. Values for the hydraulic mean depth vary for different degrees of fullness of flow. For a drain flowing half full, the cross-sectional area of flow is equal to $3.14D^2/8$, the wetted perimeter is equal to $3.14D/2$, and the hydraulic mean depth is equal to $D/4$. Similarly, for a drain flowing full, the cross-sectional area of flow is equal to $3.14D^2/4$, the wetted perimeter is equal to $3.14D$, and the hydraulic mean depth is equal to $D/4$. Thus, it can be seen that these values coincide for half-full and full flow.

Tabulated in Table 9-1 for half-full flow and in Table 9-2 for full flow in sloping horizontal drains are values for (1) m, the hydraulic mean depth of flow, in feet; (2) corresponding values for the two-thirds power of m; and (3) A, the cross-sectional area of flow, in square feet.

Values for S, the hydraulic slope of the surface of flow (in feet per foot), and for $S^{1/2}$ are tabulated in Table 9-3 for several of the common slopes used for horizontal drains in plumbing systems.

Appropriate values for n were determined experimentally by Kutter in 1869 and were found to vary with the roughness of the pipe surface

Table 9-1
VALUES OF *m* AND *A* FOR HALF-FULL FLOW

Pipe size, in	$m = D/4$ ft	$m^{2/3}$	*A* (area of flow), ft²
1¼	0.0288	0.0940	0.00520
1½	0.0335	0.1040	0.00706
2	0.0417	0.1200	0.01090
2½	0.0521	0.1396	0.01704
3	0.0625	0.1570	0.02455
4	0.0833	0.1910	0.04365
5	0.1040	0.2210	0.06820
6	0.1250	0.2500	0.09820
8	0.1670	0.3030	0.17460
10	0.2080	0.3510	0.27270
12	0.2500	0.3970	0.39270
15	0.3125	0.4610	0.61350

Table 9-2
VALUES OF *m* AND *A* FOR FULL FLOW

Pipe size, in	$m = D/4$, ft	$m^{2/3}$	*A* (area of flow), ft²
1¼	0.0288	0.0940	0.01040
1½	0.0335	0.1040	0.01412
2	0.0417	0.1200	0.02180
2½	0.0521	0.1396	0.03408
3	0.0625	0.1570	0.04910
4	0.0833	0.1910	0.08730
5	0.1040	0.2210	0.13640
6	0.1250	0.2500	0.19640
8	0.1670	0.3030	0.34920
10	0.2080	0.3510	0.54540
12	0.2500	0.3970	0.78540
15	0.3125	0.4610	1.22700

and the diameter of the pipe. The degree of fouling of the pipe interior in service must also be considered in plumbing system drains. Recommended values to apply for the coefficient n are as follows:

> For sanitary drains of galvanized-iron pipe and cast-iron soil pipe, well aligned and supported, and with normal amount of fouling in service, assume n equals 0.012 for 1¼- and 1½-in sizes; 0.013 for 2-, 2½- and 3-in sizes; 0.014 for 4-in size; 0.015 for 5- and 6-in sizes; and 0.016 for 8-in and larger sizes.
>
> For storm drains of ordinary sewer pipe after average uneven settlement and fouling, assume n equals 0.0145 for all sizes.

Table 9-3
VALUES FOR SLOPE

Slope in/ft	S, ft/ft	$S^{1/2}$
⅛	0.0104	0.102
¼	0.0208	0.144
½	0.0416	0.204
¾	0.0625	0.250

Applying Manning's formula and appropriate recommended values for *n*, the uniform flow velocity and capacity of horizontal galvanized iron or cast-iron drains of the sanitary system, laid at a slope of ¼ in/ft and flowing full at no pressure, may be found to have the rates shown in Table 9-4.

Similarly, for storm drainage systems, the uniform flow velocity and capacity of horizontal galvanized-iron and cast-iron drains, laid at a slope of ¼ in/ft (20.8 mm/m) and flowing full at no pressure, may be found to have the rates shown in Table 9-5.

The flow capacities shown in Tables 9-4 and 9-5, for drains flowing full when laid at a slope of ¼ in/ft (20.8 mm/m), may be used to establish the quantity rates of flow corresponding to other degrees of pipe fullness and for other pipe slopes. For example, for half-full flow the quantity rate of flow for each size would be just one-half as much as those tabulated for full flow. For slopes of ⅛, ½, and ¾ in/ft, the respective quantity rates of flow would be 0.707, 1.414, and 1.73 times as much as those shown in the tables for ¼ in/ft slope.

Flow capacities and velocities under conditions of uniform flow are those which will prevail where flow has sufficient time to adjust itself and achieve a state of equilibrium. Such capacities and velocities may be considered the minimum which will occur in drains regardless of the velocity at which flow enters. High entrance velocity, such as occurs at the base of stacks and leaders, produces surges in flow for a considerable distance downstream. Under conditions of surging or unsteady flow in horizontal drains of relatively short length, flow capacities and velocities may be appreciably higher without creating any hydrostatic head

Table 9-4
UNIFORM FLOW VELOCITY AND CAPACITY OF GALVANIZED-IRON AND CAST-IRON DRAINS OF SANITARY SYSTEMS AT FULL FLOW
¼ in/ft (20.8 mm/m) Slope

Pipe diameter		Velocity		Capacity	
in	**mm**	**fps**	**m/s**	**gpm**	**L/s**
1¼	31.3	1.67	0.51	7.82	0.49
1½	37.5	1.85	0.56	11.7	0.74
2	50	1.98	0.60	19.4	1.22
2½	62.5	2.30	0.70	35.2	2.22
3	75	2.59	0.79	57.2	3.61
4	100	2.91	0.88	114	7.19
5	125	3.15	0.96	193	12.2
6	150	3.58	1.09	315	19.9
8	200	4.07	1.24	637	40.2

Table 9-5
UNIFORM FLOW VELOCITY AND CAPACITY OF GALVANIZED-IRON AND CAST-IRON DRAINS OF STORM DRAINAGE SYSTEMS AT FULL FLOW
¼ in/ft (20.8 mm/m) Slope

Pipe diameter		Velocity		Capacity	
in	mm	fps	m/s	gpm	L/s
2	50	1.72	0.52	17.4	1.10
2½	62.5	1.99	0.61	31.5	1.99
3	75	2.25	0.69	51.3	3.24
4	100	2.74	0.84	111	7.00
5	125	3.16	0.96	201	12.7
6	150	3.58	1.09	327	20.6
8	200	4.35	1.33	705	44.5

in the drain. Hence, any increase in capacity or velocity of flow due to surging or unsteady flow conditions provides an added factor of safety when design is based upon uniform flow conditions.

VELOCITY OF FLOW FOR SCOURING ACTION

Drainage piping should be designed so as to afford scouring action in the piping when conveying sewage at the maximum rates of discharge which may be anticipated from fixtures connected thereto. The most important factor in achieving scouring action is sufficient velocity of flow. A velocity of 2 fps (0.61 m/s) is recommended as the minimum necessary to produce scouring action in piping conveying sewage. This velocity has the minimum amount of traction or erosive force required for lifting or scrubbing from the pipe surface loose particles, including sand, grit, and small pebbles and carrying them along with the stream. However, a velocity of at least 4 fps (1.22 m/s) is recommended for drainage piping conveying greasy wastes, for such piping is subject to deposits of grease which may accumulate as a lining of congealed solids. Higher velocity is needed in this case to produce increased erosive action and to reduce the amount of time the greasy wastes are in contact with relatively cold drainage piping.

In Tables 9-4 and 9-5, which state the velocity and capacity attained under uniform flow conditions in horizontal drains laid at a slope of ¼ in/ft, it may be seen that several of the pipe sizes yield velocities less than 2 fps (0.61 m/s) at full flow. These velocities will prevail where the pipe length is sufficiently long that flow has a chance to adjust itself and become uniform. Hence, long runs of small-size pipe should be avoided. For short runs of fixture drains, the velocity attained under uniform flow need not be observed, as higher velocity generally will occur owing to the additional velocity gained in dropping from the fixture outlet to the horizontal portion of the fixture drain. Using the equation, $V = (2gh)^{1/2}$, the velocity attained by a body in a free fall of 1 ft (0.3 m) is equal to $(2 \times 32.2 \times 1)^{1/2}$, or 8.03 fps (2.45 m/s). Owing to friction in flowing from the waste outlet of a fixture, through the outlet tailpiece and fixture trap, the actual velocity of flow entering the horizontal portion of the fixture drain after a fall of 1 ft is only a fraction of that which may be attained in free fall, generally about one-third as much. But this is more than the recommended minimum of 2 fps (0.61 m/s) and may be considered adequate for short runs of fixture drains.

Similarly, in stacks, the minimum velocity required for scouring action is most always attained with ease, including that required for greasy wastes. In vertical drains, the velocity induced by gravity is limited only by the friction developed between the interior surface of the pipe and the flowing liquid and is more than adequate for all practical purposes. Hence, the incidence of stoppage of drainage stacks is relatively nonexistent.

GRAVITY FLOW IN VERTICAL DRAINS

When a small amount of water is discharged into a large vertical drainage pipe, flow tends to cling to the pipe wall and may descend with a slightly spiral motion. As the amount of water discharged into the vertical drain is increased, flow forms a uniformly thick sheet of water on the pipe wall, encloses a core of air in the center of the pipe, and descends without any spiral motion. This condition of flow prevails for volume rates up to a rate where flow occupies from one-fourth to one-third of the cross-sectional area of the vertical drain.

Beyond this rate, flow tends to diaphragm across the cross section of the pipe and form slugs of water, which in turn break up as increased air pressure develops in the lower section of the pipe. Frequent and persistent formation and breakage of slugs of water in flow down vertical drains produce rapid oscillations in air pressure and accompanying objectionable noise. This poses a practical limit on the rate of flow which may be accommodated by a vertical drain, such as a drainage stack in a building, without producing objectionable effects.

Flow from a horizontal drain, upon entering a vertical drain through a branch or other fitting, is subjected to gravitational acceleration and gains velocity in falling in the vertical drain. Velocity increases until the frictional resistance developed by the water, flowing in contact with the pipe wall and air core, becomes equal to the accelerating force of gravity. Thereafter, velocity remains constant as flow descends further down the vertical drain.

The distance below the fitting through which flow enters the vertical drain to the point where flow reaches constant or *terminal velocity* depends upon the quantity rate of flow, the diameter of the vertical drain, and the vertical velocity resulting from the type of fitting through which flow enters. Some vertical velocity results from the downward curvature of the entrance fitting, long sweeps and Y-branch fittings yielding more vertical entrance velocity than quarter bends and sanitary T branches.

Terminal velocity achieved by water in flowing down a stack in the form of a uniformly thick sheet of water and the distance below the stack-branch fitting to the point where terminal velocity prevails have been investigated by several research agencies. The most recent reports are those rendered by Messrs. Wyly and Eaton in the National Bureau of Standards Monograph No. 31. They recommended that terminal velocity and terminal length, for flow in stacks of cast-iron soil pipe, may be computed by means of the following two formulas:

$$V_t = 3.0\left(\frac{q}{d}\right)^{2/5} \qquad [9\text{-}3]$$

and

$$L_t = 0.052\,V_t^2 \qquad [9\text{-}4]$$

where V_t = terminal velocity of flow in drainage stack, fps
L_t = terminal length below stack-branch fitting through which flow enters, ft
q = quantity rate of flow in drainage stack, gpm
d = diameter of drainage stack, in

These authorities state that flow capacity of stacks can be expressed in terms of the stack diameter and a specific fraction of the stack's cross-sectional area which may be permitted to be occupied by water at terminal velocity. This is expressed in the following formula:

$$q = 27.8\,r_s^{5/3} d^{8/3} \qquad [9\text{-}5]$$

where q = flow capacity of the drainage stack, gpm
r_s = specific fraction of the stack's cross-sectional area permitted to be occupied by water at terminal velocity
d = diameter of drainage stack, in

Suggestions have been advanced regarding a limit which should be applied as the specific fraction of a stack's cross-sectional area permitted to be occupied by water flowing at terminal velocity in a drainage stack. One recommendation is to limit water area to $\frac{6}{24}$ that of the stack, in which case flow in the stack relatively matches the uniform flow capacity

Table 9-6
COMPUTED DRAINAGE STACK CAPACITY

Diameter of drainage stack		Capacities at recommended fractions of drainage stack fullness			
		$r_s = 6/24$		$r_s = 7/24$	
in	mm	gpm	L/s	gpm	L/s
1¼	31.3	5.0	0.32	6.5	0.41
1½	37.5	8.1	0.51	10.5	0.66
2	50	17.5	1.10	22.6	1.43
2½	62.5	31.8	2.01	41.0	2.59
3	75	52.1	3.29	67.2	4.24
4	100	111.0	7.00	143.0	9.02
5	125	202.0	12.7	261.0	16.5
6	150	336.0	21.2	423.0	26.7
8	200	709.0	44.7	915.0	57.7

of horizontal drains, flowing full, at a slope of ¼ in/ft (20.8 mm/s) as shown in Tables 9-4 and 9-5. Another is to permit 7/24 of the cross-sectional area of the stack to be occupied by water at terminal velocity. When these limits are applied to the preceding equation, it may be reduced to the following:

If r_s is $6/24$ $\quad q = 2.76d^{8/3}$ [9-6]

If r_s is $7/24$: $\quad q = 3.56d^{8/3}$ [9-7]

These two equations may be used to compute rates of flow in drainage stacks corresponding to what may be deemed recommended flow capacities for various sizes. Such capacities are shown in Table 9-6.

HYDRAULIC JUMP AT BASE OF VERTICAL DRAINS

At the base of a vertical drain, flow enters the horizontal drain at relatively high velocity. For a 3-in (75 mm) drainage stack flowing at recommended practical capacity, the terminal velocity of flow is about 10.2 fps (3.11 m/s). As flow enters a long sweep fitting at the base of the stack and is diverted from vertical to horizontal direction, the water is subjected to centrifugal force and pressed against the lower curved surface in the fitting. This results in what may be termed *shooting* flow, or high-velocity flow, having the form of a sheet of water, of relatively uniform thickness, in contact with the lower curved surface in the base fitting.

Such high velocity and sheet form of flow cannot be maintained in the horizontal drain because of the frictional resistance of the horizontal pipe surface, and the shallow slope of the horizontal drain. For a 3-in (75-mm) horizontal drain laid at a slope of ¼ in/ft (20.8 mm/m) the velocity maintained under uniform flow conditions at full or half-full flow is just 2.59 fps (0.79 m/s). Hence, the velocity of flow entering the horizontal drain is approximately 4 times as much as may be maintained therein.

The transition from high to low velocity in the horizontal drain produces a *hydraulic jump* in flow a short distance downstream from the base fitting of the stack, the distance varying with the entrance velocity, depth of water in the horizontal drain, roughness of the pipe surface, diameter of the pipe, and slope of the drain. For cast-iron drains of usual sizes, the distance from the stack base fitting to the jump in flow varies to approximately 10 times the diameter of the stack. A typical hydraulic jump is illustrated in Fig. 9-1.

At the hydraulic jump, the depth of flow rises sharply and may completely fill the horizontal pipe at flows approaching drainage stack capacity. Less jump occurs when the horizontal drain is larger in size than the stack and also when the horizontal drain is increased in slope. Downstream from the jump in flow, unsteady or surging flow conditions continue until frictional resistance gradually reduces the velocity of flow to that maintained at uniform flow conditions.

In horizontal offsets of drainage stacks, the same hydraulic jump effect occurs in the horizontal section adjacent to the base fitting for the upper section of the drainage stack. If the stack is relatively large and flow from fixtures above the offset is relatively small, the jump in flow in the stack offset may be appreciably less than that produced in the horizontal drain at the bottom of the drainage stack.

QUANTITY RATE OF FLOW IN FIXTURE DRAINS

The quantity rate at which flow should be conveyed by a fixture drain is the same as the rate at which liquid waste is discharged into the drain from the waste outlet of the fixture. Flow from a fixture waste

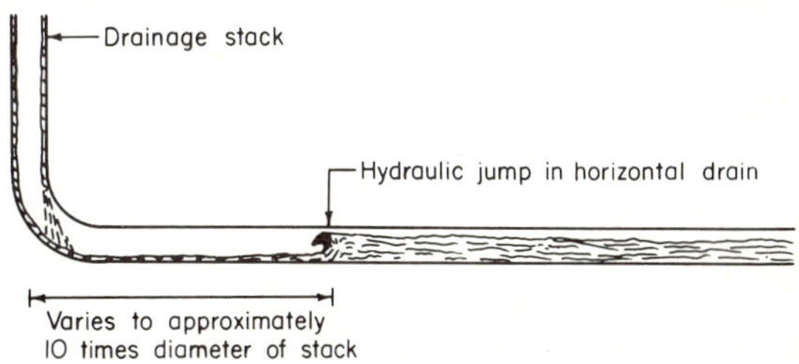

Fig. 9-1 Hydraulic jump at base of stack.

outlet is comparable with flow from a water supply orifice discharging under flow pressure. However, in this case, the orifice diameter is that of the minimum flow area existing in the waste outlet fitting and tailpiece, and the flow pressure is that due to the height of water in the fixture above the level of the minimum flow area. Thus, the quantity rate of flow discharged from a fixture into a fixture drain decreases as the water level drops in the fixture.

A formula expressing the quantity rate of flow discharged by a fixture and to be conveyed by its fixture drain is as follows:

$$q_d = 13.17 d_o^2 \sqrt{h_m} \qquad [9\text{-}8]$$

where q_d = quantity rate of flow discharged from fixture waste-outlet in practice, gpm

d_o = orifice diameter of waste outlet, in

h_m = head measured above orifice of waste outlet, ft of water column

As flow jets from the orifice of the fixture waste outlet, additional velocity is gained in dropping from the outlet through vertical sections of the fixture drain, thereby producing a corresponding decrease in cross-sectional area of flow.

QUANTITY RATE OF FLOW IN BRANCH DRAINS

The quantity rate of flow to be conveyed by a horizontal branch drain is the sum of flows received simultaneously from the various fixture drains connected thereto. For this amount of flow to be conveyed without any development of hydrostatic pressure in the horizontal drain, its design may be based upon the assumption that uniform flow conditions will prevail where the horizontal branch drain is of extended length. However, in relatively short horizontal branch drains, such as those from a group of fixtures discharging into a stack-branch fitting, the high velocities of flow from fixture drains produce surging flow conditions and appreciably higher capacities and velocities of flow than is produced under uniform flow conditions.

In no case should the quantity rate of flow conveyed by a branch drain exceed the flow capacity of the drainage stack or main drain into which it discharges. Also, where branch drains discharge into stacks, the quantity rate of flow in the branch drain should not cause excessive interference with flow in the stack so as to produce back pressure in the branch drain. Thus, the permissible flow capacity of a branch drain should be related to the probable simultaneous rate of flow in the stack. This in turn is related to the total load on the stack and to the number of branches discharging into the stack.

10

SIZING DRAINAGE SYSTEMS

SIZING FOR ADEQUACY OF PERFORMANCE

This chapter deals with the application of principles of hydraulics in sizing drainage piping so as to achieve the performance required of plumbing fixtures, fixture drains, sanitary drainage stacks, branches and building drains, storm drainage systems, and combined storm and sanitary building drains.

The specific criteria used in each sizing application are presented in detailed discussion for each section of drainage piping systems. The criteria define the limits for adequacy of performance.

FIXTURE DRAINS

In establishing proper sizes for fixture drains, consideration should be given to the diameter of the orifice in the waste-outlet fitting and tailpiece, the size of the trap, and the desired velocity of flow for proper fixture drainage and scouring action in the fixture drain. Fixtures generally have discharge orifices sized to provide suitable rates of drainage. In no case should the fixture drain size be reduced to less than that of the discharge orifice, for to do so would cause a reduction in the rate of fixture discharge. Instead, the size of the fixture trap and the fixture drain should be slightly larger than that of the discharge orifice so that they will not affect the discharge rate adversely after a moderate degree of fouling in normal service. Similarly, the size of the fixture drain should not be smaller than that of the fixture trap, so as to avoid constriction of flow and resultant increased fouling in the fixture trap.

Where fixtures discharge greasy wastes, low velocity of flow in fixture drains increases the incidence of stoppage development therein. For such fixtures, it is recommended that the size of the fixture trap and drain should be the minimum commercial size available and larger than the size of the discharge orifice. In this way, maximum scouring effect may be achieved to minimize fouling of the fixture drain.

Hence, minimum fixture drain sizes may be assumed to be the same as the minimum sizes permitted for the fixture traps of the various kinds of plumbing fixtures, as stated in the table in the section "Standard Code Regulations" in Chap. 3.

SANITARY DRAINAGE FIXTURE UNITS

A method of assigning fixture unit values to plumbing fixtures was originally proposed in 1923 by Roy B. Hunter of the National Bureau of Standards, U.S. Department of Commerce. It was presented in the publication, "Recommended Minimum Requirements for Plumbing," Department of Commerce Building and Housing Series, BH13, 1924. The suggested values were designed for application in conjunction with the probability of simultaneous use of fixtures so as to establish maximum permissible drainage loads, in terms of fixture units rather than in numbers of specific types of fixtures or gallons per minute of drainage flow, for each of the various parts of sanitary drainage systems. Since the original proposal, various changes were recommended and made by Hunter, in a 1940 National Bureau of Standards publication, BMS 65, and more recently by others participating in development of current model plumbing regulations.

Fixture unit values presently recommended for assignment to various kinds of plumbing fixtures which discharge into sanitary drainage systems are stated in Table 10-1. They are provided as a means for computing sizes of soil, waste, and vent piping based upon the loading effects produced by the discharge of many different kinds of plumbing fixtures commonly installed in buildings.

In general, the sanitary drainage fixture unit value assigned to a particular fixture is based upon the average volume discharged and the average rate of discharge for the fixture. This value is determined from the fixture's total discharge flow, in gallons per minute, divided by 7.5; or in other words, its total discharge flow in cubic feet per minute. The integer closest to the flow rate, in cubic feet per minute, is taken as the fixture unit value. Thus, the sanitary drainage fixture unit is a factor so chosen that the drainage load producing effects of different plumbing fixtures are expressed approximately as multiples of that factor.

Table 10-1
SANITARY DRAINAGE FIXTURE UNIT VALUES

Fixture or group	Trap size in	Trap size mm	Fixture units
Residential:			
Automatic clothes washer, domestic	2	50.0	3
Bathroom group consisting of a water closet, lavatory, and bathtub or shower stall:			
Flushometer valve closet			8
Tank-type closet			6
Bathtub (with or without overhead shower)	1½	37.5	2
Bidet	1¼	31.3	1
Dishwasher, domestic	1½	37.5	2
Floor drain	2	50.0	3
Floor drain	3	75.0	5
Floor drain	4	100.0	6
Food waste grinder, domestic	1½	37.5	2
Kitchen sink, domestic	1½	37.5	2
Kitchen sink, domestic, with dishwasher	1½	37.5	2
Kitchen sink, domestic, with food waste grinder	1½	37.5	2
Kitchen sink, domestic, with dishwasher and food waste grinder	2	50.0	2
Kitchen sink and wash (laundry) tray with single 1½-in (37.5-mm) trap	1½	37.5	2
Kitchen sink and wash (laundry) tray with separate 1½-in (37.5-mm) traps	1½	37.5	3
Kitchen sink and wash (laundry) tray with food waste grinder unit	2	50.0	4
Lavatory, common	1¼	31.3	1
Laundry tray (1 or 2 compartments)	1½	37.5	2
Shower stall, single head	2	50.0	2
Sink, bar, private	1½	37.5	1
Water closet, tank-type, trap arm only	3	75.0	4
Public toilet rooms:			
Urinal, pedestal, trap arm only	3	75.0	6
Urinal, pedestal, siphon jet blowout	3	75.0	6
Urinal, stall, washout	2	50.0	4
Urinal, wall [2-in (50-mm) min. waste]	1½	37.5	4
Water closet, Flushometer valve, trap arm only	3	75.0	6

Table 10-1 (*Continued*)

Fixture or group	Trap size in	Trap size mm	Fixture units
Industrial:			
Interceptors for grease, oil, solids, etc.	2	50.0	3
Interceptors for sand, auto wash, etc.	3	75.0	6
Lavatory, multiple-type (wash fountain or wash sink)	1½	37.5	2
Showers, gang (one unit per head)	2	50.0	2
Commercial:			
Dishwasher, commercial	2	50.0	2
Food waste grinder, commercial	2	50.0	3
Receptors (floor sinks) indirect waste receptors for refrigerators, coffee urn, water stations, etc.	1½	37.5	1
Receptors, indirect waste receptors for commercial sinks, dishwashers, air-washers, etc.	2	50.0	3
Sink, bar, commercial [2-in (50-mm) min. waste]	1½	37.5	2
Sink, commercial, with food waste grinder	2	50.0	3
Sink, commercial (pot, scullery, or similar type)	2	50.0	4
Sink (flushing-rim type, flush valve supplied)	3	75.0	6
Sink (service type with trap standard)	3	75.0	3
Sink (service with P-trap)	2	50.0	2
Washing machines, commercial	2	50.0	3
Medical:			
Dental unit or cuspidor	1¼	31.3	1
Dental lavatory	1¼	31.3	1
Lavatory (surgeon's, barber shop, beauty parlor)	1½	37.5	2
Sink (surgeon's)	1½	37.5	3
Miscellaneous:			
Drinking fountain	1¼	31.3	½
Mobile home park traps (one for each trailer)	3	75.0	6
Trap size 1¼ in (31.3 mm) or less	1¼	31.3	1
Trap size 1½ in (37.5 mm)	1½	37.5	2
Trap size 2 in (50 mm)	2	50.0	3
Trap size 2½ in (62.5 mm)	2½	62.5	4
Trap size 3 in (75 mm)	3	75.0	5
Trap size 4 in (100 mm)	4	100.0	6

DESIGN LOAD FOR SANITARY DRAINAGE SYSTEM

Except in the case of fixtures in which water is retained for use and discharged afterward, the rate at which water leaves fixtures and enters the sanitary drainage system is the same as the rate at which water is supplied to the fixtures. For relatively large systems where fixtures are supplied directly from public water mains, the rate at which water enters the water supply distributing system and the rate at which sewage leaves the building drainage system are the same except for a small amount of water consumed by occupants. Hence, there is relative equivalency between the water supply demand load and the sanitary drainage discharge load for which the respective systems are designed.

This ultimate equivalency is the basic premise upon which values have been assigned to express the load producing effects of fixtures, both in water supply fixture units and in sanitary drainage fixture units. Although the values assigned to fixtures for water supply purposes differ from those for sanitary drainage, the same design load in gallons per minute corresponding to a given number of fixture units is applicable to the main lines of both systems, with one proviso. The design load for sanitary drainage systems should be determined from the number of sanitary drainage fixture units and the corresponding water supply demand load, in gallons per minute as indicated in the charts (Fig. 6-3) and in Table 6-2, applicable to systems equipped predominantly with flush valves (Flushometers). The reason for this is simply that the discharge rate of water closets equipped with flush tanks is sufficiently comparable with that of flush-valve equipped water closets as to warrant use of the demand rates applicable to systems equipped with flush valves (Flushometers).

DRAINAGE STACKS AND BRANCHES

Based on the computed drainage stack flow capacity, for stacks flowing $\frac{7}{24}$ full at terminal velocity, the corresponding number of fixture units may be determined from design load charts or tables, so as to establish the total load which may be placed on a tall drainage stack. For example, the computed flow capacity of a 4-in (100-mm) stack flowing at $\frac{7}{24}$ full is 143 gpm (9.02 L/s). From design load charts or tables, it may be found that this rate of flow is equivalent to 500 fixture units.

This is the total load which may be received from all branches on a 4-in (100-mm)-tall stack. However, to avoid excessive interference between flow entering the stack and that coming down the stack, it is necessary to limit the amount of flow which may be allowed to enter the stack at each of the branches. Thus, in a building of just a few stories in height, the amount of flow entering the stack through a branch may be greater than would be permissible in a building of many stories.

Table 10-2 for sizing drainage stacks provides different permissible loadings for stacks of three stories or less in height and for stacks more than three stories in height. Included in the table are the maximum loads permitted on any horizontal fixture branch of a short stack and at any one story of a stack more than three stories in height.

SANITARY BUILDING DRAINS

Sanitary building drains should be designed to flow only partially full so as to provide an air space in the upper portion of the pipe through which air may flow and relieve pressure therein to available branches. A general recommendation in this case is to design such drains to flow at from one-half to three-quarters of the quantity rate of discharge which may be calculated for uniform flow conditions in sloping horizontal sanitary drains flowing full.

In relatively low buildings of extensive area, the condition of flow in the sanitary building drain may approach uniformity within the building, and for such buildings it is recommended that the drain be designed to convey liquid flow at one-half of its full-flow capacity under such conditions. This permits the drain to flow at one-half depth for maximum liquid load.

In buildings of the multistory type, the condition of flow in the sanitary building drain usually is different because of the high velocity at which discharges are received from drainage stacks. Considerable turbulence and surging flow occur in such cases. Under such conditions, flow is conveyed by the building drain at velocities higher than those that prevail for uniform flow conditions. For such buildings, it is recommended that the drain be designed to convey liquid flow at three-fourths of the quantity rate of flow corresponding to full-flow capacity under uniform flow conditions. The higher capacity may be permitted because of the higher velocities of surging flow, and correspondingly higher capacities for drains flowing at one-half depth.

Table 10-2 may be used for sizing sanitary building drains. The value shown therein for an 8-in (200-mm) drain laid at a ¼ in/ft (20.8 mm/m) slope corresponds to a design load, in gallons per minute, equal to the half-full flow rate for such a drain flowing at uniform flow conditions. The values shown for smaller sizes reflect an allowance made for the effects of surging flow. Similar allowances have been made for each given size of drain reflecting the effect of surging flow on the relative capacities of drains at different degrees of slope.

TABLE 10-2
MAXIMUM PERMISSIBLE LOADS FOR SANITARY DRAINAGE PIPING
In Terms of Fixture Units

Pipe diameter		Any horizontal fixture branch	One stack of 3 stories or less in height	Stacks more than 3 stories in height		Building drain, and building drain branches from stacks: Slope, in/ft (mm/m)			
in	mm			Total for stack	Total at one story	1/16 (5.2)	1/8 (10.4)	1/4 (20.8)	1/2 (41.6)
1½*	37.5	3	4	8	2	np	np	np	np
2*	50	6	10	24	6	np	np	21	26
2½*	62.5	12	20	42	9	np	np	24	31
3	75	20†	48‡	72‡	20†	np	np†	42†	50†
4	100	160	240	500	90	np	180	216	250
5	125	360	540	1100	200	np	390	480	575
6	150		960	1900	350	np	700	840	1000
8	200			3600	600	1400	1600	1920	2300
10	250			5600	1000	2500	2900	3500	4200
12	300					3900	4600	5600	6700

* No water closets permitted.

† Not over two water closets permitted.

‡ Not over six water closets permitted.

STORM DRAINAGE SYSTEM

Storm drains in buildings may be sufficiently long so that flow has ample time to adjust itself and reach a state of equilibrium or uniform flow condition. High velocity and surges in flow entering a long horizontal drain of constant shape and slope are dissipated gradually owing to pipe friction. Thereafter, flow proceeds at a rate corresponding to uniform flow conditions for the drain.

Sizing of horizontal storm drains is based upon the quantity rate of uniform flow for drains flowing full, under no pressure, after many years of service. The velocity and capacity corresponding to full flow in galvanized-iron and cast-iron storm drains, laid at a slope of ¼ in/ft (20.8 mm/m) are shown in Table 9-5. For drains laid at other slopes, appropriate multipliers are also given in the accompanying text. These multipliers may be applied to the values tabulated for drains laid at a ¼ in/ft (20.8 mm/m) slope.

The maximum rate at which storm water should be conveyed from drainage areas may be assumed to be the same as the rate at which storm water may collect thereon. This is dependent upon the maximum rainfall rate recorded in the region of the building installation and upon the amount of storm water collection area. For many regions in the United States, the maximum rainfall rate recorded over a 10-min period is 4 in (100 mm)/h. At this rate, the amount of drainage area upon which 1 gal (3.79 L) of storm water will collect in 1 min is 24 ft^2 (2.23 m^2).

By applying a value of 24 ft^2 (2.23 m^2) as the collection area equivalent to 1 gpm (3.79 L/s) of maximum storm drainage flow, tables showing uniform flow capacities in gallons per minute for sloping drains flowing full can be converted to terms of maximum permissible storm drainage areas which may be served by various sizes of drains in such regions.

Such a table for use in sizing storm drainage piping is given in Table 10-3. In the table, the values specified as maximum permissible drainage areas for horizontal storm drainage piping were derived in the manner discussed above for gravity flow in sloping drains flowing full under uniform flow conditions. For leaders, the values correspond to full-flow capacities of horizontal drains laid at a slope of ¾ in/ft (62.4 mm/m). For semicircular gutters, the values correspond to flow rates which are less than one-half of the full-flow capacities of circular horizontal drains of the same diameters. The 3-in (75-mm) gutter is assigned 29.3 percent, and the 8-in (200-mm) gutter is assigned 24.4 percent, of full-flow capacity for circular horizontal drains for such sizes. These percentages of full-flow capacity correspond, respectively, to flow depths of 37.5 and 35 percent of the diameter of semicircular gutters. Hence, flow at capacity should produce no spillage over the edge of the gutter at peak rainfall rate.

COMBINED STORM AND SANITARY BUILDING DRAINS

Where sanitary building drains are connected to storm building drains so as to form a combined storm and sanitary drain, an equivalency basis is necessary to convert the fixture unit load of the sanitary system to that of the storm drainage system in square feet of drained area. As the flow in the storm drainage system during maximum rainfall periods generally is far in excess of the flow in the sanitary drainage system at peak load, the sizing of a combined drain should be based on the storm drain design, and sanitary drainage fixture units of load should be converted to terms of equivalent square feet of drained area.

A method of converting such load units has been recommended. As a minimum, 1000 ft^2 (92.9 m^2) of drained area should be deemed equivalent to all sanitary drainage fixture unit loads that are less than 256 units. It may be seen from the water supply system demand load charts and Table 6-2 that each additional fixture unit above 256 corresponds to an additional load of 0.2 gpm (0.0126 L/s). At a maximum rainfall rate of 4 in (100 mm)/h, rain is collected at a rate of 1 gpm (0.063 L/s) on 24 ft^2 (2.33 m^2) of drained area and at a rate of 0.2 gpm (0.0126 L/s) on 3.9 ft^2 (0.36 m^2) of drained area.

Consequently, each fixture unit of sanitary drainage load in excess of 256 such units should be deemed equivalent to 3.9 ft^2 (0.36 m^2) of drained area. The total drained area equivalent to the number of sanitary drainage fixture units should be added to the total storm drainage area to determine the load on the combined system, and the appropriate size of combined drain then may be selected from the tables applicable for sizing horizontal storm drains.

STANDARD CODE REGULATIONS

Sanitary Drainage Fixture Units Fixture unit values given in Table 10-1 shall be employed in computing the total load carried by a soil or waste pipe and shall be used with tables for sizing soil, waste, drain and vent pipes, except as otherwise prescribed herein. The total discharge flow in gallons per minute for any single fixture, divided by 7.5, provides the fixture unit value for that particular fixture.

Sizing the Sanitary Drainage System The maximum permissible fixture unit load that may be connected to any horizontal fixture branch, soil stack, waste stack, building drain, and building drain branches from stacks, shall be determined from Table 10-2, except as otherwise provided herein.

Sanitary drainage systems shall have at least one main stack extending undiminished in size and as directly as possible from the building drain to the open air above the roof. Such stack shall be at least 3 in (75 mm) in size, but not larger than the size of the building drain.

Soil and waste stacks shall be at least as large as the largest branch connection thereto, except that it shall be permissible to connect the 4-in (100-mm) fixture drain of a water closet to a 3-in (75-mm) soil stack by means of a 3″ × 3″ × 4″ TY or sanitary tee.

The size of the sanitary building drain at any point shall be at least as large as the largest branch or stack located upstream from such point.

Underground waste piping shall be at least 2 in (50 mm) in size.

Where a soil or waste stack is offset at an angle of more than 45° from the vertical, and the fixture unit load conveyed by the offset exceeds the permissible load for building drains of the same size as the stack, the offset shall be sized as a building drain for that load, and the size of the stack section below the offset shall be at least as large as the offset.

Where drainage or vent piping outlets are provided for future fixture installations, they shall be considered in determining the sizes of drainage and vent piping.

Sizing the Storm Water Drainage System The size of a building storm drain or any of its horizontal branches having a slope of ½ in or less per foot shall be based upon the total drainage area served by such drains in accordance with Table 10-3.

TABLE 10-3
MAXIMUM PERMISSIBLE LOADS FOR STORM DRAINAGE PIPING AND GUTTERS
In Terms of Square Feet of Projected Drainage Area

Pipe or gutter diameter*			Horizontal storm drainage piping			Horizontal storm drainage gutters			
			Slope, in/ft (mm/m)			Slope, in/ft (mm/m)			
in	mm	Leaders	⅛ (10.4)	¼ (20.8)	½ (41.6)	⅟₁₆ (5.2)	⅛ (10.4)	¼ (20.8)	½ (41.6)
2	50	720	np	np	np	np	np	np	np
2½	62.5	1,300	np	np	np	np	np	np	np
3	75	2,200	822	1,160	1,644	170	240	340	480
4	100	4,600	1,880	2,650	3,760	360	510	720	1,020
5	125	8,650	3,340	4,720	6,680	625	880	1,250	1,760
6	150	13,500	5,350	7,550	10,700	960	1,360	1,920	2,720
8	200	29,000	11,500	16,300	23,000	1,990	2,800	3,980	5,600
10	250		20,700	29,200	41,400	3,600	5,100	7,200	10,200
12	300		33,300	47,000	66,600				
15	375		59,500	84,000	119,000				

Note: This table is based upon a rainfall rate of 4 in/h. Where greater rates frequently occur, the areas given in the table shall be adjusted by multiplying them by 4 and dividing by such greater rate, in inches per hour.

* The equivalent diameter of a square or rectangular leader may be taken as the diameter of that circle which may be inscribed within the cross-sectional area of the leader. Gutters other than semicircular may be used provided they have an equivalent cross-sectional area.

The size of subsoil drains shall be not less than 4 in (100 mm) where installed under a cellar or basement floor or where installed surrounding the outer walls of a building.

Leaders shall be sized on the basis of the horizontally projected roof areas served, except that an allowance shall be made for any additional storm water drainage load received by such roof areas from adjacent walls exposed to the elements. Leaders shall be sized in accordance with Table 10-3.

The size of a semicircular gutter shall be based upon the horizontally projected roof area served by such gutter in accordance with Table 10-3.

Sizing Combined Storm and Sanitary Building Drains The size of any combined drain shall be based on a drainage area equivalent to the total sanitary and storm drainage loads served by such drain, and shall be determined as for horizontal storm drains, using Table 10-3. The sanitary drainage load shall be converted to equivalent storm drainage area, and then added to the storm drainage area. Sanitary drainage loads shall be converted as follows:

> Where the total fixture unit load on the combined drain is less than 256 fixture units, the sanitary drainage load shall be taken as equivalent to 1000 ft^2 (92.9 m^2) of storm drainage area.
>
> Where the total fixture unit load on the combined drain is more than 256 fixture units, the sanitary drainage load shall be computed by considering each fixture unit as equivalent to 3.9 ft^2 (0.36 m^2) of storm drainage area.
>
> Where there is a continuous or intermittent flow into the drain or sewer, as from a pump, ejector, air-conditioning equipment or similar equipment, each gallon per minute (0.063 L/s) of flow at rated capacity shall be computed as being equivalent to 24 ft^2 (2.23 m^2) of storm drainage area.

11

FLOW OF AIR IN VENT PIPING

PRINCIPLES OF FLUID MECHANICS AND PNEUMATICS

This chapter deals with the principles of fluid mechanics and pneumatics that are relevant to the flow of air, including suds, in vent piping. These principles concern physical and mechanical properties of air and suds, and their application in engineering design of attendant vent piping systems.

The physical and mechanical properties of air and suds include their densities, viscosity, conditions of flow, and frictional resistance.

Energy principles include static and kinetic energy, friction between water and air, friction between air and pipe surface, pressurized flow in vent piping and from outlets, and gravity circulation of air by induced head or natural draft.

These properties and principles have specific application with regard to the flow of air and suds in vent stacks, individual vents, branch vents, and vent headers.

PHYSICAL PROPERTIES OF AIR

In any consideration of the flow of air in vent piping, several of the physical properties of air are of special interest. Those most pertinent to this subject are density, viscosity, and compressibility.

The density of air, that is, its weight per unit of volume, varies with change in temperature and moisture content. One cubic foot of dry air at atmospheric pressure and 68.4°F weighs 0.075 lb/ft^3 (1.2 kg/m^3). This is the density of what is termed *standard air,* or air at normal pressure and temperature conditions. Moisture content or humidity of the air for such conditions is an insignificant factor and may be disregarded in calculations of airflow in vent piping. The weight per cubic foot of dry air corresponding to various temperatures is given in Table 11-1.

When a fluid flows, its natural characteristics of adhesion and cohesion result in the development of internal resistance to flow. This resistance is called the viscosity of the fluid. It is a measurable property which varies greatly from one fluid to another, and in gases viscosity increases with rise in temperature. Air may be considered to be a gas at the normal temperature range existing in drainage and vent piping. The absolute viscosity and kinematic viscosity of air corresponding to various temperatures is given in Table 11-2.

Air is perfectly elastic, compressing when pressure is imposed and returning to original volume when pressure is removed. In the sanitary drainage system, only a very small amount of pressure rise may be permitted owing to the limited seal depth of fixture traps connected to the system. The attendant system of vent piping must be designed to permit

TABLE 11-1
DENSITY OF DRY AIR
At 14.7 Psia

Temperature		Density	
°F	°C	lb/ft³	kg/m³
−20	−28.9	0.09050	1.45162
−10	−23.4	0.08848	1.41922
0	−17.8	0.08656	1.38842
10	−12.2	0.08472	1.35891
20	−6.7	0.08299	1.33116
30	−1.1	0.08125	1.30325
40	4.4	0.07963	1.27727
50	10.0	0.07807	1.25224
60	15.6	0.07656	1.22802
70	21.1	0.07512	1.20492
80	26.7	0.07373	1.18263
90	32.2	0.07238	1.16098
100	37.8	0.07109	1.14028
200	93.3	0.06031	0.96737
300	149.0	0.05237	0.84001
400	204.5	0.04628	0.74233
500	260	0.04146	0.66502

Note: One pound of air at 70°F and 14.7 psia occupies 1/0.07512 or 13.31 ft³ of space.

TABLE 11-2
VISCOSITY OF AIR

Temperature		Absolute viscosity		Kinematic viscosity, ft²/s
°F	°C	Centipoises	pdl · s/ft²	
32	0	0.0175	0.00001176	0.000147
50	10	0.0178	0.00001195	0.000153
60	15.6	0.0180	0.00001208	0.000156
70	21.1	0.0182	0.00001222	0.000163
80	26.7	0.0185	0.00001242	0.000168
100	37.8	0.0191	0.00001282	0.000180
120	48.9	0.0197	0.00001322	0.000212
140	60.0	0.0203	0.00001362	0.000211
160	71.1	0.0209	0.00001402	0.000210
180	82.2	0.0215	0.00001442	0.000209

Note: One centipoise equals 0.000672 pdl · s per ft². Kinematic viscosity (in square feet per second) is equal to absolute viscosity (in poundal seconds per square foot) divided by density (in pounds per cubic foot).

admission and emission of air in all pipes so that the seals of fixture traps are subjected to an air pressure differential of not more than 1 in (25 mm) of water column above or below atmospheric pressure. For this limited amount of pressure change, the corresponding volumetric change may be determined from gas equations to be just $\frac{1}{400}$ less or more than at atmospheric pressure. Such small amounts of change in air volume in drainage piping indicate that vent piping must be designed to permit air to flow freely and without any compression other than the small amount permissible for overcoming friction head losses in air flow.

In certain parts of the vent piping system, pressure relief may occur in the form of flowing *suds* or *foam,* consisting of millions of small bubbles of air encased in films of liquid resulting from the discharge of detergents with waste water. Hence, in this discussion of the physical properties of air, it is pertinent to consider some of the important physical properties of suds. For example, manufacturers' reports show that the density of suds varies from a minimum of 0.2 to 12 lb/ft³ for newly formed suds made with many different types of detergents. For old or regenerated suds, it appears reasonable to consider their density to be approximately 1.4 lb/ft³ for the design of suds relief piping. In addition, values of the absolute viscosity may be taken as being 0.5516 centipoise, and kinematic viscosity may be taken as being 0.000264 ft²/s.

EQUIVALENT STATIC HEAD OF WATER, AIR, AND SUDS

In Chap. 5, the pressure produced by the weight of water lying above any point submerged below the water surface was discussed. It was stated that the pressure could be expressed in terms of depth below the water surface or *static head* of water above the point. Also stated was the fact that static head, in terms of feet of water depth or column, may be converted to pressure, in pounds per square inch, by means of the equation

$$p = \frac{w}{144} h \qquad [11\text{-}1]$$

where p = pressure, psi
w = density or weight of fluid, lb/ft^3
h = static head of fluid, ft

Air and suds are fluids and produce pressure or head at submerged points in accordance with the weight of overlying fluid, just as is the case with water. However, as the densities of water, air, and suds are different, the amount of pressure produced by a 1-ft head or column of each of these fluids is not the same. For equal pressure, the preceding formula may be rearranged to express equivalent static heads for water, air, and suds, as follows:

$$144\,p = w_w h_w = w_a h_a = w_s h_s \qquad [11\text{-}2]$$

where p = pressure, psi
w_w = density of water, lb/ft^3
w_a = density of air, lb/ft^3
w_s = density of suds, lb/ft^3
h_w = static head or column of water, ft
h_a = static head or column of air, ft
h_s = static head or column of suds, ft

CONDITIONS OF FLOW

In vent piping, the condition of airflow may be streamline when the velocity is relatively low, such as prevails generally during gravity circulation of air through the drainage and vent piping system. However, the condition of airflow becomes turbulent at the relatively high velocities which occur when vent piping must serve to relieve air pressure resulting from the flow of liquids in the sanitary drainage system.

The critical velocity at which airflow changes from streamline to turbulent condition may be determined from the following equation:

$$V_c = \frac{R_c}{D} \frac{\mu}{w} = \frac{2000}{D} \frac{\mu}{w} = \frac{2000\nu}{D} \qquad [11\text{-}3]$$

which is the same as Eq. (5-1). In the above equation, applying a value of 0.000163 ft^2/s as the kinematic viscosity of air at 70°F, the critical velocities of airflow in vent pipes of 1¼-, 1½-, 2-, 2½-, and 3-in (31.3-, 37.5-, 50-, 62.5-, and 75-mm) diameters are, respectively, 3.10, 2.58, 1.94, 1.55, and 1.29 fps (0.95, 0.79, 0.59, 0.47, and 0.39 m/s). They correspond in turn to quantity flow rates of 11.9, 14.2, 19.0, 23.7, and 28.4 gpm (0.75, 0.90, 1.29, 1.50, and 1.79 L/s) of airflow.

For suds flow in vent pipes, a value of 0.000264 ft^2/s as the kinematic viscosity of suds may be applied to determine the critical velocities in vent pipes of 1¼-, 1½-, 2-, 2½-, and 3-in (31.3-, 37.5-, 50-, 62.5-, and 75-mm) diameters to be, respectively, 5.02, 4.18, 3.14, 2.51, and 2.09 fps (1.53, 1.27, 0.96, 0.77, and 0.64 m/s). They correspond in turn to quantity flow rates of 19.3, 23.0, 30.8, 38.4, and 46.0 gpm (1.22, 1.45, 1.94, 2.42, and 2.90 L/s) of suds flow.

PNEUMATIC EFFECTS IN AND VENTING DESIGN CRITERION FOR SANITARY DRAINAGE SYSTEMS

When water flows in contact with air in a partially filled vertical or horizontal drain, there is friction between the water and air. This results in air being dragged along by the water and causing air to flow in the same direction in the drain. Wherever the cross-sectional area occupied by water increases sharply, such as at sudden changes in drain direction and at branch fittings through which additional flow enters the drain, the area available for airflow in the drain becomes correspondingly reduced or constricted. The effect of constricting the area available for airflow in the drain is that of a temporary stoppage or block to airflow at such points.

As air is dragged into a zone of a drain where a temporary stoppage to airflow exists, the air accumulates in the restricted volume of the drain and becomes pressurized. Highest pressure occurs at the point of airflow constriction, and pressure diminishes upstream therefrom. Consequently, when fixtures on upper floors discharge into a drainage system and water flows down a drainage stack conveying air into the lower section of the system, pneumatic effects considerably above atmospheric pressure may occur in the lower section of the sanitary drainage system. Fixture drains and branch drains connected to lower sections may be subjected to excessive back pressure sufficient to blow the water seals of fixture traps out of fixture waste outlets and into rooms.

All the air dragged along in drainage pipes by the flowing water is drawn from the upper section of the sanitary drainage system. The upper section of the system must supply air as rapidly as it is being dragged down in the system, or the volume of air in the upper section will become exhausted, and pneumatic effects considerably below atmospheric pressure may occur in that part of the sanitary drainage system. Fixture drains and branch drains connected to upper sections may be subjected to excessive aspiration sufficient to suck the water seals of fixture traps into the fixture drains.

When water seals of fixture traps are lost owing to back-pressure or aspiration effects in the sanitary drainage system, hazardous gases and objectionable odors from the system can enter rooms in which fixtures are located. Protection against such occurrence should be provided for the health, safety, and welfare of building occupants. Such protection may be afforded by the installation of an adequate system of vent piping to relieve excessive pneumatic effects in the sanitary drainage system.

The design criterion for a system of vent piping attendant upon the sanitary drainage system must be related to the strength or resistance of all trap seals connected to the drainage system. Fixture trap seals are the weakest of all the trap seals on the system, for they are permitted to have a minimum seal depth of 2 in. In view of this limiting factor, it is recommended that the venting design criterion for sanitary drainage systems be as follows: to provide an attendant system of vent piping designed to permit adequate circulation of air in all pipes and the admission and emission of air so that the seals of fixture traps are subjected to an air pressure differential of not more than 1 in of water column. This design criterion provides what is deemed a reasonable factor of safety.

QUANTITY RATE OF FLOW FROM OUTLETS

The velocity at which air flows through an outlet to the atmosphere is due to the total energy available in the vent pipe at the outlet during flow. This total energy is the sum of the potential and kinetic energies of the moving air. Potential energy, in this instance, is the pressure or head exerted by the flowing air against the inside wall of the vent pipe and is termed *flow pressure.*

In practice, the amount of kinetic energy, or velocity head, in the vent pipe during flow usually is relatively small and may be assumed to be an insignificant factor by comparison with the flow pressure. Hence, the maximum rate of air discharged from an outlet of a vent pipe may be determined with reasonable accuracy based just upon the flow pressure in the vent pipe and the diameter of the outlet.

The maximum rate at which air may discharge from an outlet into the atmosphere in practice may be expressed as follows:

$$q_d = c_d q_i = c_d(2.448\ d_o^2\ V_i) = c_d(2.448\ d_o^2 \sqrt{2gh_m}) = c_d(19.65\ d_o^2 \sqrt{h_m}) \qquad [11\text{-}4]$$

where q_d = quantity rate of air discharge from outlet in practice, gpm
q_i = ideal rate of discharge from outlet, gpm
c_d = coefficient of discharge for outlet
d_o = diameter of outlet, in
V_i = ideal velocity, fps
g = gravitational acceleration, 32 ft/s²
h_m = head measured in vent pipe during flow, ft of air column

When the value of the coefficient of discharge c_d is assumed to be 0.67, a reasonable value for ordinary pipe outlets, the equation for the maximum rate of airflow which may be obtained from an outlet in practice has the following simplified form:

$$q_d = c_d(19.65) d_o^2 \sqrt{h_m} = 0.67(19.65) d_o^2 \sqrt{h_m} = 13.17 d_o^2 \sqrt{h_m} \qquad [11\text{-}5]$$

In vent piping, the design criterion calls for limiting the air pressure differential above or below atmospheric pressure to no more than 1 in (25 mm) of water column. In terms of feet of air column, this may be established based upon the density of dry air at 70°F (21.1°C) as follows:

$$h_a w_a = h_w w_w$$

$$h_a = \frac{h_w w_w}{w_a} = \frac{1\ \text{in} \times 62.408}{12\ \text{in} \times 0.07512} = 69.23\ \text{ft (21.1 m) of air column or head [the equivalent of 1 in (25 mm) of water column or head]}$$

This conversion factor may be applied to establish a simplified equation for the quantity rate of air discharge into the atmosphere from outlets, of various nominal sizes, in a vent pipe wherein the pressure is 1 in (25 mm) of water column, as follows:

$$q_d = 13.17 d_o^2 \sqrt{h_m} = 13.17 d_o^2 \sqrt{69.32} = 13.17 \times 8.32 d_o^2 = 109.57 d_o^2 \qquad [11\text{-}6]$$

Suds discharge rates may be established in a similar manner. The feet of suds column or head equivalent to 1 in (25 mm) of water column

Table 11-3
OUTLET DISCHARGE RATES FOR AIR AND SUDS
Flow Pressure = 1 in (25 mm) Water Head

Outlet diameter d_o		Air discharge q_d		Suds discharge q_d	
in	mm	gpm	L/s	gpm	L/s
1¼	31.3	171.2	10.8	39.5	2.49
1½	37.5	246.5	15.6	56.9	3.59
2	50.0	438.3	27.7	101	6.37
2½	62.5	684.8	43.2	158	9.97
3	75.0	986.1	62.2	228	14.4
4	100.0	1753	110.6	405	25.6
5	125.0	2739	172.8	633	39.9

may be calculated, based upon the assumption of 1.4 lb/ft³ (22.5 kg/m³) as the density of old or regenerated suds, as follows:

$$h_s w_s = h_w w_w$$

$$h_s = \frac{h_w w_w}{w_s} = \frac{1\text{ in} \times 62.408}{12\text{ in} \times 1.4} = 3.71\text{ ft (1.13 m) of suds column or head [the equivalent of 1 in (25 mm) of water column or head]}$$

Applying this conversion factor, an equation may be established for the quantity rate of suds discharge into the atmosphere from outlets, of various nominal sizes, in a vent pipe wherein the pressure is 1 in (25 mm) of water column, as follows:

$$q_d = 13.17 d_o^2 \sqrt{h_m} = 13.17 d_o^2 \sqrt{3.71} = 13.17 \times 1.925 d_o^2 = 25.3 d_o^2 \qquad [11\text{-}7]$$

Comparison of the equations for suds and air discharge rates indicates that the suds discharge rate is just 25.3/109.57, or 23.1 percent of the air discharge rate for the same flow pressure. Table 11-3 shows the quantity rates at which both air and suds may discharge into the atmosphere from various sizes of vent outlets, under a flow pressure equivalent to a 1 in (25 mm) water column or head.

PRESSURE LOSS DUE TO FRICTION IN PIPING

When air flows through vent piping to relieve air pressure resulting from the flow of liquids in sanitary drainage piping, a continuous loss of pressure occurs along the piping in the direction of flow. This pressure loss is due to friction generated between the moving air and the inner surface of the vent piping in view of the fact that turbulent flow conditions prevail at the maximum rates of flow which vent pipes are designed to convey.

The amount of pressure or head loss due to friction in this case is dependent upon such factors as physical properties of air, such as its density and temperature, roughness of the interior surfaces of the pipe; length of the pipe; diameter of the pipe; and the velocity at which air flows through the pipe.

Darcy's formula for pipe friction loss may be applied to calculate the head loss due to friction developed by airflow in vent piping. It is as follows:

$$h_f = \frac{fLV^2}{D2g} \qquad [11\text{-}8]$$

which is the same as Eq. (5-21),

where h_f = head loss due to friction, ft of air column
f = coefficient of friction corresponding to the roughness of the pipe surface and diameter of the pipe
L = length of piping, ft
D = diameter of piping, ft
V = velocity of flow, fps
g = gravitational acceleration, 32.2 ft/s²

PERMISSIBLE LENGTH OF VENT PIPING

For more convenient expression of pipe friction loss in, and permissible length of, vent piping, the Darcy pipe friction formula [Eq. (5-21)] and the adapted flow formula [Eq. (5-3)] may be combined and their terms converted to those more generally used in practice, as in the following:

$$h_f = \frac{fLV^2}{D2g} \quad (5\text{-}21) \qquad q = 2.448\, d^2 V \quad (5\text{-}3)$$

$$V = \frac{q}{2.448 d^2}$$

Hence,

$$h_f = \frac{fLq^2}{(d/12)(2.448)^2 d^4 (64.4)}$$

$$= \frac{fLq^2}{(32.16) d^5}$$

$$= \frac{(0.03109) fLq^2}{d^5} \qquad [11\text{-}9]$$

and

$$L = \frac{h_f d^5}{(0.03109) f q^2} \qquad [11\text{-}10]$$

where h_f = head loss due to friction, ft of fluid column
f = coefficient of friction corresponding to the roughness of the pipe surface and diameter of the pipe
L = length of piping, ft
D = diameter of piping, ft
d = diameter of piping, in
q = quantity rate of flow, gpm
g = gravitational acceleration, 32.2 ft/s^2
V = velocity of flow, fps

The maximum permissible length of vent piping may be expressed in accordance with the design criterion for vent piping which requires that the seals of fixture traps be subjected to an air pressure differential of not more than 1 in (25 mm) of water column above or below atmospheric pressure. As the permissible head loss due to friction in this case is 1 in (25 mm) of water column, the equivalent head for air or suds may be applied in the preceding equation to determine the permissible length of vent piping for airflow or for suds flow.

For airflow, an equivalent head of 69.23 ft (21.1 m) of air column, based upon the densities of water and air at 70°F, may be applied. This results in the following formula for the maximum permissible length of vent piping for airflow:

$$L = \frac{(69.23) d^5}{(0.03109) f q^2} = \frac{2226 d^5}{f q^2} \qquad [11\text{-}11]$$

Similarly, for suds flow, an equivalent head of 3.71 ft (1.13 m) of suds column, based upon the densities of water at 70°F and that given for old or regenerated suds, may be applied. This results in the following formula for the maximum permissible length of vent piping for suds flow:

$$L = \frac{(3.71) d^5}{(0.03109) f q^2} = \frac{119.3 d^5}{f q^2} \qquad [11\text{-}12]$$

From these two equations, it can be seen that the permissible length of vent piping for suds flow is just 119.3/2226, or 5.36 percent as much as is permissible for airflow; and conversely, the permissible length for airflow is 2226/119.3, or 18.7 times as much as for suds flow.

EQUIVALENT LENGTH OF FITTINGS IN VENT PIPING

The maximum permissible length of piping L expressed in the preceding equation should be understood to mean the developed length of straight piping free of any fittings. Where fittings are included in the run of piping, consideration must be given to the fact that they impose much more frictional resistance than straight pipe of the same developed length and size. Hence, the length L should be recognized as including the developed length of the piping plus an equivalent length to be allowed in lieu of the additional resistance due to fittings.

As the types of fittings commonly used in water supply systems are also used in vent piping, the recommended equivalent lengths to be allowed for fittings in vent piping are contained in Table 5-6*a*, 5-6*b*, 5-7*a*, and 5-7*b*. The lengths stated therein may be applied as reasonably appropriate for other types of fittings used in the same capacity.

VALUE OF COEFFICIENT OF FRICTION f

In using the preceding formulas, proper selection should be made regarding the value of f, the coefficient of friction corresponding to the roughness of the pipe surface, diameter of the pipe, and pipe surface condition after a reasonable period of service. For guidance in selecting reasonable and appropriate values, one may follow the recommendations contained in research reports published by authoritative sources.

One such report is the National Bureau of Standards Monograph 31, issued July 3, 1961, in which Robert S. Wyly and Herbert N. Eaton presented a table giving values of f for use in vent stack calculations. These values are shown in Table 11-4.

The values shown in this table reportedly were computed assuming the absolute roughness of the vent stack surface to be double that given for galvanized steel pipe so as to take into account the expected effect

Table 11-4
VALUES OF f FOR USE IN VENT STACK CALCULATIONS

Diameter of drainage stack		Pipe friction coefficient
in	mm	f
3	75	0.0367
4	100	0.0330
5	125	0.0307
6	150	0.0286
8	200	0.0260
10	250	0.0242
12	300	0.0230
15	375	0.0214

Source: NBS Monograph 31.

of corrosion. In addition, each value of *f* represents the average of two values obtained from the assumption of two sizes of vent stacks for a particular size of drainage stack, in one case the vent stack and drainage stack being of the same size and in the other case the diameter of the vent stack being one-half that of the drainage stack.

QUANTITY RATE OF FLOW IN VENT STACKS

In an earlier section, capacity flow of water in drainage stacks was described as having the form of a uniformly thick sheet of water in contact with the pipe wall and enclosing a core of air in the center of the stack. At capacity, the cross-sectional area occupied by water may be from $\frac{6}{24}$ to $\frac{7}{24}$ of that of the drainage stack. In flowing down the pipe wall, terminal velocity is attained in a relatively short distance of fall, generally within one story height for stacks up to 5 in (125 mm) in size and within two stories for other sizes up to 12 in (300 mm). Velocity remains constant thereafter owing to equalization of frictional resistance and gravitational force.

Part of the frictional resistance is developed by the sheet of water in descending in contact with the core of air in the center of the stack. As a result, the water drags the air core down the stack. The average velocity at which it is dragged down by water at capacity flow has been found to be approximately equal to that of the water flowing at terminal velocity. Hence, in the design of vent stacks, it is recommended that the velocities of both air and water flowing in drainage stacks at capacity be assumed to be equal.

In the horizontal drain at the base of the drainage stack, a hydraulic jump occurs in flow. As this jump may completely fill the horizontal drain within a short distance of the base fitting, when the stack is flowing at capacity, it is reasonable for design purposes to assume that the hydraulic jump constitutes a block to airflow in the horizontal drain and that all the air dragged down the drainage stack by the water must be relieved by flowing up the vent stack to its vent terminal in the atmosphere.

Consequently, the quantity rate of airflow for which vent stacks should be designed to convey is related to the rate at which air is dragged down drainage stacks at capacity flow. When capacity flow is based upon drainage stacks flowing $\frac{6}{24}$ full at terminal velocity, the air core occupies $\frac{18}{24}$ of the cross-sectional area of the stack, and the quantity rate of airflow dragged down the drainage stack by the water at the same velocity is equal to $\frac{18}{6}$, or 3 times the quantity rate of water flow. Similarly, for drainage stacks flowing $\frac{7}{24}$ full, the quantity rate of airflow is equal to $\frac{17}{7}$, or 2.43 times the quantity rate of water flow. Computations have

Table 11-5
COMPUTED AIRFLOW CAPACITY REQUIRED BY ATTENDANT VENT STACKS

Diameter of drainage stack		Airflow rate for fraction of drainage stack area occupied by water at terminal velocity			
		$r_s = 6/24$		$r_s = 7/24$	
in	mm	gpm	L/s	gpm	L/s
1¼	31.3	15.0	0.95	15.8	1.06
1½	37.5	24.3	1.53	24.8	1.56
2	50.0	52.5	3.31	54.8	3.46
2½	62.5	95.4	6.02	99.8	6.30
3	75.0	156.0	9.84	164.0	10.3
4	100.0	333.0	21.0	348.0	22.0
5	125.0	606.0	38.2	633.0	39.9
6	150.0	1008.0	63.6	1065.0	67.2
8	200.0	2127.0	134.2	2220.0	140.0

been made to determine the respective rates of airflow required to be conveyed by vent stacks for various sizes of drainage stacks. They are presented in Table 11-5.

QUANTITY RATE OF FLOW IN INDIVIDUAL VENTS

Considerable variation occurs in the quantity rate at which air may be required to be conveyed through an individual vent for pressure relief in the fixture drain. This is directly related to the amount of air relief required in the drainage piping at the branch fitting into which the fixture drain discharges and the principal function which the individual vent is to perform in any given installation.

For example, in one installation the principal function of an individual vent may be simply to permit air to enter the fixture drain at a rate sufficient to disrupt siphonic action and thereby prevent excessive trap seal loss during discharge of the fixture. In this case, the rate of airflow into the fixture drain reaches a maximum when the drain flows approximately half full, air and water occupying equal space and moving at equal velocity. For this condition, the velocity of airflow may be considered to be approximately the same as would exist for uniform flow conditions in the horizontal section of the fixture drain at half-full flow, and consequently the quantity rates of water and airflow may be assumed to be equal. Computed airflow rates for horizontal sanitary drains flowing half full are shown in Table 11-6.

Table 11-6
COMPUTED AIRFLOW RATES REQUIRED FOR VENTING HORIZONTAL SANITARY DRAINS FLOWING HALF FULL

Diameter of drain		Slope of drain		Air and water flow rates	
in	mm	in/ft	mm/m	gpm	L/s
1¼	31.3	½	41.6	5.5	0.35
1½	37.5	½	41.6	8.3	0.52
2	50.0	¼	20.8	9.7	0.61
2½	62.5	¼	20.8	17.6	1.11
3	75.0	¼	20.8	28.6	1.80
4	100.0	¼	20.8	57.0	3.60
5	125.0	¼	20.8	96.5	6.09

In another installation, the principal function of an individual vent may be to serve as a wet vent or as a loop or circuit vent to convey air at the quantity rate of flow as may be required to enter the branch drain to prevent excessive pressure reduction therein when the branch drain flows half full. The rate of airflow required in this case is considerably higher than exists where the individual vent simply serves to disrupt siphonic action in a single fixture drain.

In still another installation, an individual vent may be required to serve in the same capacity as either a vent stack or a special relief vent. This is a much different and more demanding function, for the individual vent may be required to convey air at the same quantity rate as it is conveyed down a drainage stack, or as may be required to relieve air pressure in a drain subject to pressurization in the lower section of the drainage system. Regulations do not prohibit the connection of fixture drains to zones of drainage piping wherein relatively high air or suds relief requirements may exist. Therefore, one may conclude that the airflow rates which should be satisfied by individual vents are those which correspond to the conditions prevailing for any given installation or piping arrangement in which they are to function.

QUANTITY RATE OF FLOW IN BRANCH VENTS AND VENT HEADERS

The total rate at which air should be conveyed by a branch vent is the sum of all the airflows required to be conveyed by individual vent pipes connected thereto and served by the branch vent. As the airflow rates in individual vent pipes vary with conditions under which they are required to function, a corresponding variation occurs in the total rate of airflow to be conveyed by a branch vent. This may be determined more properly by analysis of given installations and by summing up the appropriate airflows required for each individual vent function to determine the flow rate required in a branch vent.

Similarly, the quantity rate of airflow in a vent header, which serves to convey air to the tops of a number of soil, waste, and attendant vent stacks, is the sum of the respective airflow rates required for such stacks. As the rate of airflow in such stacks is directly related to the rate of liquid flow therein, the rate of airflow in a vent header varies in accordance with the total drainage load of all the stacks served by the header.

GRAVITY CIRCULATION OF AIR BY INDUCED HEAD OR DRAFT

The force causing gravity circulation of air in a sanitary drainage and vent system is the difference in head existing between air outdoors and that in the system. This difference in head is induced by the difference in temperatures and corresponding densities of the air outdoors and the air inside the system, and by the height of air column in the system. The cooler air is more dense and produces greater head, or draft pressure, at any opening where the cool and warm air are in contact. Consequently, the cool air tends to move and displace warm air, setting up gravity circulation.

The amount of head induced or natural draft pressure produced in sanitary drainage systems for setting up gravity circulation of air may be determined for specific stack heights and air temperature conditions outdoors and inside the system, by applying the following formula:

$$\mathrm{NDP} = \frac{(2.31)(12)(w_{to} - w_{ti})H_s}{144} \qquad [11\text{-}13]$$

$$= 0.1925(w_{to} - w_{ti})H_s$$

where NDP = natural draft pressure or induced head, in of water column

w_{to} = density of air corresponding to temperature outdoors, lb/ft³

w_{ti} = density of air corresponding to temperature inside stack, lb/ft³

H_s = height of stack, ft

For any given conditions of air temperatures and stack height, gravity circulation of air will proceed at a rate such as will dissipate as friction losses in the system all the natural draft pressure or induced head.

12

SIZING VENT PIPING

SIZING FOR ADEQUACY OF PERFORMANCE IN EXTENDED SERVICE

This chapter deals with the application of principles of fluid mechanics and pneumatics in sizing vent piping so as to achieve adequacy of performance in extended service.

The specific criteria used in each sizing application are presented in detailed discussion for each section of the attendant vent piping system. The criteria define the limits for adequacy of performance.

VENT STACKS

For any given quantity rate of airflow required to be conveyed by a vent stack, the maximum permissible total length of vent piping of a given diameter may be readily computed by means of a formula [Eq. (11-11)]. The total length determined in this manner should be understood to be the sum of the total developed length of the vent stack, measured from its base connection to its terminal in the atmosphere above the roof of the building, plus the total equivalent pipe length allowable for the pipe fittings in that run of vent piping. In general, the equivalent length of pipe fittings in multistory buildings has been found to work out to be approximately 50 percent of the developed length of vent stacks. Hence, it may be assumed that the maximum permissible developed length is two-thirds of the computed total vent stack length.

It may be noted also that the maximum airflow capacity of a given diameter of vent stack of zero length is that which may be computed as the quantity rate of discharge into the atmosphere from outlets of

Table 12-1
SIZE OF VENT STACKS AND BRANCH VENTS

Size of soil or waste stack		Fixture units connected	Diameter of vent required, in (mm)								
			1¼ (31.3)	1½ (37.5)	2 (50)	2½ (62.5)	3 (75)	4 (100)	5 (125)	6 (150)	8 (200)
in	mm		Maximum developed length of vent, ft								
1¼	31.3	2	30								
1½	37.5	8	np	150							
2	50	24	np	50	150						
2½	62.5	42	np	np	100	300					
3	75	72	np	np	np	80	400				
4	100	500	np	np	np	np	180	700			
5	125	1100	np	np	np	np	np	200	700		
6	150	1900	np	np	np	np	np	np	200	700	
8	200	3600	np	np	np	np	np	np	np	250	800
10	250	5600	np	np	np	np	np	np	np	np	250

a vent pipe wherein the pressure is 1 in (25 mm) of water column. This may be calculated by means of a formula [Eq. (11-6)].

To relieve suds pressure at the base of drainage stacks, it may be appropriate to connect the base of the vent stack to a point in the horizontal drain directly adjacent to the base fitting of the drainage stack and to size the lowest two-story section of the vent stack in accordance with the size computed by means of a formula [Eq. (11-7)] for suds discharge rate from outlets to the atmosphere. The computed size for suds relief purposes may be found to be at least three-fourths of the drainage stack size.

Table 12-1 may be used for sizing vent stacks in accordance with drainage stack capacity loads. Permissible lengths of vent piping shown therein are sufficiently less than those that may be computed by formulas (in which additional allowance need be made for the equivalent length of pipe fittings) that the stated lengths may be applied directly as permissible developed length of pipe. It should be understood that this table applies exclusively to airflow loads of vent stacks attendant upon drainage stacks and is not intended for application to suds flow conditions.

VENT EXTENSIONS AND TERMINALS OF STACKS

The size of vent extensions and terminals of soil, waste, and vent stacks is directly related to their capacities for permitting air to enter the tops of stacks without causing excessive air pressure reduction in the upper section of the sanitary drainage system. In no case should the size of the vent extension and terminal of a drainage stack be reduced to less than that required for its attendant vent stack, for air must be permitted to enter the top of the drainage stack at the same quantity rate of flow as it is dragged down the stack by water at capacity load.

To achieve maximum effective circulation of air up through the sanitary drainage system during periods of relatively small load, it is recommended that the vent extension and terminal of the drainage stack which connects to the most upstream end of the sanitary building drain be of the same size as the drainage stack. If vent extensions and terminals are not to restrict gravity circulation of air in the system, the sum of the cross-sectional areas of all terminals should be equal to the cross-sectional area of a fresh air inlet through which air is supplied into the sanitary building drain of the system or, where there is no house trap and fresh air inlet on the sanitary building drain, equal to the cross-sectional area of the sanitary building drain at its point of entry into the building.

Where vent extensions and terminals are subject to frost closure conditions during frigid weather, provision should be made in sizing to ensure against the occurrence of excessive reduction of available flow area due to formation of ice within the vent extension or at the terminal. A recommendation in this regard is to provide vent extensions and terminals of at least 3-in (75-mm) size for all smaller stacks and to make such necessary changes in size by installation of a long increaser below the roof line within the building.

Similar recommendations for the sizing of vent extensions and terminals may be found in the section "Standard Code Regulations" in this chapter.

VENT HEADERS, THEIR VENT EXTENSIONS AND TERMINALS

Vent headers should be of adequate size to convey air at the quantity rates of flow required at the tops of each of the drainage stacks served thereby without causing excessive air pressure reduction in the upper section of the sanitary drainage system. When the system is operating at capacity, the quantity rate of air required to be supplied through the vent header to drainage stacks is directly related to the total drainage load of all stacks served by the vent header, rather than being just the sum of the individual airflow requirements for each of the stacks at their respective capacities. Thus, the various sections of a vent header should be sized as for a vent stack serving a drainage stack of capacity sufficient for the total drainage load, the length of the vent stack being the same as that of the longest vent stack connected to the vent header.

Vent extensions and vent terminals through which air is supplied to vent headers from the atmosphere above the building roof should be sized in the same manner as for any other section of a vent header. In this case, the vent extension and terminal convey the greatest quantity of airflow of all sections of the vent header and, hence, should be the largest in size.

A similar basis for sizing vent headers is given in the section "Standard Code Regulations" in this chapter. It provides for sizing sections of a vent header in accordance with the sum of the fixture unit loads of the stacks vented through a section and the developed length of the vent stack having the longest developed length to the open air.

INDIVIDUAL VENTS

Proper sizes for individual vents connected to fixture drains should be determined in accordance with the principal function performed by such vents in any given installation and the corresponding quantity rate of airflow required to be conveyed. This may be established by analysis of the arrangement of drainage and vent piping for any particular system.

Where individual vents are connected to fixture drains which discharge into drainage stacks, or into branch drains of stacks, above the level of the base connection of an attendant vent stack of adequate size, the principal function of the individual vent can be assumed to be that of providing sufficient airflow to the fixture drain so as to disrupt siphonic action and thereby avoid excessive trap seal loss during discharge of the fixture. The quantity rate at which air is required may be determined from Table 11-6, based upon the size of the fixture drain. A corresponding size may be computed by means of Eq. (11-11) and by reference to Table 11-3, based upon the length of vent pipe measured from the fixture drain to the branch fitting in the vent stack or to its individual vent terminal in the atmosphere.

Where individual vents are connected to fixture drains which discharge directly into drainage stacks below the level of the base connection of an attendant vent stack, or which discharge directly into the horizontal drain adjacent to the base fitting of a drainage stack, the principal function of such individual vents is that of a vent stack. In such case, the size of the individual vent should be at least as large as is required for an attendant vent stack.

Individual vents connected to fixture drains which discharge directly into the sanitary building drain or horizontal branch thereof are subject to pneumatic effects prevailing in such lower piping of the system. As these effects may vary from one extreme to another depending upon the distance at which the fixture drain connection is made from points at which airflow is constricted in the building drain, the size and load on the drain, and the presence of other building drain connections which may be available for relief of pressure therein, the quantity rate of airflow to be conveyed by an individual vent in this instance may be determined only by analysis of the given installation.

Where individual vents are connected to fixture drains which discharge into zones of the drainage system wherein suds pressure conditions may occur, the size of such vents should be determined for suds flow rather than airflow unless a special suds relief vent is connected to the suds pressure zone. Appropriate sizes for suds flow may be computed by means of Eq. (11-12) and by reference to Table 11-3, for the quantity rates of flow corresponding to conditions in the particular piping arrangement.

The minimum size recommended for an individual vent is 1¼ in. This is based upon the fact that the most critical factor with regard to the capacity of a vent pipe is its diameter. Any decrease in diameter has a seriously adverse effect upon the capacity of a vent pipe. Allowance must be made for a slight amount of fouling at the vent connection to the fixture drain. Consequently, the minimum recommended individual vent size is the same as the minimum permissible fixture drain size, 1¼ in.

In the section "Standard Code Regulations" in this chapter, the same minimum size is stated for individual vents, but the size of the fixture drain is applied as the basis for sizing the individual vent. The required vent size is one-half that of the fixture drain, except that a 1½-in vent may be provided for a 4-in fixture drain of a water closet.

BRANCH VENT DROPS TO LOWER FLOORS

Rational analysis may be applied for the sizing of branch vent drops to lower floor fixtures. If each fixture drain of such fixtures is assumed to be installed vertically and to extend downward a sufficient distance so that flow reaches terminal velocity in the vertical drop, the fixture drain may be considered to have the same performance characteristics and airflow requirements as a drainage stack.

Also, if each individual vent is assumed to be connected at the top of the vertical drain, the vent may be considered to perform the same airflow function as a vent extension at the top of a vent stack, namely, that of supplying air into the drainage stack at the same rate as it is dragged down by the liquid flowing down the vertical drainage stack.

In drainage stacks of 1¼ in (31.3 mm) through 4 in (100 mm) diameter, terminal velocity is achieved in a drop of less than one normal story height. Ordinary airflow requirements for vertical fixture drains of that size range, and conveying capacity flow, are the same as for drainage stack vent extensions.

Consequently, a branch vent drop serving a number of individual vents may be considered to perform the same function as that of a vent header or combined vent serving as the single vent pipe extending above the roof for a number of vent extensions of soil and waste stacks and their attendant vent stacks. It follows logically then that the sizing of branch vent drops should be comparable with the sizing of vent headers or combined vents in certain respects.

In the sizing of vent headers, using the standard tables developed for sizing vent stacks, the column headed "Size of soil or waste stack" is disregarded. The size is based upon the sum of the fixture units of load of all drainage stacks vented through each section of the header. The developed length applied is that of the vent stack having the greatest developed length to the atmosphere above the roof.

In essence, this amounts to considering the vent header as a vent stack which serves an imaginary drainage stack having a load equal to that of the total number of fixture units of all the connected stacks. The vent header corresponds to an imaginary vent stack with a developed length equal to that of the actual vent stack having the greatest developed length of all connected to the vent header.

The logic of this basis for sizing vent headers is established on the fact that the total amount of flow in drainage piping is related to the number of fixture units of load connected to the drainage piping, rather than to the number and size of drainage stacks into which the various fixtures discharge. If a single stack was to be used in lieu of a number of drainage stacks, and to convey the same total load, the single stack almost always would have to be larger in size than any of the drainage stacks.

In any event, a single drainage stack could theoretically be used in lieu of a number of drainage stacks. However, in that case long horizontal fixture drainage branches would be required, resulting in excessive loss of headroom due to slope in long branch drains. These branches would also be subject to excessive incidence of stoppages.

The developed length of vent stack for a single drainage stack with numerous long horizontal offsets would be comparable in length with that of the most extreme run or traverse of the drainage stack. Hence, the basis cited for sizing vent headers may be deemed reasonable and rational.

The same basis may be applied to sizing branch vent drops. But, in this case, the vent stack connection for the branch vent drop may be considered to be the equivalent of the vent terminal of a vent header or combined vent. Atmospheric pressure may be considered to prevail in vent stacks as a normal condition during service, except for relatively short periods when peak flow occurs in drainage stacks.

During peak flow periods, positive or above-atmospheric pressure conditions occur in the lower sections of drainage stacks and their attendant vent stacks. This positive pressure condition in the vent stack simply assists flow in branch vent drops, since their principal function is to supply air to the various fixture drains served when flow occurs therein.

Consequently, the rational basis for sizing branch vent drops to lower floor fixtures is the same as for sizing vent headers or combined vents, except that the maximum developed length which should be applied for branch vent drops is that measured from the vent stack branch connection to the farthest fixture drain connection served (Fig. 12-1).

Using the standard tables developed for sizing vent stacks (Tables 8-3 and 8-5), note the vent pipe sizes suitable for the developed length applicable. Also note the number of fixture units of load suitable for each vent pipe size for the developed length applicable. Then, for that developed length, select appropriate vent pipe sizes for each section of the branch vent drop corresponding to the load in fixture units. This corresponds to sizing branch vent drops on a uniform pressure loss basis for their maximum developed lengths.

BRANCH VENTS

Branch vents which serve to connect more than one individual vent to a vent stack or to a vent terminal in the atmosphere should be sized in accordance with the principal function performed thereby in any given installation and the corresponding quantity rate of airflow required to be conveyed. Just as for individual vents, proper sizes may be established

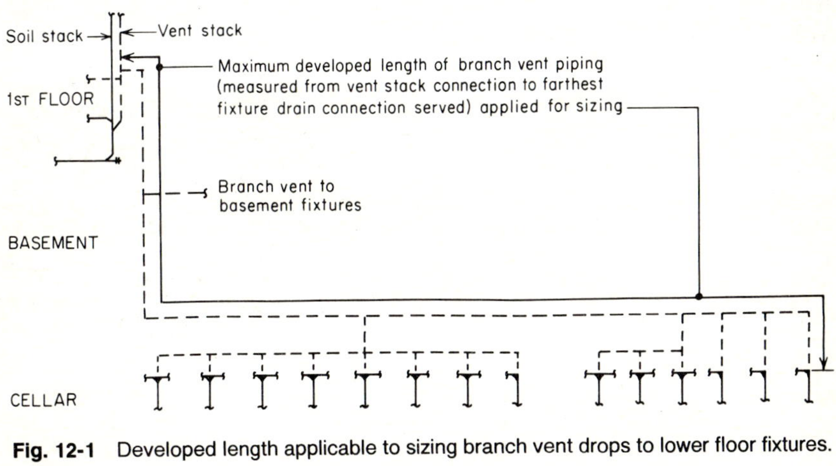

Fig. 12-1 Developed length applicable to sizing branch vent drops to lower floor fixtures.

by analysis of the arrangement of drainage and vent piping for any particular system.

Where several individual vents connect to a horizontal branch vent extending from a branch fitting in a vent stack and their respective fixture drains connect to a horizontal branch drain extending from a branch fitting in a drainage stack, a pressure relief circuit exists between the drainage stack branch fitting and the vent stack branch fitting. Each of the fixture drains and their individual vents form branch or parallel circuits for pressure relief and airflow between the drainage stack and vent stack. The proportion of the total airflow conveyed by each individual circuit is the same as would be the case in parallel circuits in a water supply system, as is discussed in Chap. 5.

In all other respects the size of branch vents may be computed in the same manner as for individual vents, that is, based upon their principal function and the quantity rates of airflow or suds flow to be conveyed. A convenient basis for sizing branch vents is given in the section "Standard Code Regulations" in this chapter. The size in this case is based upon the number of fixture units connected to the branch vent and the developed length of the branch vent measured from its vent stack or stack vent connection to the farthest fixture drain connection served by the branch vent.

CIRCUIT AND LOOP VENTS

A circuit or loop vent is little more than an enlarged individual vent, connected to the fixture drain of the most upstream fixture discharging into a horizontal branch drain and provided to permit air to flow into the horizontal branch drain at a sufficient rate so as to prevent excessive aspiration effects from occurring therein. This is the principal function of a circuit or loop vent in most ordinary installations.

However, where a circuit or loop vent serves a horizontal branch drain which connects to a pressure zone of the sanitary drainage system, such as may occur adjacent to the base of a drainage stack or in the building drain, the principal function of the vent may be to prevent excessive back-pressure effects from existing in the horizontal branch drain. Thus, the quantity rate of airflow required to be conveyed by a circuit or loop vent must be determined in accordance with its principal function and corresponding airflow requirements, just as for individual or branch vents. For any given installation, the sizes of circuit or loop vents may be computed in accordance with their air and suds flow requirements in the same manner as was discussed for individual vents.

In the section "Standard Code Regulations" in this chapter, the size of circuit or loop vents is required to be at least one-half the diameter of the horizontal soil or waste branch to which they connect. This may be adequate for most ordinary installations requiring airflow to be conveyed to prevent excessive aspiration effects in the horizontal branch drain.

RELIEF AND YOKE VENTS FOR SOIL AND WASTE STACKS

The function of a relief or yoke vent, provided in special locations in drainage stacks of extended height, is to permit air pressure to be relieved from the drainage stack directly to the vent stack by flow through a short relief or yoke vent of the same size as the vent stack. Wherever flow interference may occur in such drainage stacks, the relief or yoke vent functions as an additional elevated vent stack base connection above the point of flow interference.

Sizing in this case is related to the sizing of vent stacks and their base connections. Hence, the proper size for relief and yoke vents is the same as that of the vent stack required for the drainage stack.

SUDS PRESSURE RELIEF VENTS

Suds pressure relief vents, provided at special zones in the drainage system where suds pressure may develop, must be adequate in size to permit suds to flow out of the zone at a sufficient rate so that the pressure rise in the zone does not exceed the design criterion. Such special vents are necessary to prevent suds backup conditions at fixtures which discharge into such zones and are equipped with traps protected by vents which are inadequate to perform the function of relieving suds pressure.

An ordinary fixture trap with 1 in (25 mm) of residual trap seal depth in service has a maximum resistance of 2 in (50 mm) of water column when subjected to back pressure of suds. This resistance is equal to a column of suds 7.42 ft (2.26 m) in height, at which level suds backup through the fixture trap seal may occur.

A suds pressure relief vent need not be of extensive length, for it is only necessary to relieve suds pressure from the pressure zone to a nonpressure zone downstream from and at a lower elevation in the sanitary drainage and vent system. Hence, the principal factor involved in this case is the diameter of a relatively short vent required to relieve the quantity rate of suds flow in the pressure zone. A simple guide in this case is the quantity rate of airflow conveyed down a drainage stack into the suds pressure zone, which may be determined from rates shown in Table 11-5, and the outlet discharge rates for suds at a flow pressure of 1 in (25 mm) water head, which may be determined from the capacities for various sizes of vent outlets in Table 11-3. The maximum permissible length of suds pressure relief vents can be further computed by means of Eq. (11-12).

From these tables and by computations, it can be seen that the size of the suds relief vent in most cases is approximately three-fourths that of the drainage stack and its horizontal drain at the base of the stack. Thus, this may be taken as a recommended guide in sizing such special relief vents.

VENTS FOR BUILDING SEWAGE SUMPS AND RECEIVING TANKS

Building sewage sumps and receiving tanks, other than pneumatic ejectors, operate at atmospheric pressure and receive sewage from sanitary subbuilding drains under gravity flow conditions. The only airflow required in a vent pipe from such an airtight receptacle is that which is displaced by sewage in entering the receptacle and that which is required to enter the receptacle as the sewage is ejected by the sewage pump. Thus, the quantity rate of airflow required to be conveyed by the vent for a building sewage sump or receiving tank is the maximum rate at which sewage either may enter the tank or be pumped therefrom, whichever is the greater rate.

Based upon the quantity rate of airflow required to be conveyed and the developed length of vent piping for a given installation, a proper size may be computed for such vents by means of Eq. (11-11) and by reference to Table 11-3.

STANDARD CODE REGULATIONS

Individual Vents Individual vents shall be at least 1¼ in in diameter and no less than one-half the diameter of the fixture drain to which they connect, except that a 1½-in individual vent may be installed where a 4-in fixture drain is provided for a water closet or similar fixture.

Circuit and Loop Vents Circuit or loop vents shall be at least one-half the diameter of the horizontal soil or waste branch to which they connect.

Branch Vents Branch vents connecting more than one individual vent to a vent stack or stack vent shall be in accordance with Table 12-1. In determining the size of such piping, the column headed "Size of soil or waste stack, in inches" shall be disregarded; and the size shall be based upon the number of fixture units connected thereto, and the developed length of the branch vent measured from its vent stack or stack vent connection to the farthest fixture drain connection served by the branch vent.

Vents for Building Sewage Sumps and Receiving Tanks Vents for building sewage sumps and receiving tanks, other than pneumatic ejectors, shall be sized as branch vents.

Relief and Yoke Vents for Soil and Waste Stacks Relief and yoke vents for soil and waste stacks shall be not less than the size of the vent stacks to which they connect.

Vent Headers Sections of a vent header and its vent extension through the roof shall be in accordance with Table 12-1. In determining the size of such piping, the column headed, "Size of soil or waste stack, in inches," shall be disregarded; and the size shall be based upon the sum of the fixture unit loads of the stacks vented through such section of the header, and the developed length shall be that of the vent stack having the longest developed length to the open air.

Vent Stacks The size of vent stacks shall be determined from Table 12-1, based upon the size of soil or waste stacks served thereby, the number of fixture units connected to the soil or waste stack, and the developed length of the vent stack. Such developed length shall be that measured from the lowest connection of the vent stack with the soil stack, waste stack, or building drain, to the vent terminal in the open air.

INDEX

Acid wastes (*see* Chemical waste system)
Air:
compressibility of, 377–378
density of, 377–378
equivalent static head of, 380
viscosity of, 377–379
Air breaks:
clear water wastes, 332, 352
discharge pipes from exhaust and blow-off tanks, 332–333, 352
indirect waste pipes, 331–332, 352
relief pipes of water supply system, 332, 352
waste outlets of fixtures, 290–292, 339–340
Air chambers, 94, 117, 154
Air gaps:
fixed, 88, 114
fixture water outlets, 87, 89, 114
flush tanks, 89, 116
priming connections to pumps, 83, 114
standards for, 23
water supply tanks, 99, 119
Air pressure test, 334, 353
Airflow:
condition in vent piping, 380–381
rate of: in branch vents, 390–391
in individual vents, 389–390
from outlets, 382–384
in vent headers, 390–391
in vent stacks, 388–389
American National Standards Institute (ANSI):
formerly named American Standards Association, 33
American National Standards Institute (ANSI) (*Cont.*):
formerly named United States of America Standards Institute, 33
standards, 13, 21–24, 32–33
American Public Health Association, 32
American Society of Mechanical Engineers (ASME), 32, 108
American Society of Plumbing Engineers, 33, 95
American Society of Sanitary Engineering (ASSE), 23, 24, 31
American Society for Testing and Materials (ASTM), standards, 21–24
American Standards Association:
American Standard National Plumbing Code, 32–33
American Standard taper pipe thread, 107–108
(*See also* American National Standards Institute)
American Water Works Association (AWWA), standards, 22, 24
Anchors, pipe, 101, 104, 120, 314–315, 343
Ancient plumbing, 2–4
ANSI (*see* American National Standards Institute)
Appliances, portable household, waste connections to, 291, 340
Aquarium fixtures, connections to, 63, 76
Aspiration, 292–293, 320–321, 381–382, 400
Assembly occupancy building classification, 40, 46–48

Atmospheric pressure, 125–127, 377–378
Availability of public systems:
for sewage disposal, 284, 339
for storm water disposal, 285, 338–339
for water supply, 81–82, 113

Back pressure, 292–293, 305, 320–321, 381–382, 400
Back siphonage, protection against, 89–95, 116–117
Back vent, 324
Backflow, protection against, 89–95
Backwater valve:
location of, 296–297, 342
on subsoil drain connections, 296–297, 342
Balancing valves:
adjustment of, 271–273
equivalent length for, 146–147
in hot water recirculation system, 271–273, 277
Baptistries, 63, 76
Bar sink, waste connection to, 292, 340
Bathroom:
floors, 64–65, 71, 76
lighting, 63–64, 76
ventilation, 63–64, 76
walls, 64, 71, 77
Bathtub:
fixture supply pipe size, 229
fixture trap size, 68, 78
fixture unit rating: sanitary drainage, 366–368
water supply, 229
overflow, 56, 73
requirements for occupancies, 41, 44, 45, 48
shower above, 56, 73
walls above built-in type, 56, 73
waste outlet for, 56, 73
Bedpan washers and steamers, vapor relief pipes, 70–71, 79
Bell traps, permissible use of, 67, 78
Bernoulli's theorem, 132, 133
Bidet:
fixture trap size, 68
fixture unit rating, sanitary drainage, 366–368
water supply to, 88, 114
Blowoff pipes and tanks, steam discharge from, 332–333, 352
Boiling points of water, 123, 126–127
Branch drains:
rate of flow in, 364
sizing of, 369–371
Branch vents:
rate of flow in, 390–391
sizing of, 397–400, 402
Brass pipe, tube and fittings, standards for, 22
Briggs standard pipe thread, 107–108
Building drains:
combined: connections to, 286–287
sizing of, 370–371, 373–374
existing, use of, 345
sanitary: connections to, 301–302, 346
design capacity of, 301, 370–371
minimum size of, 374
sizing of, 370–374
storm water: design capacity of, 287, 372
sizing of, 372–375
Building occupancy classifications, 38–40
fixture requirements for, 41–52
Building Officials Conference of America, 32
Building sewers:
regulatory jurisdiction over, 285–286, 338–339
use of existing, 286, 345
(*See also* Combined storm water and sanitary drainage systems in buildings)
Building systems, required (*see* Plumbing systems for buildings)
Building trap:
cleanouts for, 297–301, 345–346
installation and location of, 297–301, 345–346
need for, 297–301, 345–346
omission of, 297–301, 345–346
Business occupancy building classification, 39, 45–46

Cast iron pipe and fittings, standards for, 21
Cast Iron Soil Pipe Institute (CISPI), 21, 24
Cavitation, 155–157
Check valves, 88, 92–94, 114–116, 146–147, 234, 270–271
Chemical closets as temporary facilities, 50

Chemical waste system:
dilution device, 294–295, 341
materials, 294–295, 341
neutralizing device, 294–295, 341
Circuit venting method, 327 329, 350, 400–401
rate of flow in, 390
sizing, 400–402
Cleanouts in drainage piping, 319–320, 344–345
building trap, 297–301, 345–346
Clear water wastes, 332, 352
ejectors for, 296, 342
Code regulations:
standardization of, 25–36
(*See also* Plumbing codes)
Combination fixture, 58, 67, 68
Combination pressure-temperature relief valve, 237, 280–281
(*See also* Relief valves)
Combination waste and vent system, 329–331, 350
Combined storm water and sanitary drainage systems in buildings:
connections to, 286–287
sizing of, 373, 375
Common vent, 321–322, 347
Compressibility:
of air, 377–378
of water, 124
Condensers and cooling jackets, 88, 114, 332, 352
Conference of State Sanitary Engineers, 32
Conservation of energy, 13
control of flow at outlets, 162
velocity limitation in pumping, 180
water temperatures, 232
Controlled-flow storm water drainage system, 287–290
conditions for use, 289–290
for flat roofs, 287
sizing table, 289
Cooling jackets and condensers, 88, 114, 332, 352
Copper pipe, tube and fittings, standards for, 22
Corrosion control:
velocity limitation for, 176–180
water supply treatment, 85–86
Corrosivity of water, 159, 167, 176, 178–180, 240–241, 248–249
Cross-connections between public and private water supplies, 87, 114
Crown vent, 321, 347

Dead ends, 316, 344
Defective drainage and vent piping systems, testing, 333–335, 353
Demand for hot water (*see* Hot water demand)
Demand load in water supply systems, 160–161, 163–167, 226–229
general method for estimating, need for, 163
load values for fixtures, 164–165
total demand including continuous flow, 167
Density:
air, 377–378
suds, 379
water, 123–124
Dental cuspidor, 68, 367
Dental lavatory, 68, 367
Design load:
sanitary drainage systems, 369–370
storm water drainage systems, 372–375
Detergents (*see* Suds; Suds pressure zones)
Direct heating equipment installations, 239–243
Dishwashing machines and fixtures, 20, 50, 59, 60, 68, 74
Disinfection methods for water supply systems, 112–113, 121–122
Draft pressure, natural, 391
Drainage piping:
air breaks, 290–292, 331–333
changes in direction, 315–316, 344–345
cleanouts for, 319–320, 344–345
fittings for, 315–316, 344–345
indirect, 290–291, 331–333
installation of, 312–316, 344–345
location of, 312–313, 342
for radioactive wastes, 316, 344
slope of, 312–313, 343–344
testing, 333–335, 353
Drainage stacks:
attendant vent stacks, 302–304, 346
branch connections to offsets, 302, 346
flow conditions, 304–306, 369–370
in offsets, 303–307
hydraulic jump at base, 362–363
noise in piping, 360

Drainage stacks (*Cont.*):
relief vents, 304–307, 350–351, 401, 403
sizing, 350–351, 369–370, 373–374
suds pressure zones, 306–311, 351
vent extensions, 302–304, 346–347
Drainage systems (*see* Sanitary drainage systems; Storm water drainage systems)
Drains, subsoil (*see* Subsoil drainage)
Drinking fountains:
design and construction of, 58–59, 74
fixture supply pipe size, 229
fixture trap size, 68
for physically handicapped, 43–44, 52
prohibited locations of, 64–65, 76
required for occupancies, 46–49
standards for, 21, 74
waste connections to, 291, 340
Drinking water (*see* Potable water supply)
Drum trap, 67, 78
Dwelling unit building classification, 39, 44–45

Early American sanitation standards, 4–6
Ejectors:
atmospheric type, 402–403
for clear water wastes, 296, 342
pneumatic type, 312, 351
for sewage, 296, 342
venting, 296, 312, 341–342, 351, 402–403
Electric water heaters, 238–243
Elevated water storage tanks, 97–100, 118–119
Emergency showers or lavatories, 50–51
Energy:
conservation of (*see* Conservation of energy)
equation for liquids, 132–133
in fluid flow, Bernoulli's theorem, 132–133
heat interchanged in friction loss, 133
kinetic, 129–130
loss due to friction, 133
potential, 129–130
Energy shutoff device, 236–237, 280–281
Equipment drains, 290–291, 339–340
Equivalent length:
of basic design circuit, 211–212, 222–223

Equivalent length (*Cont.*):
of fittings and valves, 145–150, 172
of vent piping, 386–387
Equivalent static head of air, suds, and water, 380, 383–384
Expansion and contraction of piping:
drainage and vent, 316–318
in horizontal hot water mains, 103–104
in hot water risers, 104–107
branches, 104–105
installation requirements, 103, 120
loops and swings for absorbing, 104–106
of plastic materials, 103–106
of various materials, 103–106

Faucets:
demand rates, 161–162
quick-closing, 152–154
self-closing, 94, 117
Federal Supply Service (FS), Standards Division, General Services Administration, 21, 22, 24
Filters, pressure drop through, 150
Fittings:
drainage and vent, 315–316, 344
equivalent length of, 145–150
Fixed air gaps, 88, 114
Fixture connections, 290–292, 339–340
prohibited, 87–89
Fixture drains:
rate of flow in, 363–364
sizing of, 365–366, 371
vent piping connections to, 320–323, 347–348, 396–397
Fixture traps:
for common fixtures, 66–69, 77–79
intercepting, 61, 68–70, 78–79
minimum sizes, 68
Fixture units:
conversion to equivalent storm drainage area, 373–375
sanitary drainage, 366–369, 373–375
water supply, 163–167, 184–196, 226–229
Fixtures:
for children's use, 55, 71
development of, 13–20
installation of, 63–66, 76–77
intercepting traps for, 68–70, 78–79
overflows for, 60–61, 74

Fixtures (*Cont.*)
- for physically handicapped, 13, 42–44, 51–52
- quality of, 53–54, 71
- religious, ornamental, and aquarium, 63, 76
- required for occupancies, 41–52
- special use, 61, 75
- standards for, 21, 53–54
- strainers for, 60–61, 74–75
- traps for, 66–69, 77–79
- vacuum breakers for, 89–94, 115–117
- vapor relief pipes for, 70–71, 79

Flammable oil wastes, 295, 340–341
Flange connections, 66, 77
Floor drain:
- fixture unit rating, 368
- location of, 60, 74
- objections to use, 60
- required for occupancy, 51
- strainer for, 60
- trap for, 60, 68, 74

Floors, bathroom, 64–65, 71, 76
Flow pressure at outlets, 59, 60, 134–138, 363–364, 382–384
Flush tanks, 55–56, 72, 91–92, 115–116
Flush valves (Flushometers), 55–56, 72, 91–92, 115–116
Food waste grinders, 58, 73–74
Fountains (*see* Drinking fountains)
Fresh air inlets, 297–301, 345–346
Friction head of water, 133
Friction loss:
- for airflow in vent piping, 384–385
- coefficient *f*: for vent piping, 387–388
  - for water supply piping, 140–144
- design limit: for vent piping, 381–382
  - for water supply piping, 169–170, 177–181
- equivalent length of fittings, 386–387
- in horizontal sloping drains, 355–359
- in hot water recirculation piping, 276–280
- permissible length of vent piping, 385–386
- for suds flow in vent piping, 382–384
- under turbulent flow, 127, 141, 384
- in vertical drains, 360–362
- in water meters, 138–139, 171
- in water supply piping, 133, 139–151

Galvanized iron pipe and fittings, standards for, 21
Garbage can washer, 61, 75
Gas water heaters, 239–243
Gate valves:
- equivalent pipe length for, 145–150
- in water supply systems, 110–112

Globe valves:
- equivalent pipe length for, 140–150
- in water supply systems, 110

Gravity circulation (thermal) head in hot water supply systems, 270, 275–278, 391
Grease-intercepting traps, 69–70, 78, 295, 341
Grease wastes, 295, 341
Gutters:
- rate of flow in, 372–373
- sizing of, 374–375

Hangers:
- for drainage and vent piping, 314–315, 343–345
- materials, 101, 120, 314–315, 343–344
- for water supply piping, 101, 120

Hardness of water:
- adjustment of pipe sizes for, 172–176, 230
- clogging of heater coils, 233, 239, 245, 249
- maintenance of hot water heaters, 245, 249
- pressure relief valve location for, 234–235, 280–281
- softening to eliminate, 238–239, 241, 245

Head:
- equivalency for air, suds, and water, 380, 383–384
- friction, 133
- static, 130–131
- thermal (gravity circulation), 270, 275–278, 391
- velocity, 131–132

Heat losses from hot water piping:
- estimating method for, 274–275
- recirculation pump capacity based on, 275–276
- return pipe sizes based on, 273–280

Heaters and heating methods, water, 238–239

History of plumbing, 1–2
Hoistway:
  pit drain, 292, 340
  prohibited location for piping, 313, 342
Hose bib:
  demand rate at, 161
  vacuum breaker installation for, 90–92, 115–116
Hot water demand:
  at 140°F, 257–261
  at 165°F, 257
  at 180°F, 256
Hot water heater capacity, 256–261
Hot water storage tanks:
  drain valves or cocks, 242, 254, 281
  energy shutoff devices, 237, 280
  minimum recommended capacities, 260
  pressure markings, 238, 281
  relief valves, 233–238, 280–281
  vacuum breakers for, 266, 281
Hot water supply system:
  circulation types, 261–273, 281
  combination pressure-temperature relief valve, 237, 280–281
  demand for (*see* Hot water demand)
  design objectives of, 231, 233
  indirect storage tank heater installations, 248–256
  indirect tankless heater installations, 244–248
  method for determining: circulation pump capacities, 273–279
    circulation rates, 273–279
    size of return piping, 273–279
  pressure relief valve, 233–235, 280
  relief valve outlet connection, 237–238, 281
  temperature control, 231–233, 240, 243–244, 247, 255–256
  temperature relief valve, 236–238, 280–281
  vacuum relief valve, 266, 281
  water-heating methods and heater types, 238–239
Hydraulic jump in horizontal drains, 307, 362–363
Hydraulic mean depth or radius, 356–359
Hydraulic slope, 356–359

Ice box waste connection, 291, 340
Indirect storage tank heater installations, 248–256
Indirect tankless heater installations, 244–248
Indirect waste pipes:
  equipment and fixtures connected to, 290–292, 339–340
  fixtures receiving discharge from, 61, 64, 68, 75–78
  installation of, 331–332, 352
  sizes, 331–332, 352
  traps for, 66, 77
  venting, 332, 352
Individual vents:
  quantity rate of flow in, 389–390
  sizing of, 396–397, 402
  use of, 320–323, 347–348
Industrial occupancy building classification, 39, 46
Industrial wastes, 294, 341
Inlets, fresh air, 297–301, 345–346
Institutional occupancy building classification, 40, 48–50
Intercepting fixture traps, 61, 68–70, 78–79
Intercepting strainers, 61, 75
Interconnections of water supply systems, 87, 114–115

Jet flow, 134–136

Kinetic energy, 129–130
Kitchen equipment waste connections, 292, 340
Kitchen sink (*see* Sinks)

Laundry machines, automatic:
  for dwelling units, 20, 44–45
  for physically handicapped, 52
  portable, 292, 340
  requirements for occupancies, 44–45
  water supply to, 44, 226–227
Laundry trays:
  combination fixture, 44–45
  demand rate, 161
  fixture supply pipe size, 229
  fixture trap size, 68
  fixture unit rating: sanitary drainage, 367
    water supply, 226–227
  requirements for occupancies, 44–45
  waste outlet size, 57, 73

Lavatories:
demand rate, 161
fixture supply pipe size, 229
fixture trap size, 68
fixture unit rating: sanitary drainage, 367
water supply, 226–227
multiple type, 56, 72–73
for physically handicapped, 51
requirements for occupancies, 44–50
waste outlet size, 56, 72
Lead bends, pipe, and traps, standards for, 22
Leaders:
flow in, 360–362, 372–375
inlet strainers, 286, 352
protection of exterior, 314, 343
sizing of, 372–375
traps, 286, 352
Lighting for fixture locations, 63–64, 76
Local vents for fixtures, 70–71
Loop venting:
method of, 327–329, 350, 400–401
rate of flow in, 390
sizing of, 400–402

Manning's formula, gravity flow in sloping drains, 356
Materials:
for chemical waste system, 294–295, 341
durable service life of, 38
manufacturers' recommendations for, 38, 44
performance requirements for, 37–38, 44
for pipe anchors and hangers, 101, 120, 314–315, 343–344
for piping, expansion and contraction allowance for, 103–106
selection of, 38, 167
standards for, 20–25
used, 38, 87, 114
Mercantile occupancy building classification, 40, 50
Meters (*see* Water meters)
Miscellaneous occupancy building classification, 40, 50
Mortuary table, water supply to, 88, 114
Multiple dwelling building classification, 39, 44–45

National Association of Master Plumbers, 30–31
National Association of Plumbing-Heating-Cooling Contractors, 33, 95
National Bureau of Standards, U. S. Department of Commerce, 28–32, 164, 366, 387
National Standard Plumbing Code, 33, 95
Noise in piping:
cavitation, 155–157
in drainage stacks, 360
quick-closing valves and faucets, 152–154
velocity, 151–155
water hammer, 94, 117, 151–154
water pressure regulation, 94–97, 117
Nonpotable water, restriction of use, 82–83, 113
Nuclear wastes (*see* Radioactive wastes)

Objectionable wastes, 293–295, 340–341
Occupancy classifications of buildings, 38–40
fixture requirements for, 41–52
Oil-fired water heaters, 239–243
Oil separators, 295, 341
One-family dwelling building classification, 38–39, 44
Operating table, water supply to, 88, 114
Orifices:
rate of discharge from, 134–138
rate of flow through submerged, 138
Ornamental fixtures, connections to, 63, 76
Overflows for fixtures, 60–61, 74

Parallel pipe circuits, division of flow in, 151
Peppermint vapor test, 334–335, 353
Physically handicapped facilities:
accessibility of, 13, 42–44, 51–52
door width, 42–44, 51–52
location of fixtures, 42–44, 51–52
standard for, 13, 41–42
usability of fixtures, 42–44, 51–52
wheelchair, standard, specifications and function, 42–43
Piers, 101, 120, 314–315, 343
Pipe anchors, 101, 104, 120, 314–315, 343

Pipe threads, American Standard taper, 107–108
Piping:
expansion and contraction of (*see* Expansion and contraction of piping)
installation of, 100–102, 119–121
materials, standards for, 21–24
noise in (*see* Noise in piping)
(*See also* Drainage piping; Vent piping)
Piping offset calculations:
common offsets, 108–109
rolled offsets, 109–110
uniform span between parallel offsets, 109–111
Plastic plumbing materials, standards for, 21–24
Plumbing:
ancient, 2–4
history of, 1–2
Plumbing codes:
early regulations, 25–28
engineering design basis for provisions, 28–31
model code developments, 31–33
performance requirements, 34–36
standardization of regulations, 28–33
state, 33–36
Plumbing and piping materials (*see* Materials)
Plumbing systems for buildings, 6–13, 37, 44
facilities required, 37, 41–52
performance required of materials, 37–38, 44
for physically handicapped, 13, 42–44, 51–52
testing, 112, 122, 333–335, 353
Pneumatic ejectors:
air pressure relief pipes for, 312, 351
discharge pipe connections for, 312
rate of air flow in relief pipes for, 312
venting of, 296, 342
Pneumatic water storage tanks, 95, 117
Portable household appliances, waste connections to, 291, 340
Potable water supply:
fixtures requiring, 82–83, 113
prevention of contamination of, 89–94, 114–115
protection of, 86–89, 115–117
source of, 82, 113
treatment for corrosion control, 85–86
Potential energy, 129–130
Preheater equipment, vacuum breakers for, 92–93, 115
Pressure:
atmospheric, 125–127, 377–378
back, 292–293, 305, 320–321, 381–382, 400
flow (*see* Flow pressure at outlets)
loss of, due to friction (*see* Friction loss)
natural draft, in sanitary drainage and vent systems, 391
vapor, in pumping, 155–156
water pressure regulations, 94–97, 117
in water supply systems, 94, 117, 159–160, 169, 226
Pressure regulation valve, 22, 94, 117
maximum pressure allowable at fixtures, 94, 117
in reducing water hammer and noise, 94, 117, 152
Pressure relief valve, 233–235, 280
for refrigeration machinery, 88, 114
standards for, 24
Pressure-temperature (combination) relief valves, 237, 280–281
(*See also* Relief valves)
Private water supply systems, 82, 87, 113–114
Privies as temporary facilities, 50
Process water, 88–89, 114–115
Prohibited direct connections for potable water supply, 87–89, 114–115
Projection room, plumbing facilities for, 47
Public sewer system:
availability for sewage disposal, 283–285, 339
availability for storm water disposal, 285, 338–339
regulations of district authorities, 283, 338–339
Public water system (*see* Water supply systems)
Pumps:
capacity required for hot water recirculation systems, 275–276
circulation, for swimming pools, 63, 292, 340
power delivered by, 157
suction lift of, 155–157
work done by, 157

Quick-closing valves and faucets, 152–154

Radioactive wastes:
disposal of, 284–285, 339
federal safety standards for, 285
treatment of, 295, 341
warning signs on equipment and piping, 316, 344
Rainfall intensity rates, 372, 374
Receiving tanks (*see* Sewage sumps and receiving tanks)
Refrigeration machinery:
check valves in piping to, 88, 114
interconnection of discharge pipes from, 87, 114
pressure relief valves in piping to, 88, 114
Refrigerator waste connections, 291, 340
Relief pipe from pneumatic ejector, 312, 351
Relief valves:
combination pressure-temperature, 237, 280–281
discharge pipes from, 89, 114–115, 237–238, 281
location of, 234–237, 280–281
pressure, 233–235, 280
for refrigeration machinery, 88, 114
standards for, 24
tank vacuum relief, 266, 281
temperature, 236–238, 280–281
Relief vent for soil and waste stacks, 304–306, 350–351, 403
offsets, 304–306, 350–351, 403
for stack-vented group of fixtures, 327, 349
for suds pressure zones, 306–311, 351, 401–402
for vertical offsets in building drains, 305, 350, 401–403
Religious fixtures, connections to, 63, 76
Reuse of process water, 88–89, 114–115
Reynolds, Osborne J., critical velocity experiments of, 125–128
Reynolds numbers, 127–128, 144
Roof:
storm drainage for, 285, 338–339
controlled-flow systems, 287–290
water supply tanks on, 97–100, 118
Roof drains:
strainers at inlets, 286, 352
Roof drains (*Cont.*):
watertightness of, 286, 352
Roof gutters:
rate of flow in, 372–373
sizing of, 374–375

Safe pans and receptors, 291, 340
Safety devices for hot water supply systems, 233–238, 266, 280–281
Salt water, use of, 83, 113
Sand intercepting trap, 70, 78–79, 296, 342
Sanitary building drain (*see* Building drains, sanitary)
Sanitary drainage fixture units, 366–369, 373–375
Sanitary drainage systems:
building sewer, 284–287, 339, 345
building trap, 297–301, 345–346
disposal system, 284–285, 339
fixture connections, 290–292, 339–340
fresh air inlet, 297–301, 345–346
future fixtures, 316, 344, 374
pneumatic effects, 381–382
prohibited connections, 315–316, 344
below sewer level, 295–296, 341–342
sizing of, 365–371, 393–403
soil stack offsets, 302, 346, 373–374
suds pressure zones in, 306–311
venting, 292–293, 340
design criterion for, 381–382
European single-stack, 335–338
waste stack offsets, 302, 346
Seals, trap (*see* Fixture traps)
Sediment-intercepting traps, 70, 78–79, 295, 341
Self-closing faucets, 94, 117
Service sink (*see* Slop sink)
Sewage disposal for building systems, 284–285, 339
Sewage sumps and receiving tanks:
atmospheric type, 402–403
pneumatic type, 312, 351
venting, 296, 312, 342–351, 402–403
Shock absorbers, mechanical, 94, 117, 154
standards for, 24
Showers:
above bathtubs, 56, 73
compartments, 56, 73
demand rate, 161
dimensions of, 57, 73

Showers (*Cont.*):
drainage of floors, 57, 73
emergency, 50, 51
fixture supply pipe size, 229
fixture trap size, 68
fixture unit rating: sanitary drainage, 367–368
water supply, 226–227
for physically handicapped, 52
requirements for occupancies, 44–52
waste outlet: size of, 57, 73
strainer on, 57, 73
Shutoff device, energy, 236–237, 280–281
Silt intercepting trap, 70, 78–79, 296, 342
Sinks:
bar, waste connection to, 292, 340
combination fixtures, 57, 58
commercial, 58, 74
demand rate, 161
fixture supply pipe size, 229
fixture trap size, 68
fixture unit rating: sanitary drainage, 367–368
water supply, 226–227
for indirect wastes, 331–332, 352
for physically handicapped, 51
requirements for occupancies, 44, 48, 50
waste opening size for food waste grinder unit, 58, 73
waste outlet size, 58, 73
Siphonage and siphonic action, 54, 292–293, 320–321, 389, 396–397
back, protection against, 89–95, 116–117
Sleeves, 101, 120, 314, 342
Slip joints, accessibility of, 66, 77
Slop sink (service sink):
demand rate, 161
fixture supply pipe size, 229
fixture trap size, 68
fixture unit rating: sanitary drainage, 368
water supply, 226–227
Slope:
of drainage piping, 312–313, 343–344
effect on flow in horizontal drains, 355–360
of vent piping, 313, 344
of water piping, 100, 121
Sloping drains, gravity flow in, 355–360
Smoke pressure test, 334, 353
Soda fountain fixtures, waste connections to, 290–292, 339–340
Softening of hard water, 238–239, 241, 245
Soil pipe and fittings, cast iron, standards for, 21
Soil stacks (*see* Drainage stacks)
Special use fixtures:
intercepting strainers, 61, 75
intercepting traps, 61, 69–70, 75
waste treatment and handling, 61, 75
Special wastes, 332–333, 352
Stack venting, 326–327, 349–350
(*See also* Vent stack)
Standard pipe threads and dimensions, 107–108
Standardization:
of plumbing materials, 20–25
of plumbing regulations, 25–36
Standards for plumbing materials, 20–25
State plumbing codes, 33–36
Static head:
of water, 130–131
for water, air, and suds, equivalency, 380, 383–384
Steam exhaust pipes, 332–333, 352
Sterilizers:
air break at waste outlet, 290–292, 339–340
air gap in water supply pipe, 87, 114
vapor relief pipe, 70–71, 79
Storage occupancy building classification, 39–40, 46
Storm water drainage systems:
areas to be drained, 285, 339
controlled-flow system for flat roofs, 287–290
disposal system, 285, 338–339
leaders and gutters, 286, 352
sizing of, 372–375
strainers at inlets, 286, 352
traps, 286, 352
valves for diked area drains, 340–341
Strainers:
for common fixtures, 60–61, 74–75
for roof drains, 286, 352
for special use fixtures, 61, 75
Streamline flow, 125–128, 141
Subbuilding drainage system below sewer level, 295–296, 341–342

Subsoil drainage:
- backwater valve, 296–297, 342
- piping installation, 296, 342, 375
- silt and sand intercepting traps, 296, 342
- size of piping, 296, 375

Suction lift of pumps, 155–157

Suds:
- density of, 379
- equivalent static head of water, air, and suds, 380, 383–384
- rate of discharge from vent outlets, 382–384
- viscosity of, 377–379

Suds pressure zones:
- drain connections to, 306–311, 351
- in sanitary drainage systems, 306–311
- suds pressure relief vents, 306–311, 351, 401–402
  - connection to nonpressure zone in house drain, 310–311
  - for high-rise stack systems, 310–311

Sumps:
- for clear water wastes, 296, 342
- for sewage, 296, 342
- venting, 296, 312, 342, 352, 402–403

Superheated water, 235

Supports:
- for drainage and vent piping, 314–315, 343–344
- for wall-hung fixtures, 66, 77
- for water supply piping, 100–102, 120–121

Swimming pools:
- circulation pumps, 63, 292, 340
- construction of, 62–63, 75–76
- drainage of, 62–63, 75–76
- equipment for, 62–63, 75–76
- foreign matter in, 62, 76
- maintenance of, 62, 75
- standard for, 62, 75
- types of, 62
- vortex reducing device, 63, 76
- waste connections, 291, 340
- wastes from, 291–292, 340

Tank vacuum relief valve or vacuum breaker, 266, 281

Tanks (*see* Hot water storage tanks; Water supply tanks)

Temperature relief valve, 236–238, 280–281

Temporary facilities during construction, 50

Testing:
- drainage and vent systems, 333–335, 353
- water supply systems, 112, 122

Toilet room:
- floors, 64–65, 71, 76
- lighting, 63–64, 76
- ventilation, 63–64, 76
- walls, 64, 71, 77
- (*See also* Physically handicapped facilities)

Torricelli's theorem, 134

Tower buildings:
- booster pumps, 95–96
- constant pressure systems, 95–97
- construction of, 12–13
- hot water circulation pumps, 95–97
- multizone water supply systems, 95–97
- observance of manufacturers' recommendations, 97
- pressure regulation valves, 96–97, 117
- separate systems for each zone, 95–97
- static pressure maximum at fixtures, 95–97, 117

Trap (*see* Building trap; Fixture traps)

Turbulent flow:
- in drainage stack offsets, 302, 306
- friction loss under, 127, 141, 384
- Reynolds' criterion for, 126–128
- streamline and, 125–128
- in vent piping, 380–381
- in water supply systems, 127–128, 141

Two-family dwelling building classification, 38–39, 44

United Association of Journeymen and Apprentices of the Plumbing and Pipe Fitting Industry of the United States and Canada, 30–31

United States of America Standards Institute (*see* American National Standards Institute)

U. S. Department of Commerce:
- National Bureau of Standards, 28–32, 164, 366, 387
- standardization of plumbing codes and regulations, 28–32

U. S. Housing and Home Finance Agency, 31–32
U. S. Public Health Service, 31
Urinals:
automatic flush tank, 55
demand rate, 161
fixture supply pipe size, 229
fixture trap size, 68
fixture unit rating: sanitary drainage, 366–368
water supply, 226–227
floors and walls of, 64, 76
flushing, 55, 71–72
installation and location of, 64, 76
for physically handicapped, 51
prohibited type, 55, 71
requirements for occupancies, 44, 50
trough type, 55, 71
Used materials and piping, 38, 87, 114

Vacuum breakers:
accessibility of, 115–116
for fixtures, 89–94, 115–117
in flush tanks, 56, 90–92, 116
for flush valves, 56, 90–92, 115–116
for hose connections, 92–93, 115
for hot water storage tanks, 266, 281
installation of, 90–92, 115–116
maintenance of, 92, 115
need for, 90–92, 115
for preheater equipment, 92–93, 115
types of, 90–91
Vacuum relief valve for hot water storage tanks, 266, 281
Valves:
accessibility of, 111, 121
backwater, 296–297, 342
balancing (*see* Balancing valves)
building, 111, 121
check (*see* Check valves)
curb, 111, 121
dwelling unit, 111, 121
equivalent length of, 145–150
gate (*see* Gate valves)
globe (*see* Globe valves)
line, 110–111, 121
pressure regulation (*see* Pressure regulation valve)
quick-closing, 152–154
relief (*see* Relief valves)
riser, 111, 121
Valves (*Cont.*):
standards for, 22
stop-and-waste, 111, 121
water heating equipment, 111, 121
water meter outlet, 85, 121
water service, 84, 111, 121
water supply tank, 111, 121
Vapor pressure in pumping, 155–156
Vapor relief pipes for fixtures, 70–71, 79
Velocity:
average, 128
critical, 125–128
erosion-corrosion effects in piping, 176–180
in free fall, 129–132
ideal, 134
limitation in sizing water piping, 176–184
noise effects in piping, 151–155
for scouring action in drains, 359–360
of sound in water, 152
Velocity head of water, 131–132
Vena contracta in jet flow, 134–135
Vent extensions:
attachment to, 346–347
flow rate of air in, 388–391, 394–396
to outside of building, 302–304, 346–347
through roof, 302–304, 346–347
sizing of, 302–304, 347, 394–396, 403
terminal of, 302–304, 346–347
through wall, 302–304, 347
weathertightness at roof openings for, 347
Vent headers:
connections to, 302–304, 346
flow rate of air in, 390–391
sizing of, 394–396, 403
Vent piping:
conditions of flow in, 380–381
connections to fixture drains, 320–323, 347–348, 396–397
equivalent length of fittings, 386–387
friction loss for airflow in, 384–385
installation of, 312–316, 343–344
permissible length of, 385–386
slope of, 313, 344
testing, 333–335, 353
Vent stack:
connections at base and top, 303–304, 346–347

Vent stack (*Cont.*):
- extensions to atmosphere, 302–304, 346–347, 394–396, 403
- flow rate of air in, 388–389
- header, 303, 346–347
- offsets in, 303, 346
- purpose of, 303–304
- sizing of, 393–394, 403
- terminal locations, 303–304, 346–347
- where required, 303–304, 346
- (*See also* Stack venting)

Ventilation:
- of bathrooms and toilet rooms, 63–66, 76–77
- of fixture locations, 63–66, 76–77
- of sanitary drainage systems, 292–293, 340

Venting:
- original theory proposed, 9
- of pneumatic ejectors, 296, 342
- research on, 9, 26, 28–31
- sanitary drainage systems, 292–293, 340
  - design criterion for, 381–382
  - European single-stack, 335–338
- (*See also* Circuit venting method; Combination waste and vent system; Stack venting; Wet venting)

Vertical drains, gravity flow in, 360–362

Viscosity:
- of air, 377–379
- of suds, 379
- of water, 123–125

Vortex reducing device, 63, 76

Wading pool, waste connections to, 291, 340

Wall-hung fixtures, supports for, 66, 77

Walls:
- bathroom, 64, 71, 77
- above built-in bathtub, 56, 73
- and floors, water closet, 63–66, 76–77

Washbasin:
- design development of, 13–20
- early types, 14–15, 18
- (*See also* Lavatories)

Washing machines (*see* Dishwashing machines and fixtures; Laundry machines, automatic)

Washtub:
- design development of, 13–20
- early types, 7, 10, 13, 18–20

Washtub (*Cont.*):
- (*See also* Laundry trays)

Waste stacks (*see* Drainage stacks)

Wastes:
- detrimental, 61, 75, 293–295, 340–341
- high-temperature, 294, 340
- indirect, 331–332, 352
- industrial, 294, 341
- objectionable, 293–295, 340–341
- radioactive (*see* Radioactive wastes)
- special, 332–333, 352
- (*See also* Sanitary drainage system)

Water:
- boiling points of, 123, 126–127
- compressibility of, 124
- density of, 123–124
- equivalent static head of air, suds, and water, 380, 383–384
- expansion with temperature rise, 233
- hard (*see* Hardness of water)
- head of, 130–133
- nonpotable, 82–83, 113
- potable (*see* Potable water supply)
- process, 88–89, 114–115
- salt, 83, 113
- superheated, 235
- viscosity of, 123–125

Water closets:
- for children's use, 55, 71
- connection to drainage system, 66, 77
- demand rates, 161
- design development of, 15–19
- early types, 7–9
- fixture supply pipe size, 229
- fixture trap size, 68
- fixture unit ratings: sanitary drainage, 366–368
  - water supply, 226–227
- flush tanks, 55–56, 71–72
- flush valve supplied, 55–56, 71–72
- flushing, 55–56, 71–72
- installation of, 63–66, 76–77
- location of, 63–66, 76–77
- for physically handicapped, 42–44, 51–52
- prohibited types, 54–55, 71–72
- for public use, 55, 71
- requirements for occupancies, 41–52
- seats, 55, 71
- standards for, 21
- walls and floors, 63–66, 76–77

Water flow test, 333–334, 353

Water fountains (*see* Drinking fountains)
Water hammer, 94, 117, 151–154
Water-heating methods and heater types, 238–239
Water meters:
friction loss in flow through, 138–139, 171
installation of, 84–85, 121
valving, 84–85, 121
Water outlets:
above-rim potable, 89, 115
below-rim potable, 89–94, 115–117
markings of nonpotable, 83, 113
minimum pressure at, 94–95
Water pressure regulations, 94–97, 117
Water pressure test, 334, 353
Water services, 84–85, 117–118
Water supply:
corrosion control treatment, 85–86
corrosivity of, 159, 167, 176, 178–180, 240–241, 248–249
demand: at individual outlets, 160, 161
standard method for estimating, 163–167, 226–229
fixture units, 163–167, 184–196, 226–229
hardness (*see* Hardness of water)
nonpotable, 82–83, 113
potable (*see* Potable water supply)
prohibited direct connections, 87–89, 114–115
quality of, 82–83, 113
requirements for, 94–95, 117–118
salt, 83, 113
source of, 82–83, 113
Water supply systems:
adjustment of, 94, 117
availability of, 81–82, 113
design of, 163–167, 228–229
Water supply systems (*Cont.*):
disinfection methods for, 112–113, 121–122
fixture supply pipes, 229
hot (*see* Hot water supply system)
noise in, 94, 117, 151–155
outlets (*see* Water outlets)
piping installation, 100–102, 119–121
piping offsets, 108–111
pressure in, 94, 117–118, 159–160, 169, 226
private, 82, 87, 113–114
quantity rate of flow through piping, 128–129
from outlets, 136–138
regulations of district authorities, 84, 119
sizing of, detailed method for buildings of any height, 208–212
illustrative problems, 213–226
sizing of, simplified method for low buildings, 184–198
illustrative problems, 198–208
tables, 185–196
source of water for, 82–83, 113
tests, 112, 122
for tower buildings, 95–97
turbulent flow in, 127–128, 141
valving, 84, 85, 110–111, 121
Water supply tanks, 97–100, 118–119
Western Plumbing Officials Association (WPOA), 31–32
Wet venting, 323–327, 348–349
Wheelchairs, standard, specifications and function, 42–43

Yoke relief vents for soil and waste stacks, 304–306, 350–351, 401, 403
offsets, 305, 350–351, 401, 403